AF344569

Practical Food Microbiology
and Technology

PRACTICAL FOOD MICROBIOLOGY & TECHNOLOGY

THIRD EDITION

George J. Mountney
Formerly Research Management Specialist, Food Science
U. S. Department of Agriculture
Washington, DC

Wilbur A. Gould
Professor Emeritus, Department of Horticulture
Ohio State University
Columbus, OH

KRIEGER PUBLISHING COMPANY
MALABAR, FLORIDA
1992

Third Edition 1988
Reprint Edition 1992

Printed and Published by
KRIEGER PUBLISHING COMPANY
KRIEGER DRIVE
MALABAR, FLORIDA 32950

Library of Congress Cataloging-In-Publication Data
Mountney, George J.
    Practical food microbiology and technology / George J. Mountney,
  Wilbur A. Gould. -- 3rd ed.
      p.   cm.
  Originally published: New York : Van Nostrand Reinhold, 1988.
  Includes bibliographical references and index.
  ISBN 0-89464-673-7 (acid free paper)
    1. Food--Microbiology. 2. Food--Biotechnology. I. Gould, Wilbur
  A., 1920- . II. Title.
  QR115.M68 1992
  664'.001'576--dc20   91-32944
                                        CIP

10   9   8   7   6   5   4   3   2

Contents

PREFACE TO THIRD EDITION / xi

1 INTRODUCTION / 1
Scope of Food Microbiology   1
Classification of Microorganisms   2
Pathogenic Organisms   3
Selected References   4

2 COMPOSITION OF FOODS / 5
Carbohydrates   6
Proteins   10
Lipids   17
Vitamins and Minerals   23
Selected References   24

3 ENZYMES / 25
Enzyme Classification and Nomenclature   26
Role of Enzymes in Living and Dead Tissues   27
Enzymes and Food Spoilage   30
Industrial Applications of Enzymes   31
Hydrolytic Enzymes   33
Oxidation-Reduction Enzymes   40
Catalytic Properties and Mechanism of Action of Enzymes   42
Enzymes in Frozen Foods   44
Phosphatase Test   45
Selected References   46

4 TYPES OF MICROORGANISMS / 49
Bacteria   49
Molds   56
Yeasts   60
Algae   61
Viruses and Bacteriophages   63
Parasitic Organisms   63
Selected References   66

5   FACTORS THAT INFLUENCE MICROBIAL ACTIVITY / 69
      Moisture   69
      Oxidation-Reduction Potential   71
      Temperature   73
      Nutritional Requirements   75
      Hydrogen Ion Concentration (pH)   76
      Inhibitors   81
      Selected References   81

6   HIGH-TEMPERATURE FOOD PRESERVATION / 83
      Effect of Temperature on Microorganisms   83
      Application of Heat in the Canning Industry   87
      Methods of Preserving Food by Heat   89
      Thermophilic and Thermoduric Bacteria   94
      Selected References   97

7   LOW-TEMPERATURE FOOD PRESERVATION / 99
      Effect of Low Temperatures and Freezing
        on Microorganisms   100
      Physical Principles of the Freezing Process   104
      Survival of Microorganisms on Frozen
        and Refrigerated Foods   106
      Effect of Freezing on Certain Food Products   109
      Quality of Frozen Foods   111
      Use of Microwave Heat in Cooking
        and Defrosting Foods   116
      Selected References   117

8   ACIDS AND ALKALIES IN FOOD PRESERVATION / 119
      Acid-Producing Organisms   120
      Acids as Germicidal Agents   120
      Alkali-Producing Organisms   127
      Alkalies as Germicidal Agents   128
      Selected References   130

9   SUGAR AND SALT IN FOOD PRESERVATION / 131
      Salt Preservation of Foods   132
      Halophilic Organisms   135
      Use of Sugar in Preserving Foods   136
      Selected References   139

10 RADIATION AND ULTRASOUND IN FOOD
      PRESERVATION / 141
      Development and Safety of Irradiated Foods   141
      Nature and Sources of Radiation   143
      Effect of Radiation on Microorganisms   144

Irradiation in the Food Industry   145
Ultraviolet Light in the Food Industry   148
Ultrasound in the Food Industry   153
Selected References   155

11 RED MEAT AND POULTRY / 157
Microorganisms Commonly Present in Meat   158
Microbial Counts of Meat Products   158
Spoilage in Meats   161
Tenderization of Meat   164
Microbiology of Poultry   165
Selected References   167

12 FISH AND SHELLFISH / 169
Common Fish Preservation Methods   169
Microbiology of Fish   170
Microbiology of Lobsters   173
Microbiology of Oysters   174
Microbiology of Other Mollusks   177
Paralytic Shellfish Poisoning   178
Selected References   180

13 MILK AND MILK PRODUCTS / 183
Spoilage of Milk and Milk Products   185
Pathogens and Quality Control   188
Dairy Fermentations   190
Survival of Organisms in Fermented Dairy Products   197
Microbial Inhibitors in Milk   198
Fermented Milks   200
Butter   204
Cheese   205
Selected References   210

14 EGGS AND EGG PRODUCTS / 210
Formation and Structure of Eggs   212
Microbial Content of Shell Eggs   213
Antibacterial Substances in Eggs   217
Enzymes in Fresh Eggs   218
Storage and Treatment of Shell Eggs   220
Contamination with *Salmonella*   221
Dried and Frozen Egg Products   222
Selected References   226

15 FERMENTED FOODS / 229
Sauerkraut   230
Pickles   236

Vinegar   238
Olives   245
Citric Acid   247
Selected References   247

16 FRUITS AND VEGETABLES / 249
Microbiology of Fruits   249
Microbiology of Vegetables   253
Survival of Microorganisms in Frozen Fruits
  and Vegetables   258
Selected References   259

17 FLOUR, BREAD, AND CEREALS / 261
Microorganisms in Grains and Flours   262
Breadmaking   263
Defects of Bread   264
Cereals and Other Grain Products   265
Selected References   266

18 SPICES / 267
Properties and Classification   268
Use of Spices in the Meat Industry   270
Germicidal Action of Spices   270
Storage and Treatment of Spices   271
Selected References   272

19 WATER / 273
Consumption and Uses of Water   274
Water Supplies   276
Principles of Water Purification   279
Causes and Treatment of Hard Water   282
Waste Treatment   286
Characteristics of Wastes   288
Selected References   295

20 FOOD SPOILAGE / 297
Microorganisms and Food Spoilage   299
Spoilage in Canned Foods   300
Spoilage in Meat and Other Unprocessed Foods   301
Microbiological Analysis of Foods   308
Selected References   311

21 FOODBORNE ILLNESS / 313
Etiological Agents   313
Foodborne Infections   315
Foodborne Intoxications   321

Chemical Food Poisoning   327
Epidemiology of Foodborne Illness   327
Selected References   335

22   MYCOTOXINS / 337
Specific Mycotoxins   338
Human Mycotoxicosis   340
Selected References   343

INDEX / 345

Preface to Third Edition

The original objective of writing a book for food technologists, restaurant and foodservice students, sanitarians, and for those in industry who need a general description of the microflora and their effects on specific foodstuffs rather than for microbiologists interested in basic metabolic reactions has been continued in the third edition.

A major rearrangement of chapters into more closely related subject areas and into a more logical development of chapters has been attempted. Several old chapters have been omitted and several new ones added. The subject matter areas have been regrouped into six general sections: characteristics of microorganisms and food; factors influencing microbiological activity; food preservation methods; the microbiology of specific foodgroups; food spoilage characteristics; and finally, the public health implications of food pathogens.

As in the first and second editions, which emphasized the practical aspects of food microbiology and technology, we continue to emphasize the fermentative, spoilage, and pathogenic microorganisms found on specific food types; the role of food processing on food contamination and control; the effects of different microorganisms on the several foods; the characteristics of the different types of food spoilage; and finally, the efficacy of different types of preservation treatments on different groups of foods.

The authors are particularly indebted to Dr. Mason E. Miller, former communications scientist, Cooperative State Research Service, United States Department of Agriculture, for his assistance in editing the revised manuscript.

1
Introduction

Scope of Food Microbiology 1
Classification of Microorganisms 2
Pathogenic Organisms 3
Selected References 4

SCOPE OF FOOD MICROBIOLOGY

The science of microbiology deals with organisms that are invisible or barely visible to the unaided eye. These microorganisms include viruses, bacteria, protozoa, algae, fungi, and certain small worms. Food microbiology deals with such organisms in and on food. Food microbiologists are concerned with the practical implications of the microflora of the food. Can the organism cause a disease in humans? Does the organism cause food spoilage? Is the presence of the organism aesthetically acceptable in human food? Does the organism change the functional properties of a foodstuff resulting in new tastes, odors, or textures?

Understanding the relationships among the various organisms making up the microflora of a food is important for food microbiologists. Relationships may be symbiotic, antagonistic, or commensurate. The implications of these relationships for safety, spoilage, or new product development are important.

Food microbiologists are primarily concerned with what microorganisms do to a particular human food or to consumers under a given set of conditions. They may also be concerned with the presence in foods of extraneous materials that may be toxic or displeasing for aesthetic reasons.

Both food microbiology and food technology deal with the handling, processing, preserving, storing, preparation, nutritional content, and safety of food. These disciplines are closely related. The roles that a microbiologist may fill in a food company include training production personnel, quality control of incoming and outgoing material, drafting and implementing new standards, sanitation, process development, trouble shooting, and dealing with regulatory agencies.

In the area of food safety it is essential for the food microbiologist to know the "critical hazard points" in a food processing operation. Since these are the points where contamination can take place, the food microbiologist needs to sample at these operational points as part of an overall quality assurance program. In the manufacture of cheese or fermented milk, the microbiologist must be aware of the possible presence in the milk supply of antibiotics that are illegal and interfere with the growth of the starter cultures.

CLASSIFICATION OF MICROORGANISMS

Microorganisms can be classified according to various criteria, such as optimal temperature for growth, peculiar growth requirements, and major nutrient source. The classification and terminology adopted by the American Public Health Association are described briefly here (Speck 1984).

Microorganisms are commonly classified based on their optimum growth temperature ranges. *Psychrotropic* microorganisms grow relatively rapidly at commercial refrigeration temperatures. Included in this group are the bacterial genera *Pseudomonas, Achromobacter, Flavobacterium,* and *Alcaligenes* and the mold genera *Geotrichum* and *Botrytis. Thermophilic* microorganisms survive and grow well at elevated temperatures. Species able to survive some significant level of heat treatment (e.g., pasteurization) for a short time are termed *thermoduric.* Included in this group are some species in the bacterial genera *Micrococcus, Streptococcus* (primarily enterococci), *Microbacterium, Arthrobacter, Lactobacillus, Bacillus,* and *Clostridium.* Molds in this group include *Byssochlamys fulva* and some species of *Aspergillus* and *Penicillium* under certain conditions. *Mesophilic* microorganisms are aerobic and anaerobic sporeforming bacteria that grow at 35°C but not at 55°C. All species of the genus *Bacillus* are aerobic sporeformers. Anaerobic sporeforming mesophiles that cause food spoilage by putrifaction include *Clostridium sporogenes, C. botulinum, C. putrefaciens,* and *C. perfringens.* Most mesophiles are heat resistant.

Another classification is based on peculiar growth requirements. *Osmophilic* microorganisms, which cause spoilage in honey, soft-centered chocolate candies, jams, molasses, concentrated fruit juices, and other highly concentrated sugar solutions, are adapted to growth in substrates of high osmotic pressure. Yeasts such as *Saccharomyces rouxii* and *S. mellis* are examples of osmophiles. *Halophilic* microorganisms require a minimum concentration of salt for active growth. Species in the genera *Halobacterium* and *Halococcus* and other bacterial species found principally in marine environments are examples. These microorganisms cause spoilage of fish, bacon, and hides preserved in sea salts. *Staphylococcus aureus* and *Clostridium perfringens* can also multiply in moderately salted foods.

A third way to classify microorganisms is based on the type of nutrients they metabolize. *Proteolytic* microorganisms break down proteins, accompanied by formation of undesirable flavors and odors. Genera that include proteolytic organisms are *Bacillus, Clostridium, Pseudomonas,* and *Proteus. Lipolytic* microorganisms cause spoilage of lipid-containing foods by hydrolysis and oxidation. These microorganisms are responsible for the formation of desirable flavors in cheese and fermented foods. *Saccharolytic,* or acid-forming, microorganisms break down sugars to form acids. The lactic acid bacteria, which are important members of this group, cause milk souring and are used under controlled conditions to produce cheese and fermented milk drinks. Various species in the genera *Streptococcus, Leuconostoc,* and *Pediococcus* are also saccharolytic microorganisms.

Pectinolytic microorganisms, which break down pectin, are responsible for some kinds of spoilage of fruits and vegetables. *Achromobacter, Aeromonas, Arthrobacter, Agrobacterium, Enterobacter, Bacillus, Clostridium, Erwinia, Flavobacterium, Pseudomonas,* and *Xanthomonas* are examples of genera containing species of this type. *Amylytic* microorganisms break down starch to sugars. *Cellulytic* microorganisms break cellulose into simpler compounds.

PATHOGENIC ORGANISMS

Food microbiologists must always be on the watch for the possible presence of human pathogenic organisms in foods. To help determine whether a food is free from human pathogens, microbiologists use *indicator* organisms. These can be the pathogen itself or another organism, such as *Escherichia coli,* that grows under the same conditions as the pathogen. The presence or absence of bacterial toxins is also used as an indicator for pathogens and as a means to help insure that a food is safe. Examples of pathogenic microorganisms of concern to food microbiologists are coliforms of the genera *Salmonella, Streptococcus, Staphylococcus, Shigella, Yersinia, Vibrio, Enterococcus, Bacillus,* and *Clostridium.*

SELECTED REFERENCES

Anon. 1963. "Microbiological Quality of Foods." Proc. Conf. Franconia, New Hampshire. Academic Press, New York.

Anon. 1964. An Evaluation of Public Health Hazards from Microbiological Contamination of Foods. Subcommittee on Food Protection, Food and Nutrition Board. National Academy of Sciences, Washington, DC.

Anon. 1975. Prevention of Microbial and Parasitic Hazards Associated with Processed Foods: A Guide for the Food Processor. National Academy of Sciences, Washington, DC.

Anon. 1985. An Evaluation of the Role of Microbiological Criteria for Foods and Food Ingredients. National Academy of Sciences, Washington, DC.

Frazier, W. C., and Westhoff, D. C. 1978. "Food Microbiology," 3rd ed. McGraw-Hill, New York.

Speck, M. L. 1984. Compendium of Methods for the Microbiological Examination of Foods. Am. Public Health Assoc., Washington, DC.

2
Composition of Foods

Carbohydrates 6
  Classification of Carbohydrates 6
  Action of Microorganisms on Sugars 8
  Biological Role and Occurrence of Carbohydrates 9
Proteins 10
  Classification of Proteins 11
  Distribution of Protein in Foods 12
  The Nitrogen Cycle 12
  Protein Decomposition 14
Lipids 17
  Fats and Oils 17
  Distribution of Fat in Foods 19
  Microbial Action on Fats 20
  Other Lipids 21
  Importance of Dietary Fat 23
  Effect of Heat and Metallic Oxides on Fats 23
Vitamins and Minerals 23
Selected References 24

Foods are composed of carbohydrates, proteins, fats, fatty acids, nucleic acids, minerals, vitamins, hormones, water, nucleotides, and other minor constituents. The composition and nutritive value of food are affected by the fertility of the soil, growth conditions, plant cultivar, animal breed, environment, and feed in the case of animals. The stage of maturity and other factors such as pH also can cause considerable variations.

The principal nutrients present in foodstuffs also differ among various types of foods. For example, butter is mainly fat, meat has a high protein content, and fruits have varying amounts of carbohydrates. The biochemical changes that foods can undergo during processing and storage depend on their composition. Fat products are susceptible to oxidative rancidity, while protein products are susceptible to proteolytic degradation. Some foods are protected by a tough hard shell or outer coating, which helps retard biochemical reactions. Nuts in the shell keep for a long time, but if the shell is removed, chemical and other changes occur in a short time unless some other provision is made for protection.

Adequate preservation of foods requires a knowledge of food technology and of the many biochemical and microbiological changes that usually take place in various foods during processing and storage. Many foods are produced over a wide range of geographical and climatic conditions. Often large volumes of fresh foods are produced within a relatively short time and cannot be consumed as they become available. Such foods must be preserved. The microflora and nutrients are altered by this processing and handling.

Another variation among groups of foods is their pH. Many fruits are high in acid; while vegetables, as a rule, are low. The acid content of a food helps determine the flavor and the microflora of the product. Because molds and yeast can tolerate fairly high acidity, they are most likely to be found on fruits. Molds not only tolerate severe acid conditions but can actually use the acid as a source of energy. Most bacteria, however, cannot survive in a highly acid environment, even though they produce many organic acids. There are some exceptions in the case of food fermentations where acid-tolerant bacteria can survive for a long time. The effect of food composition and other factors on microbial growth is discussed in detail in Chapter 5.

CARBOHYDRATES

By definition carbohydrates contain carbon, hydrogen, and oxygen. The name *carbohydrate* implies that hydrogen and oxygen are present in the same ratio as they are in water, that is, 2:1. Chemically, simple carbohydrates are aldehyde or ketone derivatives of polyhydric alcohols. The structural formulas shown in Table 2.1 illustrate this relationship. The common sugar D-glucose is an aldose, and D-fructose is a ketose.

Classification of Carbohydrates

Carbohydrates are generally classified into two major groups: simple sugars and compound sugars. Simple sugars are called monosaccharides and are classified according to the number of carbon (C) atoms they contain and whether an aldehyde (aldose) or ketone (ketose) group is present. Compound sugars are made up of two or more monosaccharide units and are further classified as oligosaccharides (two to nine monosaccharides) and polysaccharides (ten or more monosaccharides). A

Table 2.1. Simple Sugars

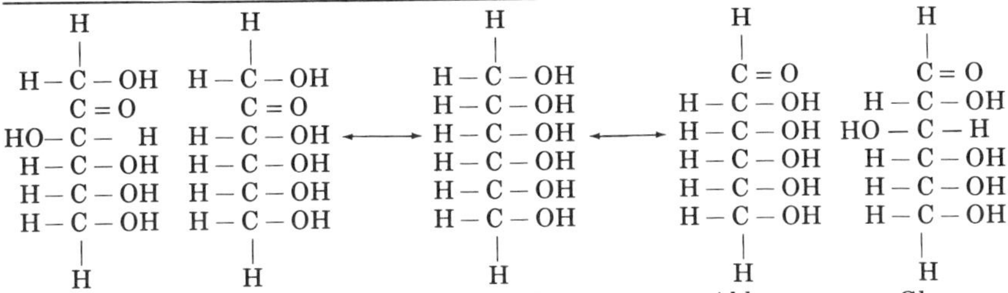

H	H	H	H	H
|	|	|	|	|
H – C – OH	H – C – OH	H – C – OH	C = O	C = O
C = O	C = O	H – C – OH	H – C – OH	H – C – OH
HO – C – H	H – C – OH ⟷	H – C – OH ⟷	H – C – OH	HO – C – H
H – C – OH	H – C – OH	H – C – OH	H – C – OH	H – C – OH
H – C – OH	H – C – OH	H – C – OH	H – C – OH	H – C – OH
H – C – OH	H – C – OH	H – C – OH	H – C – OH	H – C – OH
|	|	|	|	|
H	H	H	H	H
D-Fructose	Ketose	Hexahydric alcohol	Aldose	D-Glucose

general classification scheme with examples of specific carbohydrates (in italics) is as follows:

I. Monosaccharides
   A. Dioses
      1. Aldoses: *Glycolaldehyde*
   B. Trioses
      1. Aldoses: *Glyceric aldehyde*
      2. Ketoses: *Dihydroxy acetone*
   C. Tetroses
      1. Aldoses: *D- and L-Erythrose*
      2. Ketoses: *Erythrudose*
   D. Pentoses
      1. Aldoses: *Xylose*
                Arabinose
      2. Ketoses: *Ribulose*
                Xylulose
   E. Hexoses
      1. Aldoses: *Glucose*
                Mannose
                Galactose
      2. Ketoses: *Fructose*
                Sorbose
II. Oligosaccharides
   A. Disaccharides
      1. Reducing: *Maltose* (glucose + glucose)
                    Lactose (glucose + galactose)
      2. Nonreducing: *Sucrose* (glucose + fructose)
   B. Trisaccharides: *Raffinose* (glucose + fructose + galactose)

    C. Tetrasaccharides: *Stachyose* (galactose + galactose + glucose + fructose)

    D. Dextrins[1]:   *Amylodextrin*

           Erythrodextrin

           Achrodextrin

III. Polysaccharides $(C_6H_{10}O_5)_n$

    A. Homopolysaccharides

        1. Derived from D-glucose: *Starch*—composed of *amylose*, a long unbranched chain, and *amylopectin*, a highly branched chain

                Glycogen—more highly branched than amylopectin; found only in animal tissues; especially abundant in liver and muscle

                Cellulose—major structural component of the cell wall of plants

        2. Derived from D-fructose: *Inulin*

        3. Derived from D-xylose, D-galactose, or mannose: *Hemicellulose*

    B. Heteropolysaccharides: *Glucosides*

           Tannins

           Pectin compounds

           Gums and *mucilages*

Sugars can also be classified as reducing or nonreducing depending on the ability of their aldehyde or ketone groups to reduce Fehling's solution. All monosaccharides are reducing sugars. In the case of disaccharides, if the aldehyde and ketone groups are involved in linking the monosaccharide units together, the sugar is nonreducing. Sucrose is a nonreducing disaccharide; maltose and lactose are reducing disaccharides.

Action of Microorganisms on Sugars

Certain sugars are sometimes added to food products at concentrations of 1–10% to accelerate the growth of microorganisms. For exam-

---

[1]Dextrins are soluble gummy oligosaccharides obtained from starch and are made up of D-glucose units.

ple, the addition of sucrose or lactose to fermenting cabbage helps organisms associated with good sauerkraut production to produce acetic and lactic acids more rapidly. On the other hand, to inhibit the growth of bacteria and molds, it is generally thought necessary to have approximately 65–80% sugar. Some bacteria as well as certain molds and yeasts can tolerate a greater concentration of sugar than others. The use of concentrated sugar solutions in food preservation is discussed in Chapter 9.

Since carbohydrates are easier to utilize than proteins or fats, foods rich in sugars are readily attacked by many kinds of microorganisms. Acids and gases are the main products of carbohydrate fermentation, although appreciable amounts of many intermediate compounds also are formed. Many organisms are selective in their action on specific sugars, as the examples in Table 2.2 indicate.

The low pH of many foods retards the growth of certain microorganisms. For instance, the low pH of fruit juices acts as a preservative in some cases. The acids that develop in milk make it extremely sour. During the formation of acids, the organisms responsible for their formation are inhibited as well as any others that cannot tolerate any appreciable amounts of acid. However, some organisms are considered acid tolerant because they can survive for a long time in an acid environment.

Starch, a complex carbohydrate, is also attacked by certain microorganisms. The formation of acid from starch is much slower than from sugars because the starch must first be hydrolyzed to sugar by organisms that produce amylolytic enzymes. The amounts of sugar and acids formed depend on the amounts of enzymes produced by the microorganisms.

Biological Role and Occurrence of Carbohydrates

Carbohydrates play a major role in foods. They are formed directly in the process of photosynthesis in green plants. Carbohydrates are also found in animals in such forms as lactose in milk and glycogen in

Table 2.2. Activity of Selected Bacteria on Sugars

	Acid and gas production from		
Organism	Dextrose	Lactose	Sucrose
Staphylococcus aureus	Acid	Acid	Acid
Enterobacter aerogenes	Acid and gas	Acid and gas	Acid and gas
Alcaligenes faecalis	No acid or gas	No acid or gas	No acid or gas

liver and muscle tissue. In order for carbohydrates to be assimilated by animals, they must be converted into glucose, the sugar in blood. Surplus glucose is converted into glycogen and stored in the liver and muscle tissue.

Most of the organic compounds found in animals and plants are derived from carbohydrates. Proteins are synthesized from amino acids, which are formed in reactions between portions of carbohydrate molecules and ammonia.

Carbohydrates can be stored in animals and plants and under appropriate conditions can be converted to simple compounds and oxidized to furnish energy. Glycogen and glucose, or blood sugar, are soluble forms, which can be readily transformed into energy by the animal. In plants, starch, inulin, and hemicellulose are storage forms for various monosaccharides.

Monosaccharides are particularly important because they are the building blocks for the formation of more complex carbohydrates in plants and animals. Starch and glycogen are formed from D-glucose and inulin from fructose. Monosaccharides also are utilized in the synthesis of fat and protein. Yeasts can convert galactose, D-glucose, mannose, and fructose into alcohol. Citric and gluconic acids as well as glycogen and butyl alcohol are but a few of the biological products resulting from the action of microorganisms on monosaccharides.

The major carbohydrates found in plants are the following:

- Glucose—first product of photosynthesis
- Starch, inulin, and hemicellulose—reserve food
- Cellulose—supporting tissues
- Gums and mucilages—degradation products
- Glucosides—found in various products

In animals, the major sugars are D-glucose in the blood; lactose in milk; glycogen, the reserve food; and D-ribose and D-oxyribose, which are components of nucleic acids.

PROTEINS

Proteins are the most abundant organic molecules within cells constituting 50% or more of the dry weight. They are fundamental to all aspects of cell structure and function, and all living cells contain proteins.

The word protein is derived from the Greek term meaning holding first place. In general, they are complex nitrogenous dispersions classified as colloidal systems rather than true solutions. When proteins are hydrolyzed they break down into amino acids. All proteins conform quite closely to the following average chemical analysis: carbon, 50-55%; oxygen, 20-23%; nitrogen, 15-18%; hydrogen, 6-8%; and sulfur, 0-0.4%. Both the quality and quantity of protein present in a food affects its nutritional value. Quality depends on the number and proportion of essential amino acids present and the ease of digestion.

Proteins are composed of several amino acids. About 22 have been identified in proteins. At least ten are essential in the human diet because the body does not synthesize them in adequate amounts. Fortunately, most of the essential amino acids are found in various foods. The ten essential amino acids for human and animal nutrition are arginine, histidine, leucine, isoleucine, lysine, methionine, phenylalanine, threonine, tryptophan, and valine. Those that are not absolutely essential to human nutrition can be synthesized in the body from other amino acids.

Classification of Proteins

Simple proteins yield amino acids and no other major organic or inorganic products upon hydrolysis. They are never found in combination with nonprotein substances. The major types of simple proteins are the following:

1. Albumins are soluble in pure water, coagulable by heat, and widely distributed in plants and animals usually in small amounts. Example: egg albumin.

2. Globulins are insoluble in pure water but dissolve in dilute salt solutions and are coagulated by heat. They occur in many seeds, and can be synthesized in crystalline form. Example: edestin from hemp seed.

3. Glutelins are insoluble in water and dilute salt solutions but dissolve in dilute acids and bases. Example: glutenin of wheat.

4. Prolamines are insoluble in many common solvents but soluble in 80% alcohol. Example: gliadin of wheat and zein of corn.

5. Albuminoids are insoluble and extremely resistant to hydrolysis. Example: collagen of bones and cartilage; keratin of hoofs, horns, and feathers.

Conjugated proteins are simple proteins usually combined with nonprotein material. The term *prosthetic group* is generally used to designate the non-amino acid portion of such proteins. The following are common types:

1. Metalloproteins are simple proteins usually combined with heavy metals. In most metalloproteins the metal is very loosely bound and can be easily removed. However, with some proteins the metal is contained in a prosthetic heme group and is firmly bound. Example: hemoglobin and myoglobin.

2. Glycoproteins have carbohydrate as the nonprotein group. Example: mucin of saliva.

3. Phosphoproteins have phosphoric acid as the nonprotein group. Example: casein of milk and pepsin.

4. Nucleoproteins are complexes of proteins and nucleic acid. Nucleoproteins are known to be present in viruses and ribosomes.

5. Lipoproteins are complexes of protein and lipids. These complexes serve as transporters of lipids from the small intestine to the liver and from the liver to the fat deposit and various other tissues. Several classes of lipoproteins are found in blood plasma. Example: plasma B lipoproteins.

Derived proteins include the protein decomposition products of naturally occurring proteins. Primary derivatives are (1) coagulated proteins, which are insoluble products formed when heat or alcohol is used, and (2) metaproteins, which are the principal products formed when acids or bases are used on naturally occurring proteins. Secondary derivatives are the products of protein digestion. They include (1) proteoses (soluble in water and not coagulated by heat); (2) peptones (may be an advanced stage of protein hydrolysis); and (3) peptides (principally condensation products with very few amino acid residues).

Distribution of Protein in Foods

Cheese, eggs, meats, certain nuts, and dry legumes are rich in proteins. In these foods the protein content ranges from 10 to 35%. Cereals generally contain 5–15% protein. Fresh fruits and vegetables are usually low in protein, ranging from 1 to 5%.

The Nitrogen Cycle

Plants differ from animals in their ability to synthesize proteins. Animals depend on various foods for their protein source, whereas plants synthesize and store their protein requirements. The classical nitrogen cycle shown in Fig. 2.1 illustrates the movement of nitrogen.

The result of nitrogen fixation is the accumulation of nitrogenous compounds in the bacterial cell, which are ultimately available to the

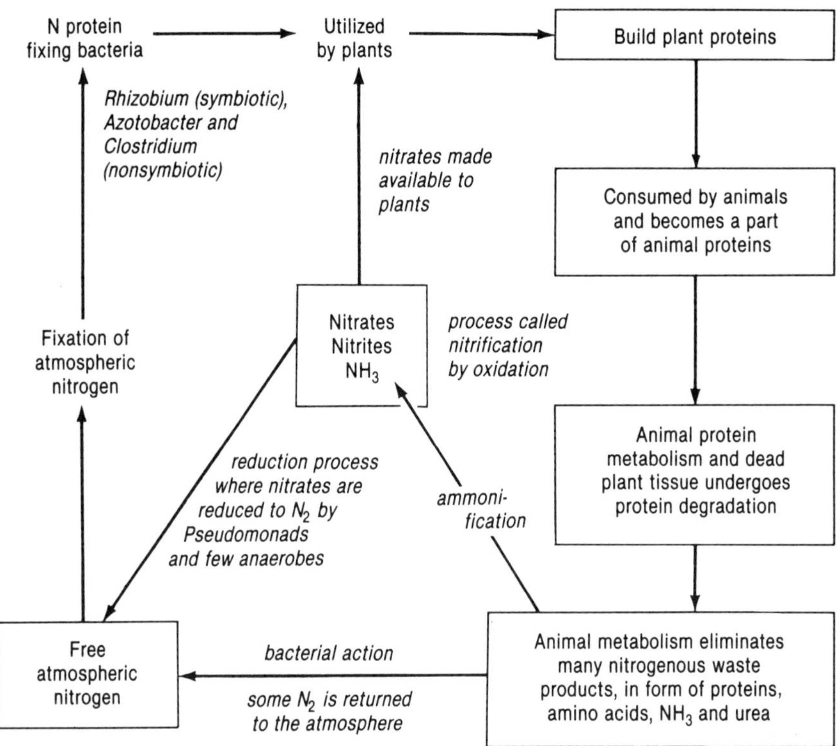

Fig. 2.1. The nitrogen cycle could not operate without bacteria. Ammonia, produced in the degradation of proteins, is oxidized by *Nitrosomonas* bacteria to nitrites. These in turn are oxidized by *Nitrobacter* species to nitrates, which can be assimilated by plants. Only specialized nitrogen-fixing bacteria can transform atmospheric nitrogen (N_2) into forms that plants can utilize.

plant. The dry residue of bacterial cells contains about 10% nitrogen and approximately 65–75% nitrogenous compounds such as globulins, albumens, and nucleoproteins. These constituents make up the protoplasm and enzyme systems of the cell. Although bacteria may contain the same amino acids as plants and animals, they differ from them in the nature of the nitrogenous materials that they can use as a source of nitrogen. Bacteria also differ from each other in nitrogen metabolism. Certain groups of bacteria can use elemental nitrogen from the air; others, simple salts; a few use amino acids only.

One of the great mysteries of biological life is how under ordinary temperatures and pressures nitrogen can be combined with other elements. In the laboratory nitrogen can be made to combine with other elements to form such well known compounds as ammonia (NH_3) and nitrates but only with difficulty.

Green plants and animals cannot use atmospheric nitrogen (N_2). However, certain bacteria can use N_2 and convert it into nitrates, which plants can use. This process is called nitrogen fixation. Plants then convert the nitrate into plant protein. By consuming plants, animals, including man, can use this protein. Through the process of metabolism, urea is excreted by animals, but it cannot be used readily by plants as a source of nitrogen until bacteria convert this urea into ammonia, nitrites, and nitrates, which can be used by plants. Microorganisms play an important role in breaking down dead organic matter and liberating complex nutrients for plant growth.

As Fig. 2.1 indicates, the nitrogen cycle depends on the activity of microorganisms. If all the microorganisms in the world were destroyed, man could not exist.

Protein Decomposition

Because protein decomposition is much more complex than the breakdown of carbohydrates, the end products are more varied. The protein molecule is quite complex; for example, the formula of gelatin is $C_{49}H_{42}O_{96}N_{105}$. Until the protein molecule is broken down into amino acids, the cell cannot use the products.

Microorganisms, through their proteolytic enzymes, break down protein into simple substances, such as amino acids or simple organic acids and ammonia. The breakdown usually follows this pattern: protein → peptones → polypeptides → peptides → amino acids → ammonia (NH_3) → elemental nitrogen (N). In addition, alcohol and gases such as carbon dioxide, methane, hydrogen, and ammonia are liberated. Ammonia is usually formed in appreciable amounts in the advanced state of protein decomposition. Odoriferous compounds such as hydrogen sulfide (H_2S) or mercaptans, indol, skatol, putrescine, and cadavarine are frequently liberated as some of the end products of protein decomposition.

Proteins are attacked by some organisms of the genus *Clostridium,* an anaerobic, sporulating organism that causes putrefaction and produces many offensive sulfur compounds (e.g., mercaptans and hydrogen sulfide) as well as hydrogen, ammonia, phenols, and carbon dioxide. Decomposition of protein by aerobic organisms is called decay. Proteins containing amino acids with sulfur, such as cystine and methionine, can be broken down with no unpleasant odor because the end products are completely oxidized and stabilized. In the case of anaerobes, the end products are not completely stabilized and foul odors are present.

The decomposition of proteins by microorganisms is very important in the processing of many kinds of foods. For instance, the ripening of cheese by microbiological activity is widely known. The tenderization of meat is another form of protein breakdown.

In food processing, the texture of food may be altered by denaturation of the protein, which may alter the metabolic behavior of the endogenous organisms and enzyme systems. This may be desirable or undesirable, and the food technologist should be aware of this phenomenon.

Milk Proteins

The proteins in milk can be used to illustrate what happens to animal or plant proteins under selected conditions when attacked by proteolytic microorganisms. Milk contains the following protein fractions: casein, 3%; lactoalbumin, 0.5%; and lactoglobulin, 0.05%. The proteins in milk are very complex chemically. For example, casein is an amphoteric protein having both acid and basic properties, but it usually possesses acidic properties. The protein molecules are broken down into the 18 or more amino acids present in milk.

Tryptophan is a good example of an amino acid that can undergo a complex series of chemical reactions. It occurs in lactalbumin in amounts ranging from 2 to 8% and is one of the essential amino acids required by humans. It is also required by several different groups of bacteria. *Escherichia coli* and certain other bacteria have the ability to produce indole, a metabolic end product of tryptophan. Because of a specific enzyme, tryptophanase, tryptophan and water break down to form indole, pyruvic acid, and ammonia, according to the following equation:

This reaction, called the indole test, is used to help identify certain species of bacteria, especially among the coliforms. The organisms are grown in a tryptophan broth for several days. Then a few drops of sulfuric acid and 1 ml of a 0.1% solution of sodium nitrite are added to the incubated culture. If indole is present in the medium, a red compound (nitroso-indole) is formed.

The amino acid tryptophan is also responsible for some of the disagreeable odors found in spoiled foods. Organisms containing the enzyme tryptophanase break down tryptophan into indole, pyruvic acid, and ammonia. Both indole and methyl indole have disagreeable odors associated with putrefaction.

The kind of organisms and amino acids present determine very largely the end products of protein decomposition. Some of these products are poisonous. Isoleucine is one of the many amino acids found in milk. When conditions are favorable and the right kind of yeasts and bacteria are present, one of the end products of this protein breakdown is fusel oil, which is poisonous.

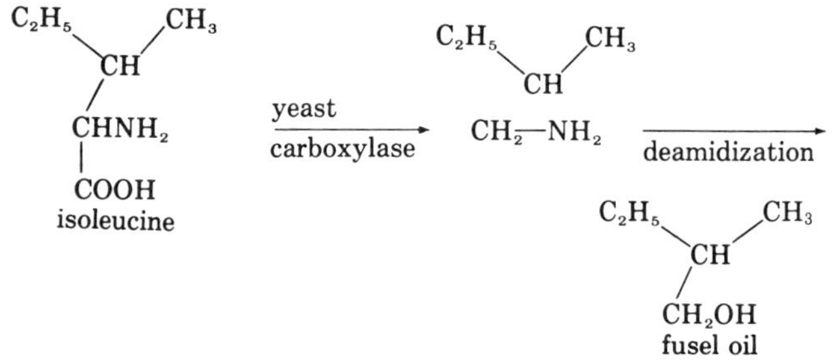

$$
\begin{array}{ccc}
\underset{\substack{| \\ \text{CHNH}_2 \\ | \\ \text{COOH} \\ \text{isoleucine}}}{\overset{\text{C}_2\text{H}_5 \diagdown \quad \diagup \text{CH}_3}{\text{CH}}}
& \xrightarrow[\text{carboxylase}]{\text{yeast}}
& \underset{\text{CH}_2\!-\!\text{NH}_2}{\overset{\text{C}_2\text{H}_5 \diagdown \quad \diagup \text{CH}_3}{\text{CH}}}
\quad \xrightarrow[\text{deamidization}]{} \quad
\underset{\substack{\diagup \\ \text{CH}_2\text{OH} \\ \text{fusel oil}}}{\overset{\text{C}_2\text{H}_5 \diagdown \quad \diagup \text{CH}_3}{\text{CH}}}
\end{array}
$$

Ptomaines

Ptomaines constitute another group of compounds resulting from the breakdown of certain proteins by protein-splitting microorganisms. Ptomaines may be defined as decarboxylated amino acids and may be represented by the following general formula:

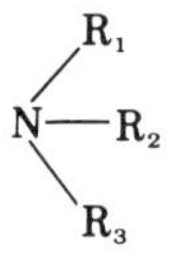

$$
\text{N} \diagup^{\!\!R_1} \!\!-\!R_2 \diagdown_{\!\!R_3}
$$

If each R group is a CH_3, one obtains trimethyl amine:

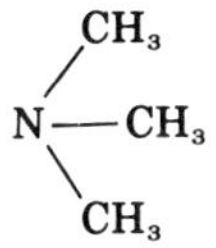

$$CH_3$$
$$N\!-\!CH_3$$
$$CH_3$$

trimethyl amine

Some but not all ptomaines are poisonous; for example, those in Limburger cheese are nonpoisonous. Nonetheless, in the past, outbreaks of food poisoning were called "ptomaine poisoning." Our present-day knowledge of bacteriology has eliminated this idea. Bacteriologists now agree that food poisoning can occur when certain microorganisms present in food produce exotoxins harmful to man.

LIPIDS

Lipids comprise a wide variety of organic compounds that are insoluble in water and soluble in solvents such as ether and chloroform. Lipids include liquid oils and solid fats, waxes, sterols, and related compounds.

Fats and Oils

The most common lipids are fats and oils, which are widespread in nature, especially in animal and vegetable matter. These compounds are esters of fatty acids with glycerol, in the ratio of one molecule of glycerol to three molecules of fatty acids. There are many different fatty acids. The characteristics of various fats are determined by the fatty acids that combine with the glycerol.

With the exception of a few fatty acids from unusual sources, naturally occurring fatty acids are straight-chain carbon compounds with an even number of carbon atoms and only one acid group (the — COOH group) in each molecule. This is illustrated by the formula for stearic acid.

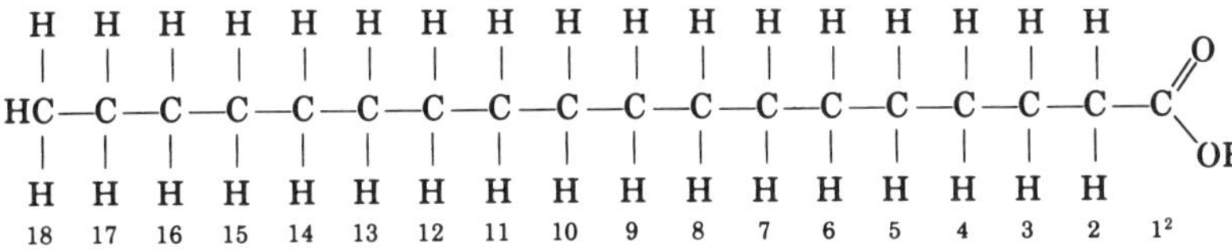

stearic acid

Fats that contain the maximum number of hydrogen atoms are known as saturated fats. Suet and tallow are examples. Unsaturated fats, which are easily attacked by microorganisms, do not contain all the hydrogen atoms possible. Various oils are good examples. However, it is possible to add hydrogen to an unsaturated fat and make it a solid saturated fat, which can be used in shortenings. The common saturated fatty acids are as follows:

Butyric acid	C_3H_7COOH	Lauric acid	$C_{11}H_{23}COOH$
Caproic acid	$C_5H_{11}COOH$	Myristic acid	$C_{13}H_{27}COOH$
Caprylic acid	$C_7H_{15}COOH$	Palmitic acid	$C_{15}H_{31}COOH$
Capric acid	$C_9H_{19}COOH$	Stearic acid	$C_{17}H_{35}COOH$

Many of the lower fatty acids have an offensive odor. Consequently, milk fat, which has a high percentage of lower fatty acids, especially butyric, has a strong offensive odor on hydrolysis.

$$
\begin{array}{c}
\quad\;\; H \\
\quad\;\; | \\
C_3H_7COO\!-\!C\!-\!H \\
| \\
C_3H_7COO\!-\!C\!-\!H \;+\; 3HOH \\
| \\
C_3H_7COO\!-\!C\!-\!H \\
| \\
H
\end{array}
\longrightarrow
\begin{array}{c}
\quad\;\; H \\
\quad\;\; | \\
HO\!-\!C\!-\!H \\
| \\
HO\!-\!C\!-\!H \;+\; 3C_3H_7COOH \\
| \\
HO\!-\!C\!-\!H \\
| \\
H
\end{array}
$$

glyceryl butyrate   water   glycerol   butyric acid

Those fats, such as lard and certain vegetable fats, which contain fatty acids of higher molecular weights (e.g., stearic and palmitic) yield very little offensive odor on hydrolysis.

$$
\begin{array}{c}
\quad\;\; H \\
\quad\;\; | \\
C_{17}H_{35}COO\!-\!C\!-\!H \\
| \\
C_{17}H_{35}COO\!-\!C\!-\!H \;+\; 3HOH \\
| \\
C_{17}H_{35}COO\!-\!C\!-\!H \\
| \\
H
\end{array}
\longrightarrow
\begin{array}{c}
\quad\;\; H \\
\quad\;\; | \\
HO\!-\!C\!-\!H \\
| \\
HO\!-\!C\!-\!H \;+\; 3C_{17}H_{35}COOH \\
| \\
HO\!-\!C\!-\!H \\
| \\
H
\end{array}
$$

stearin   water   glycerol   stearic acid
(glyceryl stearate)

The formation of fats occurs by a reaction that is the reverse of hydrolysis. That is, the –OH groups of glycerol react with the –COOH groups of fatty acids with the elimination of water. Since glycerol has three –OH groups and fatty acids have only one –COOH group, three fatty acid molecules join with one glycerol molecule to produce one molecule of pure fat.

If a mixture of fatty acids is present in a fat, it is unusual for the fatty acids in any one molecule of fat to resemble each other. If the three radicals, represented by R, are different, then further differences are introduced according to their position:

$$
\begin{array}{ccc}
& H & \\
& | & \\
R_1COO{-}C{-}H & R_2COO{-}C{-}H & R_1COO{-}C{-}H \\
| & | & | \\
R_2COO{-}C{-}H & R_1COO{-}C{-}H & R_3COO{-}C{-}H \\
| & | & | \\
R_3COO{-}C{-}H & R_3COO{-}C{-}H & R_3COO{-}C{-}H \\
| & | & | \\
H & H & H
\end{array}
$$

The diagrams represent the three possible arrangements. The three fats, although consisting of exactly the same components, will differ in melting point and other properties. This has been proved by the synthesis of fats from mixtures of known fatty acids.

A possible explanation for the relative abundance of the low-molecular-weight fatty acids in milk fat is that they originate from fermentations in the rumen of cows. This idea is supported by a limited number of analyses of milk fats from various animal species that show large amounts of low-molecular-weight fatty acids occur only in ruminants. Nonruminants, including humans, produce milk fats with much less butyric acid and other low-molecular-weight fatty acids.

Distribution of Fat in Foods

Vegetable oils and lard contain nearly 100% fat, whereas fresh fruits and vegetables contain less than 1%. Avocados with a fat content of 25% and ripe olives with about 19% fat are exceptions. Dry vegetables and soybeans have 18% fat. The fat content of nuts is quite high, varying from 33% in coconuts to 70% or more in pecans. Meats, of course, vary considerably in fat content depending upon the condition of the animal and the way the meat is cut and trimmed.

The wide variety of fatty acids present in milk fat gives this fat one of its outstanding characteristics, i.e., a gradual transition from hard (fully solidified) to liquid (fully melted) over a wider temperature range than is true for other fats. The fatty acids of low-molecular-weight and unsaturated fatty acids contribute low-melting-point tendencies, while the high-molecular-weight saturated fatty acids tend toward high melting points.

In many fats one saturated fatty acid and one unsaturated fatty acid are present in large amounts with other fatty acids dropping off in quantity in either direction, with respect to more or less carbons and lesser or greater unsaturation, from the predominant fatty acids.

Microbial Action on Fats

The oxidation of fat by microbial activity involves a breakdown of the fatty acid portion of the fat by enzymes forming short-chain acids, aldehydes, ketones, and peroxides. Lipase, an enzyme that hydrolyzes fats to free fatty acids and glycerol, is present in many kinds of foods. Because milk contains an appreciable amount of this enzyme, milk fat often undergoes lipase-catalyzed hydrolysis with the production of free fatty acids, diglycerides, monoglycerides, and in extreme cases free glycerol.

Fats are more difficult for microorganisms to attack than carbohydrates and proteins, although a few organisms produce lipase. Molds produce an appreciable amount of lipase and usually attack foods high in fats. Cheese is a good example. Some varieties of cheese are ripened by mold. Microorganisms providing lipases and hydrolyzing fats vary considerably in their fat specificity and optimal pH. Table 2.3 lists some of these organisms and the specificity of their action.

In addition to hydrolysis, fatty foods such as butter, fish, and chilled meats may undergo oxidation caused by oxidase-producing bacteria. The oxidized flavor in milk is caused largely by the formation of aldehydes and ketones as the result of the oxidation of the unsaturated fatty acids such as oleic.

Table 2.3. Specificity and pH Optimum of Common Fat-Attacking Organisms

Organism	Optimum pH	Carbon attacked
Candida lypolytica	5.5	C-1
Penicillium notalum	7.0	C-1
Mucor sufu	7.0	C-1
Rhizopus oligosporus	7.0	C-1
Chaetostylum fresenii	7.0	C-1
Aspergillus niger	5.6	C-1
Staphylococcus aureus	7.0	C-1, C-2, C-3
Aspergillus flavus	7.0	C-1, C-2, C-3
Geotrichum candidum	6.0	C-1, C-2, C-3

Bacteria are capable of causing more than one type of reaction; in fact, one reaction may mask another. Many oxidase-producing bacteria not only hydrolyze fat but may initiate the decomposition of proteins. For example, fishy flavor in butter results from the formation of trimethylamine $N(CH_3)_3$, a compound having a fish odor. It is caused by the hydrolysis of lecithin, which is present in fats and fatty materials. Many microorganisms are capable of forming trimethylamine, although there may be other substances in fats that catalyze the hydrolysis of fats.

Oxidase-Producing Organisms

Pseudomonas and *Achromobacter* genera are oxidase positive. Members of the *Alcaligenes* and *Brucella* genera are also positive, while the *Enterobacter, Escherichia,* and *Proteus* show only a weak oxidase production. Strong oxidase producers are gram-negative, while gram-positive organisms produce little oxidase.

Since *Pseudomonas* and *Achromobacter* species produce an appreciable amount of oxidase and can grow at storage and refrigerator temperatures, they may cause many off-flavors in addition to rancidity in butter, lard, chilled beef, and other fatty products.

Oxidase-producing organisms have been shown to be widely distributed in many foods. Numerous samples of milk from different cows drawn aseptically from the udder failed to show any oxidase-formers. Molasses yielded many of these organisms. In corn silage both positive and negative oxidase producers were found in about equal numbers. Plates exposed to laboratory dust did not reveal many of the organisms, although the dust in dairy barns showed numerous organisms that were positive. Many were found in decayed plant tissue as well as in soil and surface waters. Various types of cheese examined failed to show many of these organisms. Greater numbers of oxidase-positive bacteria were found in butter than in any other food product, especially when the butter was off-flavor, cheesy, putrid, rancid, or fruity.

Other Lipids

Fats and oils are classified as *simple* lipids. They are esters of fatty acids with glycerol. Waxes, also classified as simple lipids, differ from fats in that they contain alcohols other than glycerol. They serve as a protective covering on various fruits and leaves, and some have marked germicidal properties. Perhaps this is nature's way of protecting certain plants from harmful microbial activity.

Compound lipids contain other groups in addition to an ester of a fatty acid with an alcohol. The main compound lipids are as follows:

- Phospholipids (phosphatides)—Esters containing fatty acids, phosphoric acid, and other groups usually containing nitrogen
- Cerebrosides (glycolipids)—Compounds containing fatty acids, a carbohydrate and a nitrogen moiety, but no phosphoric acid
- Sphingolipids and sulpholipids.

In general, compound lipids are too complex for extensive microbial activity. However, the phospholipids in milk may be significant in contributing an unusual flavor to the product.

Derived lipids are substances having the general properties of lipids and are derived from simple or compound lipids. This group includes the fatty acids; sterols, which are solid high-molecular-weight alcohols; and hydrocarbons.

Sterols occur in both animals and plants. The sterol cholesterol is present in the animal body especially in brain and nervous tissue. It is also found lining the arteries of individuals suffering from arteriosclerosis or "hardening of the arteries," a common metabolic disease in the present generation. Ergosterol, another steroid found in certain plants, can be a very important accessory food substance. When subjected to ultraviolet light, it is converted into vitamin D. The role of sterols in food has not been fully evaluated, but many nutrition investigators believe these substances play an important part in our diet.

Phospholipids and sterols are compounds commonly associated with fats from various sources. A typical phospholipid is lecithin, which occurs in milk fat, egg yolk, soybeans, and in nearly all vegetables. The nitrogen base in lecithin is choline, a substance that is part of the vitamin B complex. It plays an important role in fat metabolism. Lecithin from egg yolk or soybean is used in the manufacture of mayonnaise, chocolate, and oleomargarine. It aids in emulsifying fat.

Although true fats are triglycerides of fatty acids (simple lipids), the term "fat," as used in the analysis of dairy products and of foods and feeds in general, includes everything that is ether soluble. In this sense, milk fat must be considered as including true fat, phospholipids (principally lecithin), sterols (principally cholesterol), carotene, xanthophyll, and the fat-soluble vitamins A, D, E, and K. There may also be traces of essential oils from weeds or feeds. (e.g., allyl sulfide from garlic and allyl thiocyanates from onions).

Importance of Dietary Fat

Fats and oils are the most concentrated sources of energy, providing 9 kcal of energy per gram, or approximately twice the energy provided by proteins and carbohydrates. They are carriers of the fat-soluble vitamins A, D, E, and K, which are essential parts of body tissues. Fats help the body to utilize carotene and calcium, reduce the need for vitamin B_1, and aid in the metabolism of milk sugar. Triglycerides, phospholipids, cholesterol, and cholesterol esters are important to the structure, composition, and permeability of membranes and cell walls. Seeds, fruits, plants, and animals store lipids which when eaten provide a source of energy. Lipids are a major component of adipose tissues, which are thermal insulation for the body, protection against shock to internal organs, and a contributor to body shape.

Fats can be classified as *saturated* and *unsaturated*. Saturated fats are generally solid at room temperature and unsaturated ones liquid. Chemically, unsaturated fats contain a double bond between some carbons, and fewer hydrogen atoms than saturated ones. Unsaturated fats are more reactive than saturated ones. Some unsaturated fatty acids (e.g., linoleic acid) cannot be synthesized in the body. For this reason, they must be included in the diet and are called essential fatty acids. The absence of these essential fatty acids may cause such pathological conditions as kidney lesions. It has been demonstrated many times that the amounts of saturated and unsaturated fats and sterols, especially cholesterol, in the diet play a major role in the development of atherosclerosis.

Effect of Heat and Metallic Oxides on Fats

Heat can cause changes in the flavor of milk through the formation of certain sulfur compounds (sulfides) associated with the fat and milk protein formed from the milk constituents.

Metallic oxides of copper and iron from food equipment may cause off-flavors in fatty foods. Copper and copper alloys, even in minute traces (1 ppm), may cause a perceptible oxidized flavor to develop in milk.

VITAMINS AND MINERALS

In addition to the three main components of foods, other factors called vitamins are essential for man. These compounds cannot be synthesized by the body but are needed in low concentrations to act with enzymes, or active portions of enzymes, to perform essential chemical reactions in the cells. Minerals are needed in trace amounts in the diet also. These nutrients are important for the activation of many enzymes. Without them cells could not function.

SELECTED REFERENCES

Bloor, W. R. 1926. Biochemistry of fats. *Chem. Rev.* 2, 243–300.
Dack, G. M. 1982. "Food Poisoning," Rev. Ed. University of Chicago Press, Chicago, Illinois.
Deatherage, F. E. 1975. "Food for Life." Plenum Press, New York.
Fennema, O. R. 1976. "Principles of Food Science, Part I, Food Chemistry." Marcel Dekker, New York.
Gray, W. D. 1959. "The Relation of Fungi to Human Affairs." Henry Holt, New York.
Green, D. E., and Stumpf, P. K. 1944. Biological oxidation and reductions. *Annu. Rev. Biochem.* 13, 1–24.
Lehninger, A. L. 1982. "Principles of Biochemistry." Worth Publishers, New York.
Neurath, H., and Hill, R. L. 1982. "The Proteins," 3rd ed. Academic Press, New York.
Oginsky, E. L., and Umbreit, W. W. 1959. "An Introduction to Bacterial Physiology," 2nd ed. W. H. Freeman Co., San Francisco, California.
Solms, J. 1973. "Functional Properties of Fats in Foods." Forster Verlag, AG/Forster Publ. Ltd. Zurich.
Stacy, M., and Barker, S. A. 1960. "Polysaccharides of Microorganisms." Oxford University Press, London.
West, E. S., Todd, W. R., Mason, H. S., and Van Bruggen, J. T. 1976. "Textbook of Biochemistry," 11th ed. Macmillan, New York.

3
Enzymes

Enzyme Classification and Nomenclature 26
  Standard System of Enzyme Nomenclature 26
  Specialized Terms 27
Role of Enzymes in Living and Dead Tissues 27
  Peroxidase 29
  Factors Affecting Enzyme Activity in Foods 29
Enzymes and Food Spoilage 30
  Autolytic Enzymes 31
Industrial Applications of Enzymes 31
  Heat of Fermentation 32
  Preparation of Enzyme Extracts 33
Hydrolytic Enzymes 33
  Esterases 34
  Carbohydrases 35
  Proteolytic Enzymes 39
Oxidation-Reduction Enzymes 40
Catalytic Properties and Mechanism of Action of Enzymes 42
  Carboxylase—An Example of Enzyme Action 43
Enzymes in Frozen Foods 44
Phosphatase Test 45
Selected References 46

Enzymes, called the catalysts of life, are complex proteins produced by living cells to perform specific biochemical reactions in cellular metabolism. One eminent scientist has said that life is just one enzyme reaction after another. The catalytic efficiency of enzymes is very high. For example, one enzyme molecule can catalyze the reaction of 10,000 to 1,000,000 molecules of substrate per minute. Enzymes are affected by acids and bases, and have maximum, minimum, and optimum temperatures for activity.

The high degree of specificity of enzymes is amazing. Each enzyme has a particular job and usually cannot do another job. For example, enzymes will catalyze a reaction of one of a pair of chemical isomers, but not the other. The only difference between such isomers is that one is a mirror image of the other. Frequently, enzymes are named by adding the suffix *-ase* to the name of the substrate, e.g., proteinase.

Enzymes play an important role in the food industry. In the baking, brewing, and confectionery industries, they are used to liquefy and saccharify starches, to convert sugars, and to modify proteins. In fruit juice and wine making, they improve the yield and flavor of juice, clarify it, and speed filtration.

ENZYME CLASSIFICATION AND NOMENCLATURE

Intracellular enzymes *(endoenzymes)* never leave the living cell and play an important role in transforming the food absorbed by the cell into forms that can be utilized in cellular metabolism. In this process large amounts of energy are liberated and made available to the cell.

Extracellular enzymes *(exoenzymes)* are enzymes produced and secreted by the cell. They diffuse from the cell into the surrounding medium and act independently of the cell in breaking down organic substances in the medium, such as proteins, starches, and fats. In this way, the insoluble materials are reduced to a form that can be absorbed through the differentially permeable cell membrane and used. The energy liberated by enzymes outside the cell is not important to the cell. In fact, exoenzymes liberate very little heat energy.

Carbohydrases are enzymes that hydrolyze polysaccharides or oligosaccharides. Enzymes acting on carbohydrates are the most investigated as well as the most widely used enzymes for laboratory and industrial applications. *Proteases* break down proteins, polypeptides, and peptides by hydrolyzing the peptide linkages. *Lipases* hydrolyze fats and oils, giving rise to free fatty acids and partial glycerides. These are essential for metabolic processes such as oxidation and resynthesis of glycerides and phospholipids. These and other types of enzymes are described in more detail later in this chapter.

Standard System of Enzyme Nomenclature

The International Union of Biochemistry has developed a system of enzyme nomenclature that has become standard and should be used in enzyme work. Each enzyme is assigned a code number of four numerals, each separated by periods and arranged according to certain principles. The first numeral indicates the main division to which the enzyme belongs: 1 = oxidoreductases; 2 = transferases; 3 = hydrolases; 4 = lyases; 5 = isomerases; and 6 = ligases. The second numeral is the subclass, which identifies the enzyme in more specific terms. The third numeral precisely defines the type of enzymatic activity, and the fourth numeral is the serial number of the enzyme in its sub-subclass. Thus, the first three numerals clearly designate the nature of the enzyme. For example, the designation EC 1.2.3.4 denotes an oxidoreductase with an aldehyde as the donor and oxygen as the acceptor, and it is the fourth numbered enzyme in a particular series. The 1978 recommendations of the International Union of Biochemistry on nomenclature and classification of enzymes catalogued over 1700 enzymes; the oxidoreductases, transferases, and hydrolases each included over 400 enzymes.

Specialized Terms

Several specialized terms that are often used in discussions on enzymes are explained in this section.

Holoenzyme refers to the protein portion of an enzyme plus the coenzyme needed for catalytic activity, that is, the total enzyme structure. *Apoenzyme* refers to the thermolabile protein component of an enzyme that determines its specificity. The terms *coenzyme, cofactor,* and *prosthetic group* are often used interchangeably to describe cocatalysts that act in conjunction with an apoenzyme to catalyze a reaction. A prosthetic group usually is a cocatalyst that is very tightly bound to the protein.

Isoenzymes, or *isozymes,* are multiple forms of an enzyme occurring in the same species. They catalyze the same reaction and arise from genetically determined differences in the primary structure of an enzyme. All proteins possessing the same enzymatic activity and occurring naturally in a single species are considered *multiple forms* of the particular enzyme. For example, lactate dehydrogenase occurs in animal tissues as five different isozymes separable by electrophoresis.

ROLE OF ENZYMES IN LIVING AND DEAD TISSUES

Enzymes are normally present in all living cells—animal, vegetable, and microbial. The metabolism of living cells is usually carried on in a well-defined order by independent enzyme reactions linked in a complex enzyme system. For example, the aerobic conversion of glucose in respiration to carbon dioxide and water takes place through a series of enzymatic reactions. In the first phase, pyruvate is formed from glucose. Then the process proceeds through 12 separate steps. In each step an intermediate product is formed as a product of the preceding reaction in which one or more enzymes take part.

Ordinarily, these intermediate products do not accumulate within the cell. The reactions are confined to the cell and are regulated. Some enzymes (e.g., catalase, peroxidase, and phenolase) that are present in the living cell play an unknown role in normal cell metabolism. However, if the cell becomes damaged by some physical or chemical means, then these or other enzymes may accelerate biochemical changes, which may result in decomposition of the cell wall and its protoplasm.

Enzymes are similar in some respects to certain normal intestinal bacteria. If death occurs, these bacteria leave the intestinal tract and invade other parts of the body. Similarly, enzymes can react on the tissues that produced them if these cells are killed or damaged.

Because enzymes are naturally present in living tissues, they are important components of most food stuffs. After the living tissues are harvested or slaughtered, many enzymes continue to function. It is in this stage that the food substances may undergo the disintegration commonly associated with food spoilage, a process known as *autolysis*.

Fruit that is removed from the plant before maturity continues to carry on internal biochemical changes catalyzed by endogenous enzymes. For example, green, mature bananas and tomatoes will continue to ripen after being harvested because of enzymatic activity. Fruit is still alive when stored and will continue to respire, taking in oxygen and giving off carbon dioxide. This process will go on until the fruit is fully "ripe." Because of the effect of temperature on enzyme-mediated reactions, fruit usually is precooled to slow down respiration before shipping or storage of products. On the other hand, removing the skin of an apple or potato accelerates enzyme activity. A pea in a pod removed from the vine may continue to undergo all the changes associated with normal ripening. If the pea is removed from the pod, these biochemical changes are greatly altered. Carbon dioxide and other gases are sometimes used to slow respiration while food products are in transit.

Alteration in the flavor and tenderness of meat long after an animal is dead also results from continued enzymatic activity. The tenderization of meat is largely a controlled enzyme process and resembles protein digestion. However, if allowed to continue, a true spoilage is the result.

Microorganisms, like all living cells, owe their activity to enzymes. Certain molds liberate lipase, an enzyme that plays an important role in the ripening of Roquefort cheese. Many bacteria possess enzyme systems that produce profound changes; some are desirable and others undesirable in foods.

Peroxidase

Peroxidase activity plays a more important role than catalase in the production of off-flavors in foods. Peroxidase is a ubiquitous enzyme occurring in all higher plants that have been investigated and in leukocytes. Milk is one of the best known sources of this enzyme. Peroxidase is very resistant to heat inactivation. Since very sensitive and simple colorimetric tests are available to measure peroxidase activity, it has been used as an indicator for the effectiveness of substerilization heat treatments. If peroxidase is destroyed, so are other enzymes of concern.

Peroxidase also appears to be important from the standpoint of nutrition, color, and flavor. For example, peroxidase activity can lead to the oxidative destruction of vitamin C. Peroxidase catalyzes the bleaching of carotenoids in the absence of unsaturated fatty acids, and also catalyzes the decoloration of anthocyanins.

Peroxidase activity varies in different vegetable tissues and with the indicator used. It is essential to select the proper indicator for each vegetable tested. Several investigators have confirmed this principle by using benzidine as the indicator in studying peroxidase activity in spinach, peas, and lima beans. They concluded that this enzyme may not be entirely responsible for deterioration in unblanched vegetables.

Four peroxidases (horseradish peroxidase, verdo-peroxidase, milk peroxidase, and cytochrome-*c* peroxidase) have been isolated in pure crystalline form, but the chemical structure has not been determined. Cytochrome-*c* peroxidase catalyzes the oxidation of ferrocytochrome-*c* only. The other peroxidases use a variety of compounds such as aromatic amines, phenols, diamines, ascorbic acid, iodides, and ferrocytochrome-*c* as substrates.

Factors Affecting Enzyme Activity in Foods

The activity, and hence the effects, of enzymes in foods is often determined by the conditions within the food. Control of those conditions is necessary to regulate enzymatic activity during food preservation and processing. These factors include pH, temperature, moisture, ionic strength, and concentrations of substrate and inhibitors.

Extreme pH values will generally inactivate enzymes. Enzymes usually exhibit maximum activity at a particular pH value called the pH optimum. Most enzymes show maximum activity between pH 4 and 8; maximum activity is usually, but not always, confined to a rather narrow pH range. Since enzymes are heat labile, heat denaturation of the enzyme protein results in a gradual loss of its catalytic properties. Low temperature delays or retards enzymatic activity. For example, frozen meat may keep for several months without any appreciable deterioration. Food preservation methods depend largely on creating an unfavorable environment for enzyme activity.

ENZYMES AND FOOD SPOILAGE

All foods are derived from living cells, which produce enzymes, and contain microorganisms, which are a good source of enzymes. Deterioration of food does not take place in normal healthy tissue. However, when injury occurs or the tissues reach maturity, the enzymes may start a series of biochemical changes, many of which are undesirable and bring about spoilage.

For example, meat is stored in a nonliving state in which the respiratory enzymes cease to function, but others carry on and induce undesirable changes in the meat. Microorganisms also play an important role in spoilage. Low temperatures and carbon dioxide are employed to create an unfavorable environment for microorganisms and enzymes.

Under practical storage temperatures, autolysis goes on at a very slow rate. Studies made by chemists in the U.S. Department of Agriculture have established that protein in stored food products is broken down into simpler substances such as peptones, peptides, amino acids, and ammonia. Phosphate, normally present in the tissues, is separated from organic compounds. Fats undergo hydrolysis with liberation of fatty acids. If these acids are members of the lower fatty acid series, offensive odors are usually evident.

The blanching of vegetables before storage is necessary in order to inactivate enzymes. The chemical changes that take place in frozen vegetables are not very well understood.

Enzymes play a minor role in spoilage of canned foods because they are usually destroyed by heat treatment during processing. However, if enzymes in canned foods are not completely destroyed or inactivated, they may give rise to undesirable flavors, color, and texture, thus lowering the quality; transform starch to sugar, which tends to increase toughness; and destroy ascorbic acid (vitamin C).

In milk, the fat–water interface is the site of chemical reactions that lead to off-flavors. Certain forces at interfaces greatly accelerate such changes. Other "pro-oxygenic" factors that stimulate production of off-flavors are increased temperature, sunlight or ultraviolet light, and time (aging).

Autolytic Enzymes

Autolytic means self-destruction. Autolytic enzymes may be an inherent part of a food or present in microbial cells. Regardless of their origin, these enzymes play an important role in foods. Although all living cells contain autolytic enzymes, their activity is not well understood. When a cell dies, these enzymes act very promptly and digest the cell.

It is believed that most enzyme actions are reversible. Perhaps some of the synthetic enzyme processes of cells during life reverse their action after death. The optimal growth temperatures for an organism favor enzyme action. An excellent example is the tenderization of meat. Leafy vegetables undergo the same enzyme action until spoilage is apparent. Bananas as well as other fruits become very soft during autolysis, followed by loss of weight, color, and flavor. When extensive autolysis has taken place, the resulting environment is most favorable for growth of bacteria, molds, and yeasts.

Enzymes are also used to clarify wine and fruit juices. The use of enzymes in clarification of fruit juices increases the yield, shortens the processing times, and results in clear juice and good color. In the closely related jelly industry, other enzymes are used to improve gel strength by reducing the sugar content.

One of the most interesting uses of enzymes in foods is in the manufacture of chocolate-covered, liquid-center candies. If cherries are used, they are first covered with a semisolid sugar solution to which the enzyme sucrase has been added. After the chocolate covering is added, the enzyme hydrolyzes the sugar to the more soluble glucose and fructose, thus producing the liquid center.

INDUSTRIAL APPLICATIONS OF ENZYMES

Enzymes have several distinct advantages for use in industrial processes. They are of natural origin and are nontoxic. They have great specificity of action; hence, they can bring about reactions without unwanted side reactions. They work best under mild conditions of moderate temperature and near neutral pH and do not require drastic conditions such as high temperature, high pressure, high acidity, and the like, which necessitate special expensive equipment and may cause undesirable side reactions. Enzymes act rapidly at relatively low concentrations, and the rate of reaction can be readily controlled by adjusting temperature, pH, and the enzyme concentration. Finally, enzymes can be inactivated easily when the reaction has gone as far as is desired.

Enzymes used in industrial processes are isolated from a variety of animal and plant tissues and microorganisms. Because of the relative ease with which microorganisms can be grown on a large scale, enzymes from microbial sources are commonly used in industrial processes.

One of the oldest and most important uses of enzymes in food technology is in various fermentation processes, but they have many other applications as well. For example, the production of clear beer depends on the use of enzymes. Ordinarily, beer becomes cloudy when cooled because of the presence of complex proteinaceous materials that form a haze at lower temperatures. Beer that is treated with proteolytic enzymes remains clear when it is cooled.

Heat of Fermentation

The heat evolved in an active fermentation is the result of energy liberated in the individual enzymatic reactions. In large-scale fermentations, billions of cells are active and large amounts of heat may be generated. If this heat is not removed, the active cells are affected. For instance, in vinegar making, the heat generated during the fermentation, if uncontrolled, may be sufficient to inhibit biological activity. Table 3.1 shows some typical energy values of various enzymatic reactions.

All fermentative processes are exothermic. It has been calculated

Table 3.1. Energy Liberated From 1 Gram of Substrate by Enzymes

Exoenzymes	Calories	Endoenzymes	Calories
Pepsin	0	Lactic	82
Trypsin	0	Dehydrogenase	149.5
Rennent	0	Urease	239
Lipase	4	Oxidase	2530
Invertase	9.3		
Maltase	10		
Lactase	23		

that 1 g (net weight) of *Micrococcus ureae* decomposes 180–1200 g of urea per hour. Also, 1 g (net weight) of lactose-fermenting organisms can break down 178–14,890 g of lactose per hour.

These observations indicate the ease and rapidity with which microorganisms may obtain energy from food and their ability to maintain themselves under a variety of environmental conditions.

Preparation of Enzyme Extracts

Most enzymes are soluble in water. Some enzymes diffuse out of the cell and can be obtained by purification of the extracellular material. Other enzymes are held in the cell. To extract these enzymes, it is necessary to destroy the cells. Extraction by water, dilute glycerine, or dilute acid or alkali is facilitated if cells have previously been dried; frozen and allowed to thaw; plasmolyzed; or allowed to autolyze. These processes disintegrate cell structures and render membranes more permeable. The first enzymatic fractions extracted from living cells always consist of a mixture of several enzymes and a host of accompanying inactive substances.

Since the amount of enzyme present in a given product cannot yet be measured by ordinary quantitative methods of analysis, activity is taken as the measure of concentration. This is done by measuring the amount of substrate undergoing a change in unit time under defined conditions, such as pH, temperature, and initial concentration of substrate when acted on by a unit mass of the material processing enzymatic activity. When crude enzyme preparations are purified, their enzyme activity increases, but their mass decreases.

HYDROLYTIC ENZYMES

Among the most important hydrolytic enzymes *(hydrolases)*, which catalyze the hydrolysis of various types of substrates, are esterases, carbohydrases, proteolytic enzymes (proteinases and peptidases), and amidases. Other hydrolytic enzymes are cellulase, hemicellulase, and pentosanase; these may be useful in recovering cellulosic wastes and tenderizing cellulosic food products. For example, pentosanase increases the yield of high-quality starch from wheat by hydrolyzing insoluble pentosans. In Table 3.2, some microbial extracellular hydrolytic enzymes are listed, along with their substrates and reaction products.

Table 3.2. Microbial Extracellular Hydrolytic Enzymes

Enzyme	Substrate	Catabolic products
Esterases		
Lipases	Glycerides (fats)	Glycerol + fatty acids
Phosphatases		
Lecithinase	Lecithin	Choline + H_3PO_4 + fat
Carbohydrases		
Fructosidases	Sucrose	Fructose + glucose
Alpha-glucosidases (maltase)	Maltose	Glucose
Beta-glucosidases (cellobiase)	Cellobiose	Glucose
Beta-galactosidases (lactase)	Lactose	Galactose + glucose
Amylase	Starch	Maltose
Cellulase	Cellulose	Cellobiose
Cytase	—	Simple sugars
Nitrogen-carrying compounds		
Proteinases	Proteins	Polypeptides
Polypeptidases	Proteins	Amino acids
Desamidases		
Urease	Urea	CO_2 + NH_3
Asparaginase	Asparagine	Aspartic acid + NH_3
Desaminases	Amino acids	NH_3 + organic acids

Esterases

The hydrolysis of substances containing an ester linkage with the formation of free acids and alcohols is catalyzed by esterases.

Lipase is widely distributed in nature. The enzyme is physiologically important because it hydrolyzes fats and oils, giving rise to free fatty acids and partial glycerides, which are essential for such metabolic processes as fatty acid transport, oxidation, and resynthesis of glycerides and phospholipids. In addition, this enzyme is of considerable economic significance in the food industry.

If not appropriately controlled, lipase can hydrolyze lipids producing undesirable rancid flavor in milk products, meat, fish, and other food products containing fat. On the other hand, lipase is essential for the production of desirable characteristic flavors in certain foods. For example, lipase is necessary for the development of specific flavors in blue (Roquefort), Camembert, and several Italian cheeses. Also, the action of lipase is required to develop the characteristic flavor of milk-chocolate candy and buttermilk pancakes. Lipase is also found in resting and germinating seeds containing oils as reserve food. The properties of lipase in resting seeds are different from those in germinating seeds. Some maintain that the enzyme then is in the proenzyme stage, or zymogen, and is activated by H^+ ions. Some enzymes are active in the neutral or even weakly alkaline stage, while others are active only in acid stages.

Phosphatase is probably present in all living cells. Yeast serves as a useful source of this enzyme. It attacks hexophosphates, yielding hexose and phosphoric acid. It may also act on other organic phosphates, nucleotides, and phytin. The presence of phosphatase in milk is used as an indicator in quality control to determine whether the milk has been adequately pasteurized.

Chlorophyllase occurs in all plants and microbial tissues that contain chlorophyll. Chlorophyllase hydrolyzes the phytyl group from chlorophyll to give phytol and chlorophyllide. The enzyme may be quite important in determining the stability of the green color of vegetables prior to and during processing.

Carbohydrases

Among the carbohydrases, it is convenient to distinguish polysaccharases, which hydrolyze polysaccharides, from glycosidases, which hydrolyze glycosidic linkages in oligosaccharides.

Glycosidases

The glycosidase *invertase* was first isolated from yeast in 1860. Since then the enzyme has been extensively studied and purified from a variety of sources, using different extraction methods. Only the enzymes from *Saccharomyces cerevisiae* and *S. carlsbergensis* have industrial importance.

Invertase hydrolyzes cane sugar (sucrose) to reducing sugars. Cane sugar is dextrorotatory. The hydrolysis product contains equal amounts of glucose and fructose and is levorotatory since fructose is more levorotatory than glucose is dextrorotatory. Because of the change of rotation, this mixture is termed *invert sugar*. Since the hydrolysis is described as an inversion, the enzyme received the name invertase. Invertase can also act on trisaccharides containing the same linkage between glucose and fructose as occurs in cane sugar. For example, raffinose is hydrolyzed to fructose and melibiose, a disaccharide.

Invertase has several important applications in the food industry. It can be employed in manufacturing artificial honey. Invert sugar, which is much more soluble than sucrose, is also produced. A large use of crude invertase is to prevent crystallization in concentrated molasses.

The solubility of invert sugar is important in the manufacture of confectioneries, liqueurs, and frozen desserts where high sucrose concentration would lead to crystallization. Invertase is also used in the preparation of chocolate-coated liquid-center candies. Molding and coating are carried out while the centers are firm, after which the invertase that has been added to the cast centers acts to yield a smooth, stable cream.

An interesting example of enzyme specificity is displayed by invertase. Although yeast and fungal invertases both hydrolyze sucrose, the nature of their action differs. Yeast invertase is a fructosidase, attacking the fructose end, whereas fungal invertase is a glucosidase, attacking the glucose end of the sucrose molecule. This may be demonstrated by comparing their activities against raffinose. Yeast invertase hydrolyzes raffinose into fructose and melibiose, but fungal invertase causes no reaction since glucose is not terminal in the raffinose molecule:

$$\text{Raffinose (fructose-glucose-galactose)} \xrightarrow{\text{yeast invertase}} \text{Fructose + Melibiose (glucose-galactose)}$$

$$\text{Raffinose} \xrightarrow{\text{fungal invertase}} \text{No reaction}$$

Polysaccharases

Conversion of starches to sugars by use of polysaccharases is an important commercial application of enzymes. Probably amylases have the greatest commercial use.

$$\text{Starch} \xrightarrow[\text{(liquefying amylase)}]{\alpha\text{-amylase}} \text{Maltose + Dextrin}$$

$$\text{Starch} \xrightarrow{\beta\text{-amylase}} \text{Maltose + Dextrin}$$

$$\text{Starch} \xrightarrow{\text{glucoamylase}} \text{Glucose}$$

Amylases, also called diastases, are most likely present in all living cells that contain starch. Amylases probably consist of several enzymes. In the presence of a thermostable substance of unknown composition, known as a complement, amylases show extended activity and readily hydrolyze dextrin. That dextrins are never found in the free state in the cell may be accounted for by the association of amylases with a complement in living cells. Autolyzing yeast serves as a source of complement.

Amylases act on starch, glycogen, and derived polysaccharides to hydrolyze the 1,4-α-glucosidic linkages. The amylases may be divided into three groups. α-Amylase is an endoglucosidase that hydrolyzes 1,4-α-glucosidic linkages randomly to produce a mixture of glucose and maltose. α-Amylase is often referred to as the "liquefying" enzyme because of its rapid action in reducing the viscosity of starch solutions. β-Amylase is an exoglucosidase with specificity for a 1,4-α-D-glucosidic linkage involving the penultimate glucose on the nonreducing end of the molecule. The enzyme successively removes maltose units from the nonreducing end of 1,4-α-D-glucosidic polymers until it comes to a branched point in the molecule. There action stops, and glucoamylase, which hydrolyzes glucose units from the nonreducing terminal end of the substrate, starts reacting.

The amylases are used extensively in the baking industry. They increase the sugar content for yeast fermentation through greater starch conversion, which in turn increases gas formation and improves crust color. Softening of bread dough on the other hand is largely caused by protease action.

Amylases have many uses in the food processing industries. An extremely important use of amylases is in the production of sweet syrups from very low DE[1]-high dextrin syrup to very high DE-high sugar syrup to meet the demand of different food industries. A recent extensive application of amylolytic enzymes is the use of fungal glucoamylase for the production of glucose from starch.

Production of glucose from starch by conversion with glucoamylase was proposed before 1950. However, commercialization was not possible until some ten years later when glucoamylase essentially free from transglucosylases became available at economical cost. This was achieved by the discovery of very high glucoamylase-yielding strains of fungi that produce minimal amounts of transglucosylases. The presence of transglucosylases is undesirable because they produce glucose polymers having 1,6-α-linkages, which reduce dextrose yields and interfere with dextrose crystallization.

The amylases in malt are important in the alcoholic fermentation of grains. Amylases help to liquefy the starch to produce a free-flowing slurry. This is probably the oldest of all enzyme applications and has been practiced since antiquity.

Other uses of amylases in food industries are quite minor with respect to the amounts of enzymes involved, but are quite important and frequently almost indispensible in the manufacture of certain products. Amylases are used in processing cereal products for food dextrin and sugar mixtures, in processing baby foods, and in making some breakfast foods. In the preparation of free-flowing chocolate and lico-

---

[1]DE means dextrose equivalent. Dextrose is another name for glucose.

rice syrups, amylases are used to liquefy the starch present, which keeps the syrups from congealing. They also are used for recovering sugars for reuse from scrap candy of high starch content.

In addition to their important applications in food processing, large amounts of bacterial amylases have other industrial uses: textile desizing, preparing modified starch coatings for paper and fabrics, and making cold water dispersible laundry starches, for example.

Another polysaccharase with important industrial applications is *cellulase.* The term cellulase has been applied both to pure, well-characterized enzymes and to mixtures of enzymes produced by organisms that degrade cellulose. These mixtures are sometimes called cellulase complexes or systems. Cellulases are formed by many bacteria, fungi, higher plants, and some invertebrate animals. The physiological roles of these enzymes include the extracellular formation of soluble oligosaccharides and glucose for use as a carbon source by microorganisms and for modifying cell walls during growth and morphogenesis in higher plants. The available commercial cellulases have generally been obtained from *Aspergillus niger, Trichoderma viride,* and *Trichoderma koningii.*

Microbial cellulases are of tremendous importance in the degradation of cellulose waste products. The potential commercial uses of cellulase to convert waste paper and sawdust to food, to tenderize plant materials, to disrupt microbial cell walls for release of protein, and to elucidate the structure of cellulosic materials should not be minimized.

The enzymes that act upon cellulose and its derived products can be divided into three groups. C1, a factor whose action has not been clarified, is required for the breakdown of highly crystalline cellulose. The second group is the β-glucanases, which are of two types: exo-β-1,4- and endo-β-1,4-glucanase. The exoenzymes resemble glucoamylase in that they remove successive glucose units, effecting inversion of configuration at the anomeric carbon. The endoenzymes randomly split cellulose chains into glucose. This activity is also referred to as Cx-activity. The third group includes β-glucosidases, which show the highest affinity toward small-molecular-weight substrates. They are responsible for the cleavage of oligosaccharides ranging from cellobiose to cellohexose.

Pectinases, another type of polysaccharase, hydrolyze pectic substances, which are constituents of plant cell walls and, in the middle lamella, serve as cementing materials to hold cells together. A linear chain of anhydro-D-galacturonic acid units connected by 1,4,-α-linkages is the basic structure of pectic substances. Approximately 75% of the carboxyl groups of pectin are esterified with methanol. Pectinesterase removes methoxyl groups from methylated pectic substances. This enzyme has often been referred to as pectase, pectin methoxylase, pectin demethoxylase, pectolipase, and pectin methylesterase. Polygalacturonase splits the glycosidic bonds between galacturonic acid molecules.

$$\text{Pectin} \xrightarrow{\text{pectinesterase}} \text{Methanol} + \text{Polygalacluronic acid}$$

$$\text{Polygalacturonic acid} \xrightarrow{\text{polygalacturonase}} \text{Mono- and Oligogalacturonic acids}$$

Pectic enzymes are of interest to food scientists because of their role in ripening fruits, in maintenance of viscosity in processing fruit products, and in textural changes. They are used to facilitate filtration and clarification and to increase yields of fruit juices and beverages, and in the production of low-methoxyl pectins, which are used for low-calorie gels. Pectinases may cause excessive softening of many fruits and vegetables and "cloud" separation in such products as tomato and citrus juices.

Proteolytic Enzymes

All enzymes that catalyze hydrolysis of proteins to peptones, polypeptides, and amino acids are called proteolytic enzymes. These enzymes hydrolytically cleave the peptide linkage with the formation of a free amino and carboxylic acid group.

Animal proteinases include such enzymes as pepsin, rennet, trypsin and chromotrypsin, and cathepsin. Plant proteinases include such enzymes as papain, chymopapain, ficin, and bromelin. All of these enzymes are endopeptidases; that is, they hydrolyze internal peptide linkages along the protein chain, do not usually attack terminal units, and produce peptides. In contrast, exopeptidases hydrolyze the peptide linkages that join terminal amino acid residues to the main chain, thus producing amino acids. Aminopeptidases act upon the ends of protein chains having terminal amino groups, and carboxypeptidases act on the ends having a terminal carboxyl group. A greatly simplified picture of the action of a protease may be illustrated as follows:

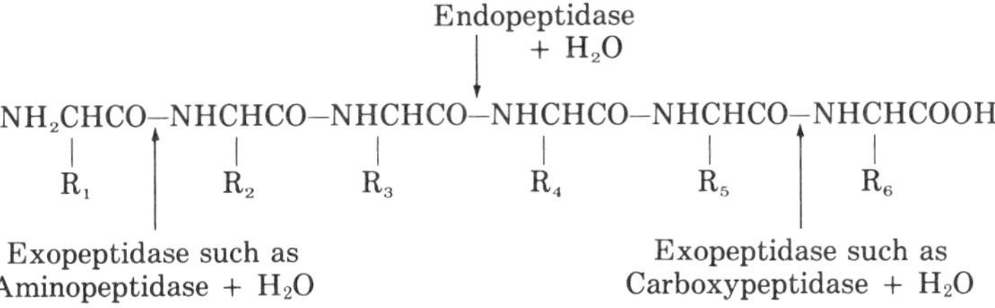

Peptidases act on short-chain oligopeptides. Some are quite specific and will only hydrolyze peptide bonds involving specific amino acids; examples of such specific enzymes are leucyl-peptidase and iminopeptidase.

OXIDATION–REDUCTION ENZYMES

Many substances that are stable in the presence of molecular oxygen rapidly undergo aerobic oxidation in living cells. It appears that every living cell contains enzymes that catalyze oxidation of various substrates by activating oxidizing substances. In the course of an oxidation reaction, the oxidizing agent becomes reduced. Thus, the same enzyme system is involved with both oxidation and reduction reactions. For example, potato tissue contains an enzyme that oxidizes certain aldehydes in the presence of nitrates; in this reaction nitrate is reduced to nitrite. Enzymes that catalyze such oxidation–reduction reactions are classified as *oxidoreductases*.

The experimental investigations of certain oxidation systems has been greatly helped by the use of methylene blue (MB) as an oxidizing agent. Upon reduction this dye is converted into a colorless leuco compound (MBH_2). Methylene blue also is rapidly reduced by certain living cells under aerobic conditions. It is supposed that such cells contain oxidizable substances (AH_2) with labile hydrogen atoms, which are activated by *dehydrase* enzymes. In donating hydrogen to a hydrogen acceptor (methylene blue), these substances are themselves oxidized.

$$AH_2 + MB \xrightarrow{\text{dehydrase}} A + MBH_2$$

Generally, oxidizable substances can be removed from cells and tissues by washing with water. Then, by adding back potential hydrogen donors to the washed cells or tissues, the dehydrases present may be determined. For example, the seeds of species of mallow, orange, and plum decolorize methylene blue in the presence of oxalates. It is believed that these seeds contain an oxalic dehydrase. Likewise, the presence of malic, formic, and succinic dehydrases in the seeds of the runner bean has also been demonstrated. In living cells, two common hydrogen acceptors in dehydrase reactions are the oxidized forms of nicotinamide-adenine dinucleotide (NAD) and nicotinamide-adenine dinucleotide phosphate (NADP).

Molecular oxygen (O_2) acts as a hydrogen acceptor in reactions catalyzed by *oxidases* in which hydrogen peroxide (H_2O_2) is formed. Hydrogen peroxide is toxic to most cells and is rapidly broken down by the enzyme *catalase,* which is invariably present in the cells of aerobic organisms. Glucose oxidase and catalase have numerous industrial applications. For example, they are used to remove glucose from dried eggs, meats, and potatoes, thus preventing "browning," preserving the color and flavor, and extending the shelf life of such products. The reactions involved are as follows:

$$C_6H_{12}O_6 + O_2 + HOH \xrightarrow{\text{glucose oxidase}} \text{Gluconic acid} + H_2O_2$$

$$2H_2O_2 \xrightarrow{\text{catalase}} 2H_2O + O_2$$

Other applications of glucose oxidase include the following:

1. Formation of hydrogen peroxide
   - To form "nascent" oxygen in the presence of catalase
   - In conjunction with a secondary system for peroxide (e.g., peroxidase and chromogen) as a test for glucose
   - In the treatment of flour, by forming peroxide with catalase-free glucose oxidase
   - In biological chlorination
2. Formation of gluconic acid and salts thereof as a means obtaining higher purity salts
3. Removal of glucose
   - In the analysis of mixed sugars by determining the reducing power before and after removing glucose
   - To prevent Maillard browning in dried eggs, meats, and potatoes
4. Removal of oxygen
   - To stabilize against the deteriorative effect of light in citrus-juice-based soft drinks
   - To costabilize ascorbic acid and cyanocobalamin (B_{12}) in aqueous vitamin preparations
   - To remove occluded air from oil in water emulsions to prevent the development of rancidity
   - To prevent oxidation of beer
   - To prevent wine from going to vinegar
   - To prevent enzymatic browning of fresh frozen fruit, including cherries and peaches
   - To prevent corrosion of cans, such as soft drinks, and maintain color

Some cells that cannot directly stimulate the aerobic oxidation of catecholic compounds such as gum guaiacum contain the enzyme peroxidase. For example, neither horseradish root alone nor hydrogen peroxide alone can oxidize guaiacum, but a blue color develops if both guaiacum and hydrogen peroxide are applied to the cut surface of horseradish root. Catechol oxygenase is considered to be the main enzyme in the system that causes the color changes. Several theories have been given on this reaction. This enzyme is a dehydrase that activates cellular substances containing the catechol group, which then is converted by dehydrogenation into an orthoquinone. It is supposed that molecular oxygen is also activated and that it then functions as a hydrogen acceptor and is reduced.

Orthoquinone is a very strong oxidizing agent and can itself effect all the color changes by which direct oxidases are characterized. It may be that peroxidase immediately acts on the hydrogen peroxide and so contributes to the oxidation powers of direct oxidases. Accordingly the color changes would be caused by mixtures of a catechol compound, oxygenase, molecular oxygen, and peroxidase.

Another type of oxidoreductase has been demonstrated in yeasts and higher plants. This enzyme, which converts methyl glyoxal into lactic acid, is called methly-glyoxatase.

$$CH_3COCHO + H_2O \xrightarrow{\text{methyl-glyoxatase}} CH_3CHOHCOOH$$

CATALYTIC PROPERTIES AND MECHANISM OF ACTION OF ENZYMES

Enzymes that have been separated from living cells appear to fulfill the principal requirements of a catalyst. Without themselves serving as the sources of the end products, they can accelerate reactions that occur spontaneously but slowly or promote reactions that could not occur in their absence. An enzyme may enter into chemical combination during the course of the reaction, but the end products are entirely derived from the substrate.

For reactions of short duration, at moderate temperatures and constant and favorable pH, with a limited amount of substrate rather than the enzyme, the concentration of enzyme will be the same at the end as at the beginning of the reaction. The initial catalytic powers of the enzyme system remain unimpaired. However, enzymes are readily inactivated under unfavorable conditions such as sometimes arise during the course of a reaction. For example, trypsin is only active in alkaline solution and is gradually inactivated by the hydrogen ions set free when this enzyme cleaves proteins into amino acids in unbuffered solutions. This particular inactivation is reversible, and lytic activity can be restored by making the solution alkaline.

The velocity of an enzymatic reaction, like that of any other catalytic reaction, is proportional to the concentrations of the substrate and enzyme. The diverse and often complex kinetics of enzymatic reactions depend on these concentrations, temperature, pH, and other factors.

With few exceptions, enzyme preparations do not pass through dialysis membranes. Estimation of their molecular weights by physical methods have given figures ranging from 20,000 to 300,000. They are regarded as colloidal catalysts, which develop extensive active surfaces when dispersed in a reaction medium. It is supposed that enzymatic reactions take place on these surfaces. However, certain enzymes (e.g., lipase) can actually bring about chemical changes when dispersed in liquids in which they are quite insoluble. Such enzymic reactions resemble surface catalysis induced by certain metals such as colloidal platinum. The degree of dispersion of enzymes appears to exercise an important influence on surface activity.

Chemical forces also play an important part in the action of enzymes. The whole surface of enzyme particles is probably chemically active. These chemically active regions possess affinities for and the power to effect changes in the appropriate substrate or substrates. It is thought that during the course of a reaction the enzyme combines chemically with and activates the substrate, after which it decomposes, yielding the products. Since the products of a reaction generally have little affinity for the enzyme, they diffuse from the enzyme surface, thus making room for further activation of substrate.

Certain substances that inactivate enzymes may be absorbed by enzymes. Others may combine chemically at the active centers, as happens when oxidation enzymes containing iron are inactivated by hydrogen cyanide or hydrogen sulfide.

Carboxylase—An Example of Enzyme Action

The mechanism of action of carboxylase illustrates certain aspects of how enzymes work. Carboxylase catalyzes the reversible decarboxylation of acetaldehyde to form pyruvic acid.

$$CH_3COCOOH \underset{}{\overset{\text{carboxylase}}{\rightleftharpoons}} CH_3CHO + CO_2$$

Many yeasts and a few bacteria produce ethyl alcohol (C_2H_5OH) when grown on glucose. After the bacteria have grown, if pyruvic acid is added to the medium, large amounts of acetaldehyde will be formed. Or if the bacteria are dried at low temperature under conditions that kill the cells and the cells are ground into a fine powder, the powder will produce acetaldehyde from pyruvic acid. Although the cells are dead, the enzyme involved in converting pyruvate to acetaldehyde is still active.

Yeast cells behave very much in the same way as bacteria in converting pyruvate to acetaldehyde. The enzyme carboxylase can be obtained from yeasts by grinding, filtration, and precipitation as acetone powder. In this method, yeasts are ground to destroy the cell wall and membrane, filtered, and then precipitated as acetone powder. The powdered yeasts are suspended in 1% KCl and centrifuged. The liquid supernatant contains the enzyme and will catalyze the reaction of pyruvate to acetaldehyde.

If carboxylase or many other enzymes are dialyzed all biological activity is apparently lost. But, with the addition of concentrated dialysate, activity is restored. Thiamine pyrophosphate and magnesium have been identified in the dialysate of carboxylase. The dialyzed protein, thiamine pyrophosphate, or magnesium acting separately seems to have no action on pyruvate, but when all three are added to the substrate, the reaction proceeds. Oginsky and Umbreit (1959) state that both magnesium and thiamine pyrophosphate are usually associated with carboxylase coenzyme activity.

ENZYMES IN FROZEN FOODS

Enzymes continue to function very slowly in foods held at low temperature. If the quality of the frozen food is to be maintained indefinitely, the enzymes must be inactivated. Otherwise, they will cause off-flavors and discoloration. Even at $0°F$ $(-17.8°C)$ enzymes are not inactivated, but only slowed down. After a week or month of frozen storage, a food may show abnormal odors and taste.

Many fruits become dark after the skin is removed because of oxidizing enzymes. The enzymes are successfully inactivated by excluding atmospheric oxygen. Chemical inhibitors such as sulfurous acid and the addition of sugar or syrup also can suppress the activity of these enzymes.

It has been well established that active enzyme systems can spoil fruits and vegetables even at subzero temperatures and low moisture levels. Consequently most vegetables and some fruits are given a blanching treatment in boiling water or steam to inactivate these enzyme systems. Improvement of the keeping quality of the product during cold storage results. The following are accomplished during the blanching process: (1) destruction of large numbers of microorganisms on the surface of the food, which decreases the microflora in the product during processing; (2) inactivation of many enzymes that would otherwise contribute to the deterioration of the product during storage; (3) removal of many mucilage-like substances, which contribute to off-flavors; (4) "color set," which prolongs the life of the natural color of the food; and (5) removal of an appreciable amount of extraneous substances, which would otherwise contribute to many undesirable flavors in frozen foods.

PHOSPHATASE TEST

As noted earlier, the phosphatase test is a reliable method for measuring the efficiency of pasteurization of milk and cream. The test is based on the presence in milk of an enzyme (phosphomonoesterase or phosphatase) that catalyzes the hydrolysis of monoesters of phosphoric acid. The enzyme is totally destroyed by pasteurization at 62°C (143°F) for 30 min or at 71°C (160°F) for 15 sec. To determine if enzyme destruction has occurred (i.e., whether the milk has been subjected to proper pasteurization), a phenyl phosphoric ester is added to the milk. If the enzyme is present, the ester will be hydrolyzed and free phenol will be released. The indicator 2,6-dibromoquinone-chloroimide is then added to the milk. This compound reacts with phenol to give a colored product, an indophenol. The intensity of the resulting color is an index of the amount of phenol released and, therefore, of the activity of the enzyme and the degree of pasteurizing efficiency. The heat required to inactivate phosphatase is slightly higher than that required to destroy *Mycobacterium tuberculosis*. Thus, a negative phosphatase test is taken to indicate that any pathogenic bacteria present also have been destroyed in the pasteurization process.

The chemical equations involved in the phosphatase test follow.

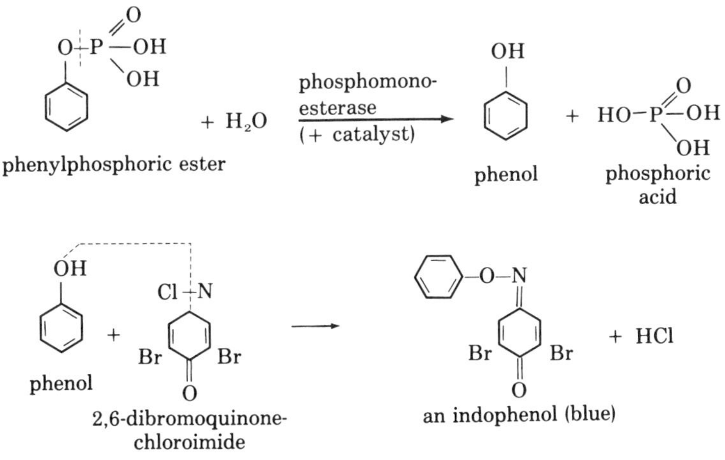

It should be noted that the phenylphosphoric ester undoubtedly exists as a salt of the acid, but for simplicity, the reaction is written with the ester in its acid form.

The appearance of a blue color in the finished test is an indication of improper pasteurization. Commercially, however, a certain amount of leeway is permitted. Standards may be obtained indicating the limits of this tolerance. For more accurate work, butyl alcohol (high-grade, neutralized) may be used to dissolve the colored compound and separate it into a distinct layer, which eliminates any error caused by opaqueness of the milk.

Although the phosphatase test is used by regulatory and quality control laboratories to detect the efficiency of pasteurization, several conditions may alter the accuracy of the test. For example, reactivation of phosphatase may occur after several days. Milk and cream heated to high temperatures for short periods give negative phosphatase results. When the same samples are held for 3 or 4 days at 40°–50°F (4°–10°C), a positive reaction sometimes is observed.

One explanation for this phenomenon may be the presence of bacterial phosphatase, which seems to be more heat stable than milk phosphatase. A temperature of 170°F (77°C) for 30 min is required to inactivate bacterial phosphatase, whereas milk phosphatase is inactivated at 143°F (62°C) for 30 min. In one study, pasteurized cream gave a negative phosphatase test, then upon standing at 40°–50°F (4°–10°C) it again gave a positive reaction. The positive test was caused primarily by *Bacillus cereus* and *B. mesentericus* enzymes. Since both organisms are sporeformers, they were able to survive the pasteurization temperature. Thus, bacterial phosphatase production may partially explain the reactivation of phosphatase in the pasteurization of selected food products.

SELECTED REFERENCES

Anon. 1939. Enzymes in Food and Food Preservation; Food and Life. Yearbook of Agriculture. U.S. Dept. of Agriculture, Washington, DC.

Barron, E.X. 1943. Mechanism of carbohydrate metabolism. An essay on comparative biochemistry. *Adv. Enzymol.* **3,** 149–189.

Bartolome, G. L., and Hoff, J. E. 1976. Firming of potatoes: Biochemical effects of preheating. *J. Agric. Food Chem.* **20,** 266.

Bernhard, S. 1968. "The Structure and Function of Enzymes." W. A. Benjamin, New York.

Denault, L. J., and Underkofler, L. A. 1963. Conversion of starch by microbial enzymes for production of syrups and sugars. *Cereal Chem.* **40,** 618–629.

Green, L. A., and Stumpf, S. 1944. Biological oxidations and reductions. *Annu. Rev. Biochem.* **13**, 1–24.

IUB. 1978. "Enzyme Nomenclature," Recommendations of the Nomenclature Committee of the International Union of Biochemistry. Academic Press, New York.

Miller, B. S., and Johnson, J. A. 1955. Fungal enzymes in baking. *Bakers' Dig.* **29**, 95–100.

Oginsky, E. L., and Umbreit, W. W. 1959. "An Introduction to Bacterial Physiology." W. H. Freeman Co., San Francisco, California.

Reed, G. 1975. "Enzymes in Food Processing," 2nd ed. Academic Press, New York.

Suzuki, S. 1964. An overall look at the dextrose industry in Japan. *Starke* **16**, 285–293.

Underkofler, L. A. 1966. Manufacture and uses of industrial microbial enzymes. *Chem. Eng. Prog. Symp. Ser.* **62**, 11- 20.

Underkofler, L. A., Barton, R. R., and Rennert, S. S. 1958. Production of microbial enzymes and their applications. *Appl. Microbiol.* **6**, 212–221.

Underkofler, L. A., Denault, L. J., and Hou, E. F. 1965. Enzymes in the starch industry. *Starke* **17**, 179–184.

Whitaker, D. R. 1971. *In* "The Enzymes," 3rd. ed., Vol. V, pp. 273–290. P. Boyer (Editor). Academic Press, New York.

4
Types of Microorganisms

Bacteria   49
  Pseudomonadaceae   50
  Achromobacteriaceae   50
  Brevibacteraceae   51
  Bacillaceae   51
  Corynebacteriaceae   51
  Enterobacteriaceae   52
  Propionibacteriaceae   54
  Lactobacillaceae   54
  Streptococcae   55
Molds   56
Yeasts   60
Algae   61
Viruses and Bacteriophages   63
Parasitic Organisms   63
Selected References   66

Microorganisms include bacteria, molds, yeasts, algae, viruses, parasitic worms, and protozoa. These organisms differ in size and shape and in their biochemical and cultural characteristics.

The smallest common bacteria are 0.15 μm wide and 0.3 μm long. A large bacillus may range from 0.8 to 2.5 μm in diameter and from 8 to 10 μm in length. Most of the important bacteria in foods are 0.5–2.5 μm in diameter and 2–10 μm in length.

Yeast cells are much larger than bacterial cells, usually ranging from 2.0 to 7 μm in diameter. Molds are multicellular, and their components vary with the type of mold. Viruses are the smallest units of living matter and have the simplest structures.

BACTERIA

Various morphological types of bacteria are shown in Fig. 4.1. The class Schizomycetes includes most of the bacteria that are associated with foods. Eubacteriales and Pseudomonales are the two principal orders. The bacterial families of most concern in foods are described in this section.

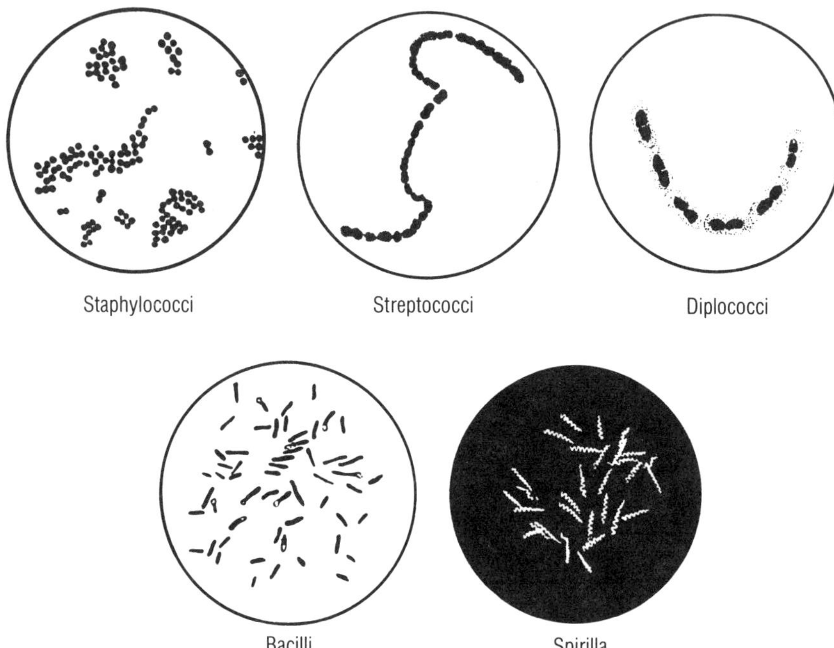

Staphylococci   Streptococci   Diplococci

Bacilli   Spirilla

Fig. 4.1. Morphology of bacteria. Top row: cocci in clusters, chains, pairs (encapsulated); bottom row: bacilli (rods) and spirals.

Pseudomonadaceae

The principal genus in this group is *Pseudomonas*. These bacteria are gram-negative and usually motile. Organisms in this group are responsible for many kinds of food spoilage. Their action on carbohydrates is minor; however, in foods containing proteins and fats they demonstrate considerable proteolytic and lipolytic activity. These organisms grow rapidly at refrigerator temperatures and in many cases produce slime on the surface of meats. *Pseudomonas fluorescens* produces a greenish pigment; some species, such as *P. nigrificans*, form a black pigment on protein foods. In many cases, the biological activity of *Pseudomonas* spp. may be identified by their proteolytic and lipolytic activity and pigment production.

Achromobacteriaceae

These bacteria are gram-negative rods, aerobic to facultative and generally motile. Carbohydrates are not usually used. There are three genera that are important in foods because of their ability to grow at low temperatures. *Alcaligenes viscolactis* may cause ropiness in milk. *Achromobacter* causes spoilage of meat products by producing slime and off-odors. *Flavobacterium* produces an orange pigment on meats and dairy products. All of these organisms are widely distributed in nature.

Brevibacteraceae

Brevibacterium lineus and *B. erythrogenes* produce orange and red pigments in certain cheeses and may also have a role in the ripening processes. Their carbohydrate and oxygen requirements are variable, and their morphology ranges from gram-positive short rods to cocci. Variation also exists in motility characteristics. All members in this family are asporogenous.

Bacillaceae

Bacillus subtilis, B. polymyxa, and *B. stearothermophilus* are heat-resistant sporeformers. *Bacillus coagulans* is used to produce commercial lactic acid. *Bacillus stearothermophilus* may cause flat-sour spoilage in canned vegetables. *Clostridium* spp. are typical anaerobic bacteria. *Clostridium thermosaccharolyticum* produces a gaseous spoilage in canned vegetables. *Clostridium sporogenes* produces a stormy fermentation of milk. *Clostridium butyricum* can be a troublesome organism in foods, especially in large blocks of cheese where the center area of the cheese offers ideal anaerobic conditions. A variety of smelly gases are produced, and a favorable environment for other spoilage organisms may be created. *Clostridium botulinum* produces toxins in food which cause botulism food poisoning. The Bacillaceae are widely distributed in nature. *Clostridium perfringens* (Fig. 4.2) are typical examples.

Corynebacteriaceae

These organisms are of special interest from the public health standpoint. *Corynebacterium diphtheriae* is the causative agent of diphtheria in man. *Corynebacterium pyogenes* may cause mastitis in dairy cattle. The characteristic feature of this group is the granules in the rod, which can be easily observed by appropriate staining techniques. These organisms have no special significance in spoilage of foods.

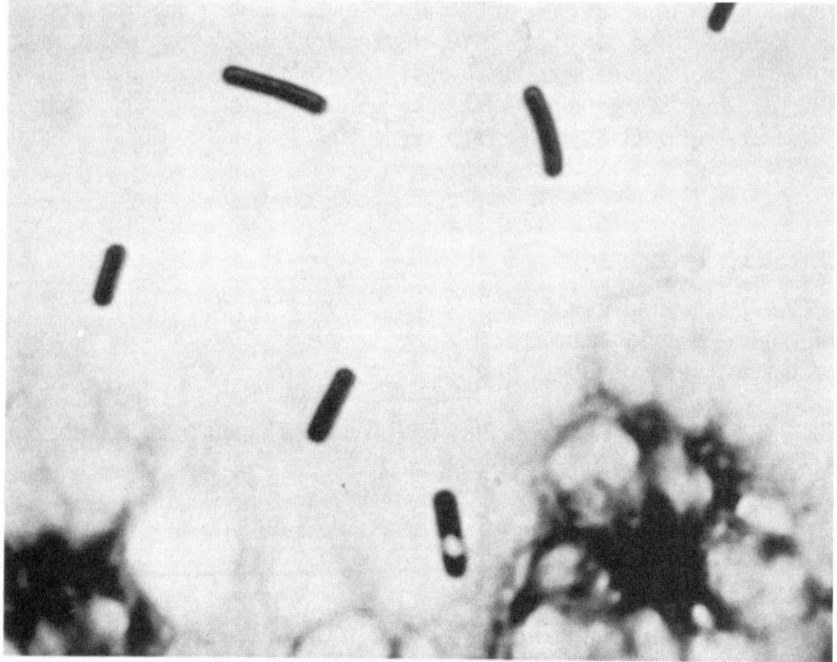

Fig. 4.2. *Clostridium perfringens* showing eccentric spore. (5000×)

Enterobacteriaceae

These organisms are gram-negative non-sporeforming rods. The genera *Aerobacter, Erwinia, Escherichia,* and *Serratia* may bring about biological changes in food spoilage, while *Proteus, Salmonella,* and *Shigella* are significant in public health. *Salmonella* especially is involved in food poisoning and will be discussed more completely in Chapter 21.

Aerobacter aerogenes and *Escherichia coli* are often called the coliform group. The two species are distinguished because of their source. *Aerobacter aerogenes* in usually of plant origin, while *Escherichia coli* is of intestinal origin in animals. They can use carbohydrates, with production of gases and other metabolic by-products that are undesirable. Coliforms are indicators of sewage pollution. When present in oysters, coliforms may indicate the presence of other intestinal pathogens that may be of major importance in foods.

Erwinia may cause wilts, soft rots, or necrosis in vegetables and fruits. *Erwinia carotovora* often causes soft rot in carrots.

Serratia is noted for its red pigment production on the surface of many foods, especially if carbohydrates are present. For example, when it occurs on bread, it is referred to as blood bread.

Proteus are gram-negative, motile rods. Morphologically they are difficult to distinguish from other members of the intestinal group. *Proteus vulgaris* is the type species. It produces acid and gas from glucose but not lactose. This species has been implicated in spoilage of eggs and meats and also suspected in food poisoning. Ammonia is readily formed from the breakdown of urea. Putrefactive odors have been associated with the breakdown of proteins.

At present there are over 1200 *Salmonella* species, generally called serotypes. Organisms in this genus include *S. enteritidis, S. paratyphi, S. schottmuelleri, S. typhimurium,* and *S. typhi,* the cause of typhoid fever. All have been involved in foodborne infections. *Salmonella pullorum,* the cause of pullorum in chickens, has also been isolated in several foodborne outbreaks involving humans.

Salmonella are morphologically similar to *Proteus* (Fig. 4.3). All are

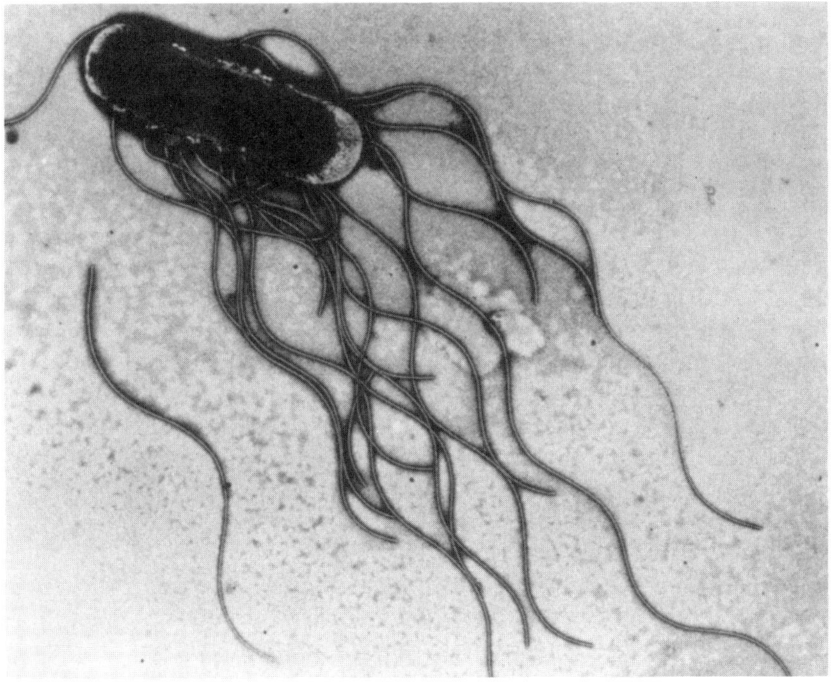

Fig. 4.3. Electron photomicrograph of *Salmonella,* a gram-negative rod that can contaminate food and cause disease and even death. Courtesy USDA.

gram-negative, nonsporulating organisms. Serological techniques are more useful than cultural features in identifying the different species. They can produce acid from glucose but not lactose. Their proteolytic properties are negligible.

The microscopic features of *Shigella* are similar to *Proteus* and *Salmonella*, except they are nonmotile and produce no gas from glucose. One species, *S. dysenteriae*, is the type species. They may be associated with foodborne infections.

Propionibacteriaceae

Propionibacterium shermanii is the best example in this group and is typical of other species. They are gram-positive, asporogenous, anaerobic to microaerophilic rods, sometimes resembling cocci in short chains. Their metabolic activity is interesting because they can produce acetic and propionic acids and carbon dioxide from lactic acid. These organisms play an important role in Swiss cheese ripening because they ferment lactates to produce gas, which contributes to the characteristic eyes in this variety of cheese. They also contribute to the fine delicate flavor of Swiss cheese. No doubt they play an important role in other desirable food fermentations and in the production of desirable flavors and odors in foods.

Lactobacillaceae

Members of this family are important because of their ability to produce lactic acid as a desirable fermentation. Some are homofermentative, producing principally lactic acid, but some acetic acid and carbon dioxide may be produced. The heterofermentative members of this family not only produce lactic acid but also several volatile compounds and small amounts of alcohol. *Lactobacillus acidophilus, L. bulgaricus, L. casei, L. pentoaceticus, L. brevis,* and *L. thermophilus,* are all gram-positive rods, nonmotile, and microaerophilic. They require for the most part complex foods for their energy requirements.

Lactobacillus leichmannii is used for microbiological assays of vitamins and antibiotics in foods.

Since most of the members of this group are microaerophilic, they grow poorly on the surface of culture media but produce colonies submerged beneath the surface. Lactobacilli predominate in nearly all types of acid fermentation in foods.

Morphologically, these organisms are quite pleomorphic, even within the species. Some species produce metachromatic granules comparable to those formed in *Corynebacterium diphtheriae*.

Streptococcae

Some of the streptococci are pathogenic to man and animals. *Streptococcus pyogenes* may cause infections in man, and *S. agalactiae* is largely responsible for mastitis in dairy cattle. (Fig. 4.4) Fortunately, it has a low virulence in man.

Streptococcus lactis and *S. cremoris* are important starter organisms in fermented dairy products. They are largely responsible for the ordinary souring of milk. They are gram-positive, ferment lactose, and produce D-lactic acid. They are easily cultivated on whey agar. The temperature for growth may range from 50°F (10°C) to 113°F (45°C).

Streptococcus faecalis has been identified in the spoilage of canned hams and sometimes in bacon. The temperature range for growth may vary from 50°F (10°C) to 122°F (50°C). They can survive pasteurization of milk, tolerate 7% salt, and grow at pH 10.0.

Streptococcus liquefaciens produces a rennin-like enzyme, which slowly coagulates milk; it is often called "sweet curdling." Later an

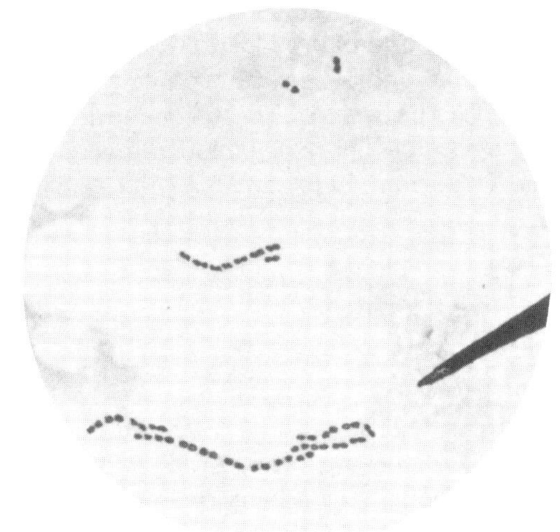

Fig. 4.4. The genus *Streptococcus* includes some species that are pathogenic to man. Shown here are the typical chains formed in liquid media. From H. V. Atherton and W. A. Dodge, University of Vermont.

actual proteolysis takes place. The acid production generally equals that produced by *S. lactis*. The organism produces D-lactic acid and acetyl-methyl-carbinol. Except for proteolysis of milk, the cultural features of *S. liquefaciens* are comparable to those of *S. lactis*.

Streptococcus faecalis and *S. liquefaciens* often are present in the intestinal tract of man and animals. Therefore, their presence in foods and on utensils may be significant from a sanitary viewpoint. However, *S. liquefaciens*, due to its rennin-like enzyme, may play an important role in the ripening of certain cheeses. The ability of this group of organisms to grow over a wide temperature range makes them all the more significant in foods.

Leuconostoc mesenteroides is associated especially with sauerkraut and pickle fermentations. This organism initiates the desirable lactic acid fermentation in these products. It differs from other lactic acid species in its ability to tolerate fairly high concentrations of salt and sugar—up to 50% sugar. These bacteria produce an undesirable slimy or gummy fermentation in syrups and ice-cream mixes. A desirable feature is the ability of *L. mesenteroides* to produce dextran, a product used in blood plasma.

Leuconostoc citrovorum (*Streptococcus citrovorus*) and *L. dextranicum* (*S. paracitrovorus*) are important organisms in dairy starter cultures. They are able to ferment citric acid with the production of diacetyl, the compound that contributes to the fine delicate flavor of butter. Diacetyl is also found in bread, coffee, and other fermented food products in varying amounts. The morphological features of these species are comparable to those of *Streptococcus lactis*. They are gram-positive and asporogenous. These organisms are considered nonpathogenic for humans. This group of organisms will be discussed more fully in Chapter 15.

MOLDS

Molds, like bacteria and yeasts, play an important role in foods. Some molds are desirable because they produce products that enhance the flavor of certain foods. For example, Roquefort and Camembert cheeses owe their characteristic flavor to the growth of molds in the cheese. Molds are useful in producing certain enzymes, such as amylase for bread making or the production of commercial citric acid. Some molds are undesirable because they contribute to certain kinds of spoilage in foods. Other molds produce toxins and thus may be significant in public health (see Chapter 22).

The order Mucorales has spores borne in sporangia. A summary of a few molds commonly found in foods is sufficient to emphasize their importance in foods and food products.

Aspergillus glaucus and *A. repens* are often responsible for undesirable changes in foods. These species grow well in high concentrations of salt and sugar. *Aspergillus niger* is frequently found in foods. *Aspergillus ochraceous* is found on growing tomatoes (Fig. 4.5). The fruiting bodies are black with a variation in colors. Some strains are used to produce citric, gluconic, and oxalic acids. Figure 4.6 illustrates two of the many mold forms.

The genus *Penicillium* is characterized by branching of the conidophore as shown in Fig. 4.7. *Penicillium camemberti* and *P. roqueforti*

Fig. 4.5. *Aspergillus ochraceous* showing mycelia and conidial heads. Courtesy of Continential Can Co.

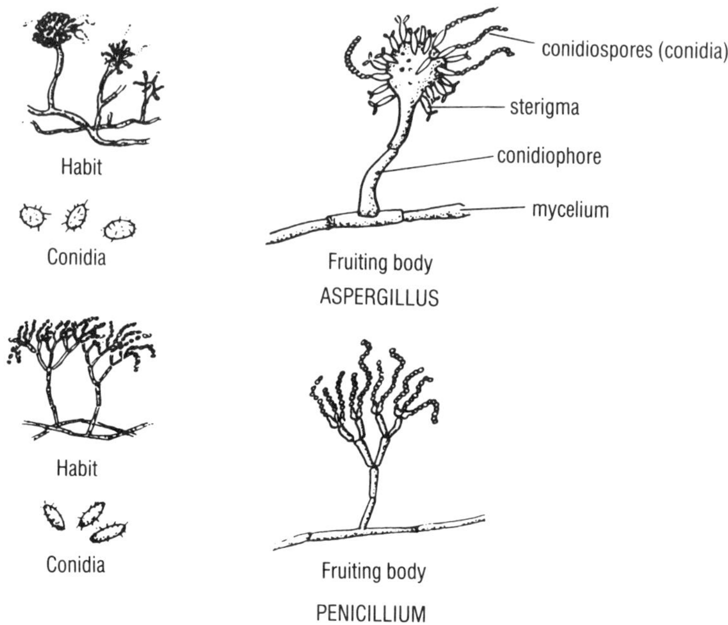

Fig. 4.6. Common mold forms.

with blue-green conidia are used in ripening Camembert and Roquefort cheeses.

Geotrichum candidum (*Oospora lactis*) or "dairy mold" gives a white cottony growth on dairy products. It also appears on tomatoes.

The genus *Neurospora* produces asexual spores. It may be characterized by a long mycelium. The aerial hyphae form branched chains and pigmented budding conidia. *Neurospora sitophila* produces a pink conidia and is sometimes called "red bread mold," or blood bread.

White spots on refrigerated meat may be caused by *Sporotrichum carnis.* The conidia are oval shape and may branch out from all parts of the mycelium.

Alternaria tenuis (Fig. 4.8) forms a mass of mycelium with dark brown conidia. It may be associated with spoilage of citrus fruits and tomatoes.

The morphological features of molds can be examined satisfactorily by the slide culture technique (Fig. 4.9). This method permits examination with all of the parts intact and thus makes identification of the mold comparatively easy. Consult the many reference books on fungi for complete identification and environmental factors for growth.

Fig. 4.7. *Penicillium* showing mycelia and brushlike conidial heads. Courtesy of Continental Can Co.

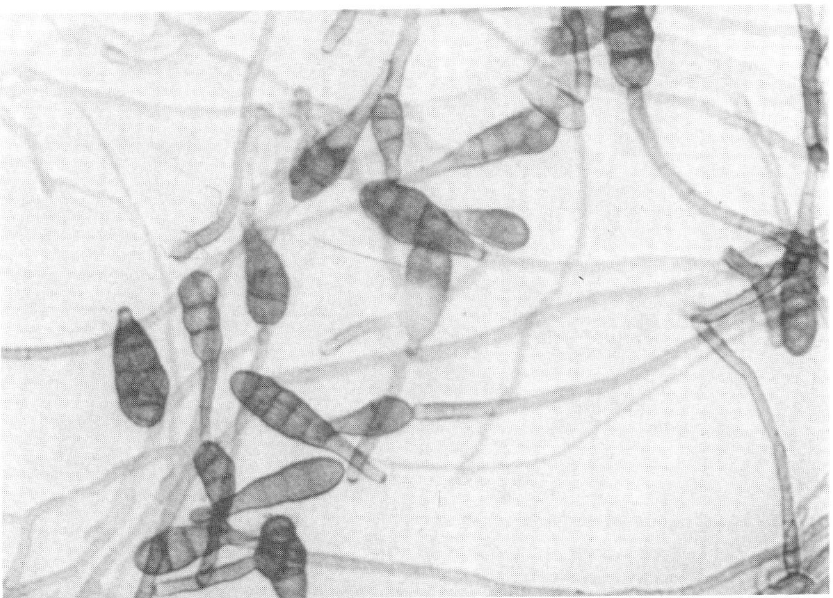

Fig. 4.8 *Alternaria,* a cause of rot in tomatoes. Spores show typical Indian-club shape; filaments are septate and of medium thickness. Courtesy of Continental Can Co.

Fig. 4.9.   Slide culture cell technique.

YEASTS

Yeasts and yeastlike fungi, which may be desirable or undesirable in foods, are widely distributed in nature. They are found in orchards and vineyards, in the soil, air, intestinal tract of animals, and in certain insects. Physiologically yeasts have no chlorophyll and are dependent on plants and animals for their energy, i.e., they are saprophytic or parasitic.

Yeasts are unicellular organisms, although they may not appear to be so. Some are cylindrical or elongated, changing at one end into a filamentous form. Others are ellipsoidal. Yeasts range in size from 2 to

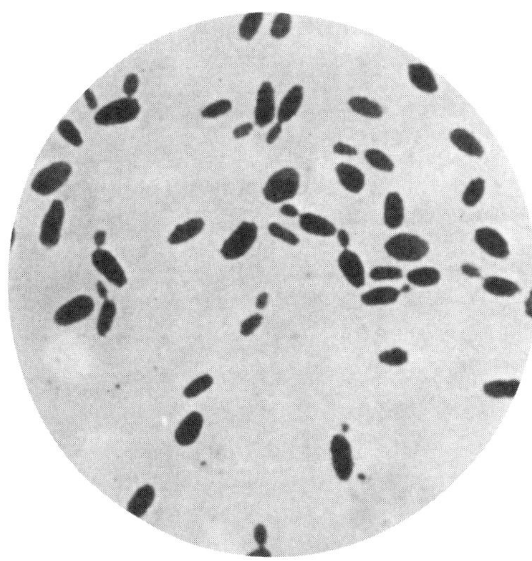

Fig. 4.10.   Yeast cells are much larger than bacteria. Normally spherical or slightly elongated, they frequently show small buds, their manner of reproduction. From H. V. Atherton and W. A. Dodge, University of Vermont.

6 μm wide and from 10 to 30 μm long. As a rule most yeasts are larger than most bacteria.

Yeasts may reproduce asexually by budding. All true yeasts also may reproduce sexually. When two yeast cells meet, a tube is formed to allow passage of nuclei; then they fuse into a single nucleus in sexual reproduction. The fused nuclei divide and ascospores are formed, which are encased in a sac called an ascus. Therefore, all true yeasts are called "sac fungi" and should be included in the class Ascomycetes. Asexual or binary fission is comparable to bacterial reproduction. False yeasts produce no ascospores and belong to the Fungi imperfecti.

Yeasts play an important role in the food industry because they produce enzymes that favor desirable chemical reactions, such as leavening of bread, and the production of alcohol, glycerol, or invert sugar.

Some yeasts are chromogenic and produce a variety of pigments including green, yellow, pink, and black. They are also capable of synthesizing certain essential B vitamins.

The type of end products produced in yeast fermentations can be controlled within certain limits by the oxygen supply. Most yeasts require an abundance of oxygen for growth, but in certain industrial fermentations, an exclusion of oxygen determines the yield and kind of end product. The temperature range for yeast growth varies from 32°F (0°C) to 122°F (50°C), although the optimum range is 68°–86°F (20°–30°C).

False yeasts (class Fungi imperfecti) include several genera that are important in foods. *Cryptococcus utilis* is a food yeast and *C. kefyr* is a lactose-fermenting yeast. *Candida mycoderma* is a film-forming yeast, which grows in acid environments such as beer, pickles, and sauerkraut. They use the acid, making conditions more favorable for other spoilage organisms to survive. *Candida krusei* is used in a starter culture along with *Lactobacillus bulgaricus* in Swiss cheese making, to stimulate a uniform acid production. Since *L. bulgaricus* is microaerophilic, the yeast may create a metabiotic state that is beneficial to *L. bulgaricus*.

Rhodotorula glutinus (Fig. 4.11) produces a pink pigment that is undesirable in sauerkraut. The water-soluble pigment accumulates in the kraut juice, giving it a pink color, and so it is called pink kraut.

ALGAE

The role of algae in food should not be underestimated. In fact, some algae can use atmospheric nitrogen and combine it with oxygen, mak-

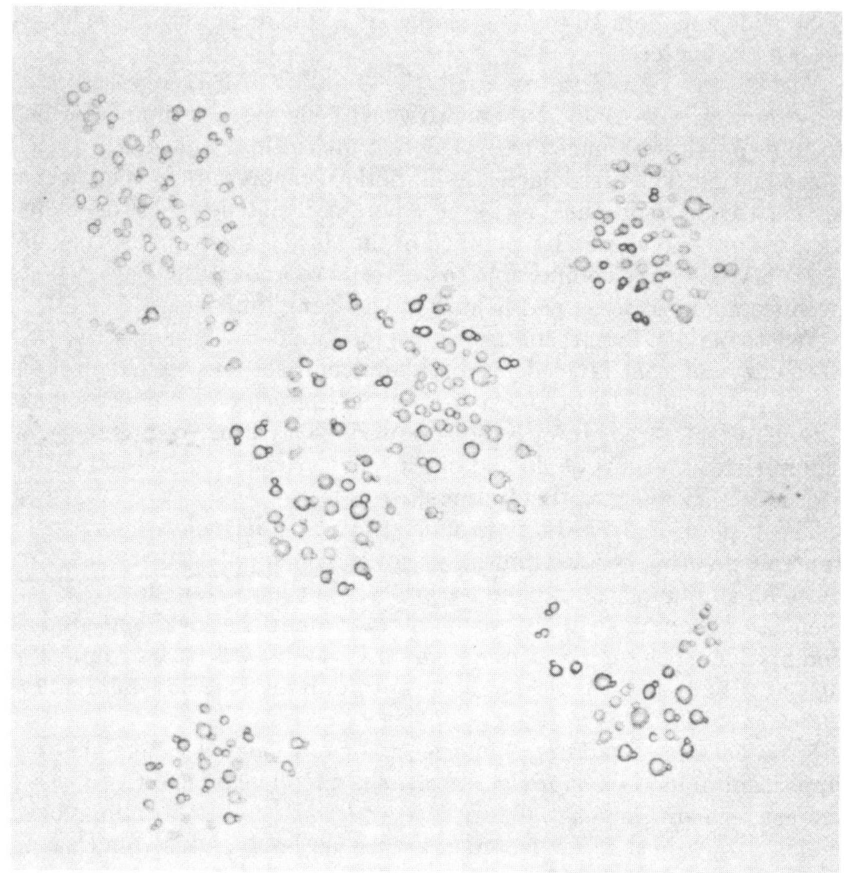

Fig. 4.11. *Rhodotorula* species.

ing it available to plants and animals in much the same way as nitrogen-fixing bacteria. Some algae form a sheath that protects nitrogen-fixing bacteria from dehydration, and at the same time the algae provide carbohydrates as a source of energy.

Some algae produce pigments that may be of interest in foods. One species produces carotene, a precursor of vitamin A, while others synthesize vitamin D. When fish eat the phytoplankton, the vitamins are stored in the liver, making it a rich source of vitamins for man.

Red algae such as Irish moss and brown algae such as kelp are harvested as a source of food. Agar, a common bacteriological culture medium, is also used in fruit juices and beverages as a clarifying agent.

Irish moss is used in chocolate milk to hold the chocolate in suspension. Algae are rich in iodine, bromine, and potassium; therefore certain algae can be used as commercial fertilizers. Farmers in France and Germany have already used algae in this way.

Unfortunately, there are some forms of algae that when eaten by shellfish make the shellfish toxic to animals that eat the shellfish.

Algae may play an important role in food waste stabilizing ponds by supplying the oxygen to oxidize the food wastes in lagoons. However, more information is needed before any definite conclusions can be drawn as to the feasibility of such a process.

VIRUSES AND BACTERIOPHAGES

Viruses are 10–450 nm in size; cannot reproduce without a living host; attack only susceptible host cell lines; infect plants, animals, and bacteria; and have the capacity to produce specific diseases in specific hosts. Transmission occurs in foods, water, and air. Viruses that infect bacteria are called bacteriophages. Viruses are included in the order Virales.

Viruses are too small to be visualized with an ordinary compound microscope. Only after the electron microscope was developed was direct observation of viruses possible. A photomicrograph of a typical virus is shown in Fig. 4.12.

Viruses consist of a DNA or RNA core surrounded by a protein coat. Because they lack all the apparatus for normal cellular metabolism, they must utilize the cellular machinery of the host cell in order to grow and divide. Once they invade a host cell, however, viruses can multiply very rapidly (Fig. 4.13).

Bacteriophages are of interest in foods, especially dairy starter cultures, because they parasitize the bacterial cells and impair their ability to produce acid. This is a serious problem with butter and cheese cultures, e.g., *Streptococcus lactis, S. thermophilus,* and *S. cremoris.*

Table 4.1 is a summary of human intestinal viruses that may occur as contaminants in foods.

PARASITIC ORGANISMS

A number of parasitic worms can also be transmitted by food to cause diseases in humans.

Cestodes are flatworms that inhabit the intestinal tract, heart, and

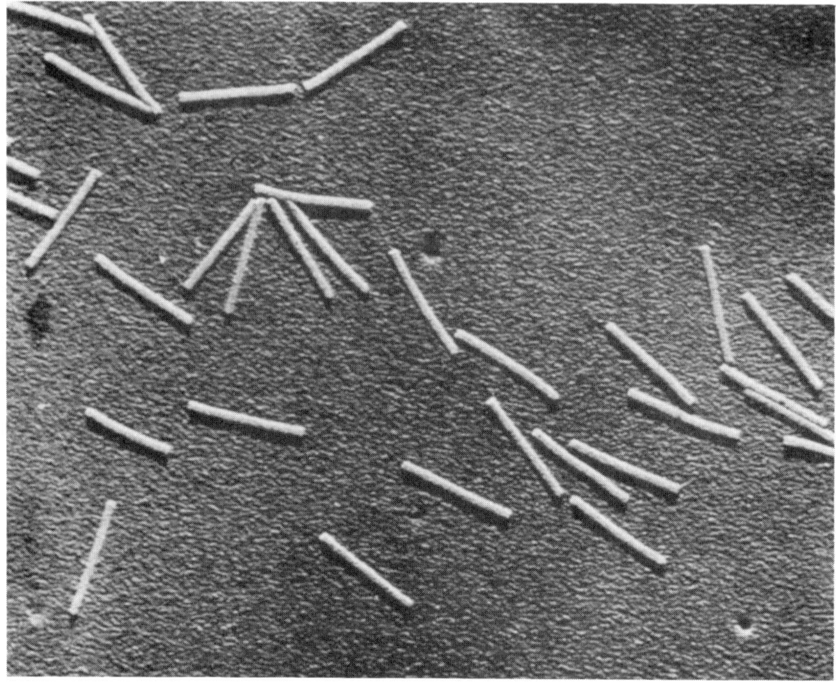

Fig. 4.12. Tobacco mosaic virus under electron microscope. Courtesy of USDA.

Table 4.1. Human Intestinal Viruses with High Potential as Food Contaminants

Picornaviruses
　Polioviruses 1-3
　Coxsackievirus A 1-24
　Coxsackievirus B 1-6
　Echovirus 1-34
　Enterovirus 68-71
　Probably hepatitis A
Reoviruses
　Reovirus 1-3
　Rotaviruses
Parvoviruses
　Human gastrointestinal viruses
Papovaviruses
　Human BK and JC viruses
Adenoviruses
　Human adenoviruses types 1-33

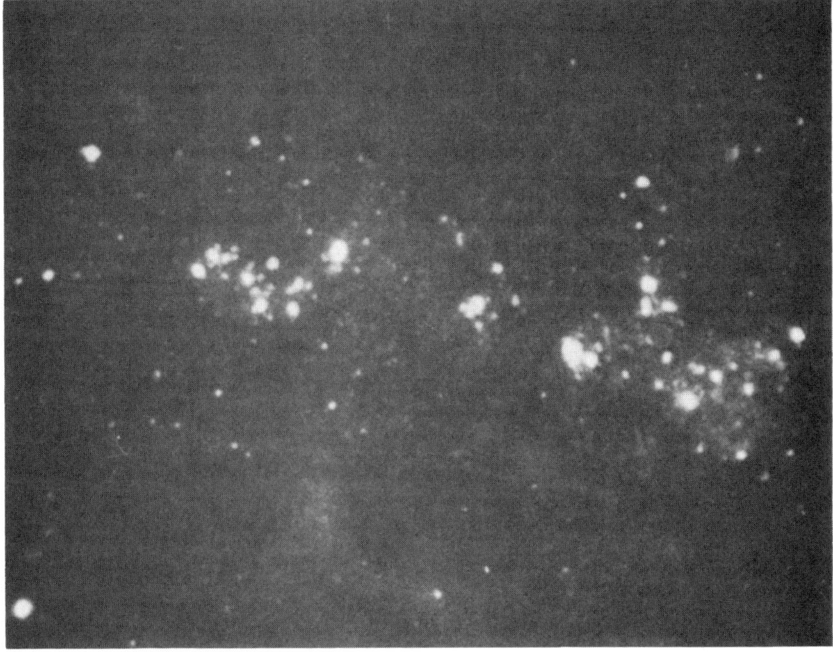

Fig. 4.13. Stained virus particles in tissue cells. (10,000×)

lungs of animals. Beef, swine, dogs and other canine species, bears, and fish can all harbor tapeworms and flatworms, which can be transmitted to and can infect humans.

Trematodes are nonsegmented flatworms that possess a mouth and oral sucker and depend on a snail as an intermediate host before infecting humans by being ingested in drinking water or aquatic plants. Intestinal flukes, pyriform worms from fish, sheep and chinese liver flukes, and oriental lung flukes are all examples of food-transmitted parasites.

Nematodes or true roundworms also can be transmitted from animals to humans. Eggs carried in excrement from roaches and dung beetles ingested by cattle, sheep and hogs contaminate humans. Trichinosis is an inflammation of the muscle tissue caused by ingesting the worm *Trichinella spiralis* (Fig. 4.14). Pork is the most common vector. Capillary worms, whipworms, and pinworms are other examples of nematode parasites.

Protozoa are microscopic single-celled animals, which can be taken in with food or water to cause human illness. *Entamoeba histolytica,*

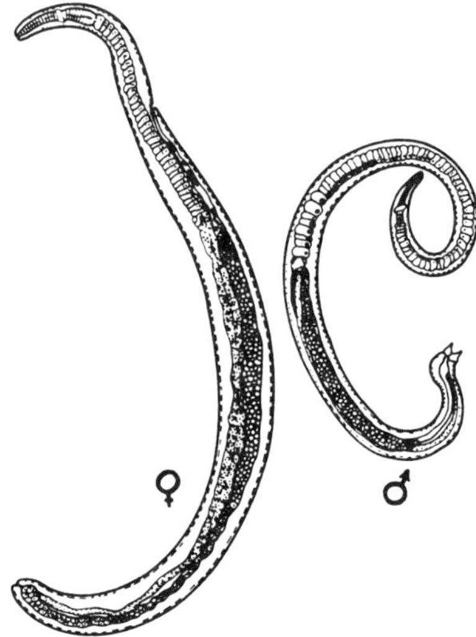

Fig. 4.14. Female (left) and male (right) of the adult worm that causes trichinosis in humans and animals (*Trichinella spiralis*).

Toxoplasma gondii, Balantidium coli, and *Giardia lamblia* are the most common foodborne protozoan parasites.

SELECTED REFERENCES

Ayres, J. C., Mundt, J. O., and Sandine, W. E. 1980. "Microbiology of Foods." W. H. Freeman and Co., San Francisco, California.

Barnett, H. L., and Hunter, B. B. 1972. "Illustrated Genera of Imperfect Fungi," 3rd ed. Burgess Publishing Co., Minneapolis, Minnesota.

Buchanan, R. E., and Gibbons, N. E. 1974. "Bergey's Manual of Determinative Bacteriology," 8th ed. Williams and Wilkins Co., Baltimore, Maryland.

Guthrie, R. K. 1980. "Food Sanitation," 2nd ed. AVI Publishing Co., Westport, Connecticut.

Hird, D. W., and Pullen, M. M. 1979. Tapeworms, meat and man: A brief review and update of cysticerosis caused by *Taenia saginata* and *Taenia solium. J. Food Protect.* 42(1), 58–64.

Larkin, E. P. 1973. The public health significance of viral infections of food animals. *In* "The Microbial Safety of Foods," B. C. Hobbs and J. B. H. Christian (Editors), Academic Press, London.

Lodder, J. 1970. "The Yeasts. A Taxonomic Study." North-Holland Publishing Co., Amsterdam.

Phaff, H. J., Miller, M. W., and Mrak, E. M. 1966. "The Life of Yeasts." Harvard University Press, Cambridge, Massachusetts.

Speck, M. L. (Editor). 1984. Compendium of Methods for the Microbiological Examination of Foods. American Public Health Association Inter-Society/Agency Committee on Microbiological Methods for Foods, Washington, DC.

Troy, V. S. 1960. Mold Counting of Tomato Products. Continental Can Company, New York.

5
Factors That Influence Microbial Activity

Moisture 69
  Significance of Free and Bound Water 70
Oxidation-Reduction Potential 71
Temperature 73
Nutritional Requirements 75
Hydrogen Ion Concentration (pH) 76
  Endogenous Acidity of Foods 77
Inhibitors 81
Selected References 81

Six major environmental factors influence the growth of microorganisms in foods: moisture, oxygen concentration, temperature, nutrients, pH, and inhibitors. The food supply available to microorganisms depends on the composition of the food on or in which they are found. Organisms differ in their ability to use carbohydrates, proteins, and fats as sources of energy.

MOISTURE

Water is essential for the growth of all cells. Most foods contain sufficient water to support growth; however, in drying or freezing, the water is removed or remains in a solid state, which makes it unavailable for the organisms to carry on normal metabolic activity.

Water in most cases carries food nutrients into the cells and removes the waste products that accumulate from metabolic activity. Moreover water helps to maintain the natural shape and turgidity of cells.

Water is present in all foods in varying amounts. For instance, the water content of fresh fruits varies from 75–90%. Leafy vegetables, such as green beans, spinach, and lettuce, have a water content of 90–95%. Tomatoes, cucumbers, and watermelons average about 95%

moisture. Milk contains about 87% moisture; fresh nuts 5–15% water. Sugar and lard are practically free of moisture.

To understand the relationship of moisture to cells, osmotic pressure and its effects must be considered. Diffusion occurs when different liquids or the same liquid at different concentrations come in contact. The two solutions flow together until both liquids have the same concentration. Osmosis is the diffusion of particles through a semipermeable membrane, such as a cell membrane. The direction of flow through the membrane is from an area of lower concentration to one of higher concentration. This diffusion occurs at different rates depending on the properties of the membrane (which may be permeable to one substance in a solution, but not to another) and on the properties of the particle (size, charge). The osmotic pressure is the force driving the diffusion of particles through a membrane.

Two solutions separated by a semipermeable membrane can be isotonic, hypertonic, or hypotonic, depending on the relative concentrations of the solutions. We are interested in the effects that the relationship of concentrations of solute and solvent outside the cell (media) have on the solution inside the cell (cytoplasm).

An *isotonic* solution is one in which the concentration of solutes and solvents are the same on both sides of the membrane. Since the concentrations are the same, there is no net movement through the membrane. An isotonic solution such as physiological saline (0.85% sodium chloride in water) is the most favorable medium for life.

In a *hypertonic* solution, the fluid outside the cells is more concentrated than the intracellular fluids. This causes a net movement of water out of the cells. As cells lose water, they lose turgor pressure and shrink. This process is called *plasmolysis*. Strong solutions of salt and sugar, or mixtures of these, are frequently used to preserve food such as meat. If salting is done properly, it is a good way to stop microbial activity.

A *hypotonic* solution has a higher solvent (water) concentration than does the cytoplasm. This means that water enters cells, causing swelling and possibly bursting. Hypotonic solutions have no practical value in food preservation. For instance, if one crushed fruit, a hypotonic condition would be attained which could be harmful to microorganisms, but the fruit itself would also be damaged.

Significance of Free and Bound Water

The form in which water exists in food is important from the microbiological standpoint. Bound water is present in the tissue itself and is a part of the living tissue just as much as proteins and carbohydrates. It is vital to all the physiological processes associated with the cell. Free water exists in and around the tissue of cells. It can be removed from cells without seriously interfering with the vital processes. Because free water is necessary for the survival and metabolism of microorganisms present in the food, dehydration is an important method of preserving food. Free water governs microbial activity, whereas bound water has very little effect upon the organisms.

Microorganisms depend on water for growth. The amount of water available for microorganisms is referred to as *water activity* (a_w). Water activity is a measure of the free moisture in a product and is the quotient of the water vapor pressure of the substance divided by the vapor pressure of pure water at the same temperature. Pure water has a water activity of 1.00 and is in equilibrium with a relative humidity of 100%. In freezing, pure water is removed from solution and the solute becomes more concentrated. Because some water exists as bound water, the water activity of a system is not exactly the same as its water content. Bacteria require more water than yeasts, which require more than molds. This is illustrated by the data in Table 5.1.

OXIDATION–REDUCTION POTENTIAL

All forms of life require oxygen to carry on metabolic activities. Free atmospheric oxygen (O_2) is utilized by certain groups of microorganisms, whereas others utilize more reduced forms of oxygen present in organic compounds such as carbohydrates.

Microorganisms may be classified according to their oxygen requirements. Aerobic organisms grow in the presence of atmospheric oxygen. Anaerobic organisms grow in the absence of atmospheric oxygen. Fac-

Table 5.1. Minimum Equilibrium Relative Humidity (RH) Permitting Development of Spoilage Organisms

Type of microorganism	Minimum RH (%)
Normal bacteria	91
Normal yeasts	88
Normal molds	80
Halophilic bacteria	75
Xerophilic fungi	65
Osmophilic yeasts	60

Source: Adapted from Mossell and Ingram (1955).

ultatively anaerobic organisms grow in either the absence or the presence of atmospheric oxygen. Microaerophilic organisms grow in the presence of a reduced supply of atmospheric oxygen.

The presence of aerobes and anaerobes may change the oxidation–reduction potential in a food to the point where the growth of certain organisms is inhibited. Many fresh foods have a low oxidation–reduction potential because of reducing substances such as vitamin C, reducing sugars, and other compounds. These compounds tend to keep the oxidation potential at a low level. In the case of the Swiss cheese, the heating of the milk may alter the texture of the food or change the reducing and oxidizing compounds so that anaerobic organisms may find a favorable environment in the center of the cheese.

The biochemical behavior of microorganisms in food is largely determined by the amount of oxygen present. If sufficient oxygen is present, most food materials are completely though slowly oxidized. Carbon dioxide and water are the products when sugar is oxidized. If amino acids are oxidized, ammonia, carbon dioxide, and water are formed. Free fatty acids, carbon dioxide, and water are the result of the oxidation of fats.

However, if the oxygen tension is lowered, more intermediate products and fewer end products are formed by microbial action. In this case the organisms may produce alcohol and lactic, acetic, and formic acids. The protein breakdown usually results in amino acids with varying amounts of ammonia, hydrogen sulfide, sulfites, carbon dioxide, and water. Fats are broken down to glycerol and fatty acids, with very little carbon dioxide and water formation. Under strictly anaerobic conditions, foul-smelling compounds are formed from proteins such as butyl alcohol, mercaptans, and hydrogen sulfide. The formation of these is frequently referred to as putrefaction.

In an alcoholic fermentation of sugar by yeasts, in which alcohol is the desired product, a small amount of oxygen is necessary for the yeasts to convert the sugar into alcohol. If yeast and vitamins are desired, then large amounts of oxygen must be supplied. In this case, the sugar is completely oxidized, thus producing more energy per gram than under anaerobic conditions and a greater yield of yeast cells.

As those examples indicate, oxygen plays an important role in the type and amount of biological product obtained from microbial action, the amount of substrate consumed, and the energy released in the breakdown. Sugar fermentation by yeasts is familiar to all biologists. With an abundance of oxygen available for yeast production, the principal reaction is

$$C_6H_{12}O_6 + 6O_2 \longrightarrow 6CO_2 + 6HOH + 674 \text{ calories}$$

Under anaerobic conditions for alcohol production, there is little increase in the yeast present and the main reaction is

$$C_6H_{12}O_6 \longrightarrow 2C_2H_5OH + 2CO_2 + 22 \text{ calories}$$

Thus, some 30 times as much energy is produced in the first process as in the second.

Molds grow very poorly under anaerobic conditions; consequently, they are not a serious factor in food spoilage under these conditions. Ensilage is an excellent example of the influence of oxygen on the byproducts of fermentation. If the ensilage is well packed in the silo, lactic acid and gassy fermentations are the result. If it is loosely packed so that more oxygen is present, then lactic acid is produced in small amounts but mold growth is pronounced.

TEMPERATURE

Temperature affects the growth and metabolic activity of all living cells. At high temperatures all organisms are destroyed, while at freezing temperatures microbial activity is slowed but not necessarily stopped. Microorganisms vary in their temperature requirements for growth and can be classified into three groups: psychrophilic, mesophilic, and thermophilic.

Each organism or group of related organisms exhibit a small range of optimum temperatures for growth (Table 5.2). The minimum temperature for growth may be several degrees below the optimum. The rate

Table 5.2. Classification of Bacteria According to Their Temperature Requirements

| Bacterial type | Temperature required for growth (°C) | | | General sources of these bacteria |
	Minimum	Optimum	Maximum	
Psychrophilic	0 to 5	15 to 20	30	Water and frozen foods
Mesophilic	10 to 25	30 to 40	35 to 50	Pathogenic and many nonpathogenic bacteria
Thermophilic	25 to 45	50 to 55	70 to 90	Many spore-forming bacteria usually from soil and water

of growth decreases as the temperature is lowered. This is the basic principle of preserving foods by refrigeration and freezing. Maximum temperature refers to the temperature slightly above the optimum at which growth will occur. At temperatures above the maximum, organisms are killed.

Microorganisms vary considerably in their rate of growth, even under optimal conditions. In the case of bacteria, a common measure of growth rate is *generation time*. This is the time it takes for the number of cells growing in a culture to double. The generation times of several bacterial species are listed in Table 5.3. The effect of temperature on growth rate is illustrated by the data on generation times for *Escherichia coli* presented in Table 5.4. A decrease in generation time corresponds to an increase in growth rate. As can be seen from the table, the highest rate of growth (lowest generation time) in this case occurred at 98°F (37°C).

The effect of temperature on various biological processes is often expressed by an index called the temperature coefficient or Q_{10}. Elliot and Michener (1965) defined Q_{10} as the ratio of activity rate or growth rate at one temperature to that at a temperature 10°C higher. The Q_{10} values for the growth of *Pseudomonas fluorescens* under various conditions are listed in Table 5.5. In general, the rate of reaction increases two- to fourfold for each 10° increase in temperature. However, the growth rate of microorganisms does not follow this pattern as closely as the reaction rates of enzyme systems.

Table 5.3. Generation Times of Various Selected Bacterial Species

Organism	Medium	Temp. (°C)	Generation time (min)
Aerobacter aerogenes	Broth or milk	37	16–18
	Synthetic	37	29–44
Bacillus mycoides	Broth	37	28
Bacillus thermophilus	Broth	55	18.3
Escherichia coli	Broth	37	17
	Milk	37	12.5
Lactobacillus acidophilus	Milk	37	66–87
Mycobacterium tuberculosis	Synthetic	37	792–932
Rhizobium japonicum	Mineral salts + yeast + mannitol	25	344–461
Salmonella typhi	Broth	37	23.5
Staphylococcus aureus	Broth	37	27–30
Streptococcus lactis	Lactose broth	37	48
	Milk	37	26
Treponema pallidum	Rabbit testes	37	1,980

Table 5.4. Effect of Temperature on Generation Time of *Escherichia coli*

Temperature		
(°F)	(°C)	Generation time (min)
68	20	60
77	25	40
86	30	29
98	37	17
104	40	19
113	45	32
122	50	No growth

NUTRITIONAL REQUIREMENTS

Almost without exception microorganisms must depend on nutrients for both energy and growth. Because of the surface to volume ratio of microorganisms, their energy requirements are high. Generally those organisms found in a particular foodstuff are there because the required nutrients are in the food. If the nutrients that a particular species requires are not present, it will not survive. Energy requirements vary from simple carbon sources to complex cellulose compounds, which can be broken down by only a very few organisms. Proteins and amino acids can also be used as energy sources. Nitrogen requirements also vary. Some microorganisms cannot use proteins; others can grow with very crude nitrogen sources. Accessory food substances also influence growth. Some species are so sensitive to a particular vitamin that they are used as an assay organism. Occasionally, trace mineral deficiencies can prevent growth of microorganisms.

Table 5.5. Effect of Temperature, Nutrient, and Aeration on the Q_{10} for Growth of *Pseudomonas fluorescens*

Media	Culture	Q_{10} values at various temperature intervals				
		4° to 10°C	10° to 15°C	15° to 20°C	20° to 25°C	25° to 32°C
Glucose	Stationary	4.04	3.10	1.90	2.31	0.018
	Shake	3.52	1.69	1.88	2.50	0.028
Citrate (Na)	Stationary	4.25	2.96	1.96	2.02	0.019
	Shake	3.89	2.10	2.59	1.66	0.024
Casamino acids	Stationary	4.49	3.77	1.69	1.49	1.27
	Shake	2.26	2.72	1.93	1.64	0.019

Source: Jezeski and Olson (1962).

A typical growth curve for bacteria growing in a culture containing nutrients that are suitable but limited in quantity is shown in Fig. 5.1. The shape of this curve is determined by the change in nutrient concentrations in the medium during culture. During the initial phase (1–2), the number of bacteria remains constant or actually decreases as the inoculated bacteria become adjusted to the environment. During the log phase (2–3), the number of bacteria increases rapidly because of the plentiful supply of nutrients. Multiplication begins to slow down during the negative growth phase (3–4) as the food supply decreases and waste products, which may be toxic to the organism, accumulate. During the stationary phase (4–5), the death rate is in balance with the birth rate and so the total number of bacteria remains constant. Finally, during the accelerated death phase (5–7), the number of bacteria decreases rapidly because the food supply becomes exhausted and cannot support additional growth.

HYDROGEN ION CONCENTRATION (pH)

Microorganisms vary in their optimum pH for growth. Most bacteria grow best at a pH near neutral (7); yeasts can grow in a pH range of 4 to 4.5; and molds can grow from pH 2 to 8.5 but favor an acid pH. Some microbiological reactions in food are controlled by pH when one group of organisms ferments sugars so the pH becomes so low that other microorganisms cannot survive.

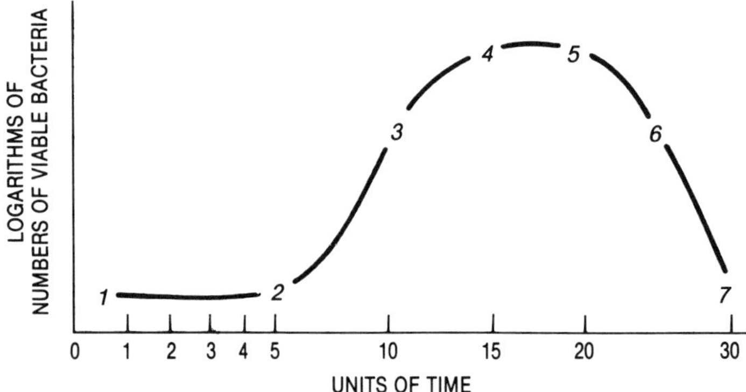

Fig. 5.1. Typical bacterial growth curve. See text for explanation of various phases.

The pH, a measure of the hydrogen ion concentration, is the key factor in determining the survival of bacterial spores. A food with a pH of 4.6 or less is termed a high-acid or acid food and will not permit spore germination. If the pH is above 4.6, the food is called a low-acid food and will not inhibit bacterial spores. A pH of 4.6 or less inhibits the growth of the most resistant bacterial spore of public health significance, i.e., *Clostridium botulinum.* If this spore is not inhibited, it may germinate, grow, and produce a toxin; if ingested, this toxin causes neurological damage and sometimes death.

The Food Processors Institute recommends five methods for acidification of food products:

1. Blanch the food ingredients in an acidified, aqueous solution. The concentration of acid, food particle size, and blanch temperature all influence the length of blanch time required to achieve a final pH less than 4.6.
2. Direct immersion of the product in an acid solution following blanching prior to filling. Sufficient hold time in the acid solution is required to achieve a pH below 4.6. The time depends on the concentration of acid and particle size.
3. Direct batch acidification. This is the usual method for small quantities and, particularly, for liquid products. After mixing, the pH should be checked to assure that the pH is below 4.6.
4. Direct addition of a predetermined amount of acid to individual containers during production runs. This method involves addition of tablets or known volumes of acid or some other means of direct acidification. Control is very critical to achieve the desired results.
5. Adding acid foods to low-acid foods in controlled portions to achieve the desired pH.

Endogenous Acidity of Foods

The optimum pH for nearly all microorganisms is near the neutral point or 7.0. Acid-tolerant organisms have an advantage in an acid medium over organisms inhibited by acid. Molds and yeasts as a rule are acid tolerant and are usually associated with acid foods, especially fruits. Since most kinds of bacteria are not acid tolerant, they are seldom found in normal healthy fruit.

There are a few species of acid-tolerant bacteria that are important in food products. *Lactobacillus* and *Streptococcus* species are capable of producing various acids in milk; however, their survival in an acid environment is usually short. These organisms are highly desirable in the production of cheeses, fermented milks, and butter. Tomato juice, an acid food product, is subject to spoilage when contaminated with *Bacillus thermoacidurans.*

A few molds are capable of utilizing the acids present in food products as a source of energy. In this case, the acidity of the food is reduced or neutralized by the basic products of metabolism. Yeasts and most bacteria as a rule cannot use acid components of food as a nutrient.

The pH values for a variety of foods are listed in Table 5.6. Foods

Table 5.6. pH and Moisture Content of Foods

Food	pH	Moisture (%)
Apples (whole)	3.4–3.5	—
Apple juice	3.3–3.5	—
Apricots (dried)	3.6–3.4	—
Asparagus (green)	5.0–5.8	—
Beans		
Baked	4.8–5.5	—
Green	4.9–5.5	95.1
Lima	5.4–6.3	80.7
Soy	6.0–6.6	—
With pork	5.1–5.8	—
Beef (corned, hash)	5.5–6.0	55.6
Beef (stew)	5.6–6.2	72.6
Beef (dried)	5.5–6.8	59.3
Beets (whole)	4.9–5.6	85.4
Blackberries	3.0–4.2	—
Blueberries	3.2–3.6	87.9
Boysenberries	3.0–3.3	—
Bread		
White	5.0–6.0	—
Date and nut	5.1–5.6	—
Broccoli	5.2–6.0	—
Carrots (chopped)	5.3–5.6	91.0
Carrot juice	5.2–5.8	—
Catfish	—	81.2
Cheese		
Parmesan	5.2–5.3	—
Roquefort	4.7–4.8	—
Cherry juice	3.4–3.6	82.0
Chicken (roasted)	6.2–6.4	47.2
Chicken (with noodles)	6.2–6.7	—
Chop suey	5.4–5.6	—
Cider	2.9–3.3	—
Clams	5.9–7.1	83.7
Cod fish	6.0–6.1	82.6
Corn-on-the-cob	6.5–6.6	76.0

Table 5.6. *(continued)*

Food	pH	Moisture (%)
Corn		
Cream style	5.9–6.5	—
Whole grain		
Brine packed	5.8–6.5	—
Vacuum packed	6.0–6.4	—
Crab apples (spiced)	3.3–3.7	—
Crackers	7.0–8.5	—
Cranberry		
Juice	2.5–2.7	—
Sauce	2.3–2.3	62.1
Currant juice	3.0–3.0	—
Dates	6.2–6.4	—
Duck (roasted)	6.0–6.1	46.2
Figs	4.9–5.0	87.4
Flour	6.0–6.5	—
Frankfurters	6.2–6.2	—
Fruit cocktail	3.6–4.0	—
Goose (roasted)	—	57.5
Gooseberries	2.8–3.1	—
Grapefruit		
Juice	3.0–3.3	—
Pulp	3.4–3.4	—
Sections	3.0–3.5	—
Grapes	3.5–4.5	—
Ham (spiced)	6.0–6.2	—
Herring	—	68.3
Hominy (lye)	6.9–7.9	82.6
Huckleberries	2.8–2.9	90.4
Jams (fruit)	3.5–4.0	—
Jellies (fruit)	3.0–3.5	—
Kidney (beef)	—	76.0
Lamb (cutlets)	—	65.3
Lemons	2.2–2.6	—
Lemon juice	2.2–2.4	—
Liver (calf)	—	66.8
Loganberries	2.7–3.3	—
Mackerel	5.9–6.2	70.1
Milk		
Cow	6.4–6.8	87.0
Evaporated	5.9–6.3	—
Molasses	5.0–5.4	—
Mushrooms	6.0–6.5	92.4
Olives (ripe)	5.9–7.3	78.0
Orange juice	3.0–4.0	—
Oysters	6.3–6.7	—
Peaches	3.4–4.2	88.0
Pears (Bartlett)	3.8–4.6	88.0
Peas	5.6–6.5	82.9
Pheasant (baked)	—	51.5
Pickles		
Dill	2.6–3.8	—

(continued)

Table 5.6. *(continued)*

Food	pH	Moisture (%)
Sour	3.0–3.5	—
Sweet	2.5–3.0	—
Pimento	4.7–5.2	—
Pineapple		
Crushed	3.2–4.0	80.0
Sliced	3.5–4.1	—
Juice	3.4–3.7	—
Plums	2.8–3.0	—
Pork (leg)	—	53.7
Potatoes		
White	5.4–5.9	—
Mashed	5.1–5.1	—
Potato salad	3.9–4.6	—
Prune juice	3.7–4.3	—
Pumpkin	5.2–5.5	—
Rabbit	—	75.0
Raspberries	3.2–3.7	—
Rhubarb	2.9–3.3	94.8
Salmon	6.1–6.3	72.0
Sardines	5.7–6.6	53.6
Sauerkraut	3.1–3.7	93.2
Juice	3.3–3.4	—
Shrimp	6.8–7.0	59.4
Soups		
Bean	5.7–5.8	—
Beef broth	6.0–6.2	—
Chicken noodle	5.6–5.8	—
Clam chowder	5.6–5.9	—
Duck	5.0–5.7	—
Mushroom	6.3–6.7	—
Noodle	5.6–5.8	—
Oyster	6.5–6.9	—
Pea	5.7–6.2	—
Tomato	4.2–5.2	—
Turtle	5.2–5.3	—
Vegetable	4.7–5.6	—
Spinach	5.1–5.8	92.1
Squash	5.0–5.3	—
Strawberries	3.1–3.5	—
Sweet potatoes	5.3–5.6	66.7
Tomatoes	4.1–4.4	—
Juice	3.9–4.4	—
Tuna	5.9–6.1	54.1
Turnip greens	5.4–5.6	80.0
Veal	—	75.0
Vegetable		
Juice	3.9–4.3	—
Mixed	5.4–5.6	—
Vinegar	2.4–3.4	—
Youngberries	3.0–3.7	—

may be classified as acid foods, acidified foods, or low-acid foods. Because of the effect of acidity on microbial growth, it is important for food technologists to undertand these three terms. The Food and Drug Administration has established the following definitions:

- *Acid foods* are foods that have a natural pH of 4.6 or below.
- *Acidified foods* are low-acid foods to which acid(s) or acid food(s) are added. These foods include, but are not limited to, beans, cucumbers, cabbage, artichokes, cauliflower, puddings, peppers, tropical fruits, and fish, singly or in any combination. They have a water activity (a_w) greater than 0.85 and have a finished equilibrium pH of 4.6 or below. These foods may be called, or may purport to be, "pickles" or "pickled." Carbonated beverages, jams, jellies, preserves, and acid foods (including such foods as standardized and nonstandardized food dressings and condiment sauces) that contain small amounts of low-acid food(s) and have a resultant finished equilibrium pH that does not significantly differ from that of the predominant acid or acid food, and foods that are stored, distributed and retailed under refrigeration, are not classified as acidified foods.
- Low-acid foods are any foods other than alcoholic beverages with a finished equilibrium pH greater than 4.6 and a water activity (a_w) greater than 0.85. Tomatoes and tomato products that have a finished equilibrium pH less than 4.7 are not classed as low-acid foods.

INHIBITORS

Many chemical compounds can inhibit the growth of microorganisms by preventing metabolism, by denaturing the protein portion of the cell, or by causing physical damage to the parts of the cell. An example is sodium benzoate.

SELECTED REFERENCES

Banwart, G. J. 1979. "Basic Food Microbiology." AVI Publishing Co., Westport, Connecticut.

Elliott, R. P., and Michener, H. D. 1965. Factors Affecting Growth of Psychrophilic Microorganisms in Foods. Tech. Bulletin *1320.* U.S. Dept. of Agriculture, Washington, DC.

Frazier, W. C., and Westhoff, D. C. 1978. "Food Microbiology," 3rd ed. McGraw-Hill, New York.

ICMSF. 1980. "Microbial Ecology of Foods." Vol. I: Factors Affecting Life and Death of Microorganisms. Academic Press, New York.

Jezeski, J. J., and Olson, R. H. 1961. The activity of enzymes at low temperatures. *Proc. Low Temperature Microbiol. Symp.* Campbell Soup Co., Camden, New Jersey.

Mossell, D. A. A., and Ingram, M. 1955. The physiology of the microbial spoilage of foods. *J. Appl. Bacteriol.* 18, 233–268.

6

High-Temperature Food Preservation

Effect of Temperature on Microorganisms 83
  Causes of Death by Heating 84
  Effect of Heat on Bacterial Spores 87
Application of Heat in the Canning Industry 87
Methods of Preserving Food by Heat 89
  Cooking 89
  Cold Pack 91
  Hot Pack 91
  Oven Process 91
  Pasteurization 92
Thermophilic and Thermoduric Bacteria 94
Selected References 97

EFFECT OF TEMPERATURE ON MICROORGANISMS

The bacterial cell has no unique mechanism for surviving at temperatures above or below its natural environment. Bacterial cultures can grow at temperatures ranging from 32°F (0°C) to 185°F (85°C), but the rate of growth varies with each species of bacteria. Biologists define the optimum temperature as that at which an organism grows most rapidly in a relatively short time (i.e., 2–24 hr). As noted in Chapter 5, each bacterial species has a relatively narrow range of temperature at which it grows best (*see* Table 5.2). At temperatures lower than its optimum, an organism may grow and reproduce but at a slower rate; at temperatures above the optimum, the rate of growth and reproduction is increased, but so is the speed of chemical reactions. These reactions may result in synthesis of protoplasm or the generation of energy or both. At sufficiently high temperatures, proteins are destroyed and enzymatic activity ceases.

Oginsky and Umbreit (1959) cited references showing that the maximum growth temperature is slightly below the minimum temperature necessary to inactivate cell enzymes. For instance, at least one *Bacillus subtilis* enzyme is inactivated at 131°F (55°C). Therefore, *B. subtilis* will grow at 130°F (54°C) but not at higher temperatures. These authors also observed that the malic dephosphorylating adenosine triphosphate enzyme of thermophilic organisms showed marked resistance to heat. It is likely that variations in the temperature sensitivity of most mesophilic organisms may be caused by variations in the heat sensitivity of their enzyme proteins.

If an enzyme that produces an essential metabolite is inactivated at the temperature at which a bacterial culture is incubated, then the cells will cease to grow and reproduce. However, the cells may continue to grow at the elevated temperature if the essential metabolite is added to the culture medium.

Lactobacillus arabinosus provides a good example of this phenomenon. This species grows at 79°F (26°C) in the absence of added tyrosine, aspartic acid, or phenylalanine. When *L. arabinosus* is incubated at 98.5°F (37°C), however, phenylalanine and tyrosine are essential nutrients and must be added to the medium for cell growth to occur; at 103°F (39°C) aspartic acid is required. If this culture is incubated at 79°F (26°C) in the absence of CO_2, then all three amino acids must be supplied. Conversely, if CO_2 is added to the medium at a high incubation temperature, *L. arabinosus* will grow without aspartic acid or phenylalanine. Oginsky and Umbreit (1959) explain this temperature dependence of nutritional requirements on the inactivation of the requisite enzyme(s) at higher incubation temperatures.

The effect of various temperatures on bacterial life is summarized in Table 6.1. The temperatures at which various foods can be safely stored depend to a large extent on the microorganisms that may be present. The conditions for safe storage of foods are presented in Table 6.2.

Causes of Death by Heating

Several theories have been advanced to explain the exact physical and chemical mechanisms that bring about the death of bacterial cells when they are heated in the presence of water, and why there is such a great variation among cultures of the same species in the time and the temperature required to destroy these organisms. The three theories most generally accepted are

Table 6.1. Effect of Temperature on Bacterial Life

°F	°C	Temperature effects
250	121	All forms including spores killed in 15–20 min
240	115	All forms including spores killed in 30–40 min
230	110	All forms including spores killed in 60–80 min
212	100	Vegetative cells but not spores killed
200	93	Growing cells of bacteria, yeasts, and molds usually killed
180	82	Thermophilic organisms grow
170	76	All important pathogenic bacteria for humans except spore-forming pathogens killed
100 } 98.6	38 } 37	Active growing range for most bacteria, yeasts, and molds
60	15.6	Growth retarded for most organisms
50 } 40	10 } 4.4	Optimum growth range of psychrophilic organisms
32	0	Usually growth of all organisms stopped
0	−18	Bacteria preserved in latent state
−420	−250	Many species survive at temperature of liquid hydrogen

Source: Adapted from Research Department Bulletin, Glass Container Association.

(1) heat inactivation of vital enzymes,
(2) cellular intoxication from metabolic products, and
(3) destructive changes in the physical state of essential lipids.

Heat causes an irreversible change in proteins including enzymes called *denaturation*. The isolation of heat-resistant enzymes (e.g., apyrase, cytochrome oxidase, dehydrogenase, and succinoxidase) from heat-resistant bacteria supports the theory that protein denaturation is responsible for the death of cells.

It is possible that when a cell is heated, the metabolism of the cell accelerates to the point where the cell is poisoned by its own metabolic products. However, it is doubtful that sufficient time elapses between heating and the death of the cell for enough toxic substances to accumulate to kill the cell.

A general relationship exists between the melting points of the fats found in an organism and the temperature range that will bring about its death. Fats from mammals are solid at room temperature, whereas fats from fish have low melting points and are liquid at room temperature. Killing temperatures are lower for fish than for warm-blooded animals. Further support for this theory is found in the fact that the heat resistance of spores of *Clostridium botulinum* is much greater when there is an accumulation of high-molecular-weight fatty acids in the medium in which the cells are grown.

Table 6.2. Temperature Range for Safe Storage of Foods

Zone I. Sub-freezing Temperatures, 0° to 15°F (−18° to −9.0°C)
  Frozen meat, fish, and vegetables
  Frozen fruits
  Ice cream
  Homemade frozen desserts

Zone II. High Humidity (85%) and Moderate Air Circulation, 34° to 37°F (1° to 3°C)
  Fresh meat, chicken, and fish
  Sliced smoked ham and bacon
  Sliced cold cuts of meat
  Leftover canned and cooked meat

Zone III. 38° to 40°F (3° to 4°C)
  Fresh milk, cream, and buttermilk
  Cottage cheese and butter (both covered)
  Fresh orange and tomato juice (covered)
  Bottled beverage (for chilling)

Zone IV. 40° to 43°F (4° to 6°C) and Moderate Humidity
  Berries, pears, and peaches
  Ripe grapefruit and oranges
  Ripe tomatoes (short time only)
  Fresh eggs
  Oleomargarine
  Custards and puddings (day or two only)
  Prepared salads (for chilling)

Zone V. 40° to 45°F (4° to 7°C) and High Humidity
  Cherries and cranberries
  Lettuce and celery
  Spinach, kale, and other greens
  Beets, carrots, parsnips, and turnips
  Peas and lima beans
  Cucumbers and eggplant (short time only)

Zone VI. 55° to 60°F (13° to 16°C), Fairly High Humidity, and Moderate Circulation
  (good fruit cellar or storage cellar well ventilated)
  Apples, cabbage, potatoes, pumpkin, squash, unripened tomatoes, and maple syrup
  (in tight container)

Zone VII. Normal Room Temperature—Dry Storage
  Ready prepared cereals
  Crackers
  Bottled beverages

Zone VIII. Normal Room Temperature Storage
  Peanut butter and honey
  Salad oils and vegetable shortenings
  Catsup and pickles
  Jelly and preserves
  Dried fruits and bananas (short time)
  Flour
  Dried peas and beans
  Sugar and salt

Effect of Heat on Bacterial Spores

A slight heat shock will stimulate the metabolic activity of bacterial spores. For example, in the presence of methylene blue, the vegetative cells of *Bacillus subtilis* dehydrogenate glucose at 104°F (40°C) much more rapidly than the endospores. However, when the vegetative cells and endospores are heat-shocked at 176°F (80°C) for 30 min and the temperature is then reduced to 104°F (40°C), the vegetative cells no longer reduce methylene blue in the presence of glucose, but the endospore activity markedly increases. If spores that have been activated are heated again to 176°F (80°C), their enzyme systems are destroyed, leaving them comparable to vegetative cells.

APPLICATION OF HEAT IN THE CANNING INDUSTRY

In 1782, Scheele suggested the application of heat to preserve vinegar in bottles. In 1810, Appert, a scientist and French confectioner, was awarded 10,000 francs by Napoleon for suggesting the best method for preserving food for his armies. He boiled the food and placed it in bottles under an airtight seal. In the same year Peter Durand secured an English patent for preserving fruits, fish, and vegetables in sealed cans. In about 1819, Ezra Daggett and Thomas Kensett introduced the process in America for packing salmon, lobsters, oysters, and other seafoods. The next year, Underwood and Mitchel built a factory in Boston where cranberries, currants, plums, and quinces were canned. It is interesting to note that glass jars were used exclusively in the early days of canning. However, glass was gradually abandoned because the glass made then was unable to withstand high temperatures. The jars were also expensive, bulky, and costly to transport. In 1825, Kensett obtained a patent to use tin cans, and built a canning factory for the use of tin cans.

The original Appert method of preserving food by heat was to cook the food in open kettles and then put it into sealed containers. The boiling point of water, around 212°F (100°C), was the highest temperature obtainable. In 1874, a closed kettle was developed that allowed for superheating water with steam. Later steam pressure was used, a forerunner of the present autoclave and more recently the pressure cooker. Relations between steam pressures and temperatures are shown in Table 6.3.

Table 6.3. Relationship of Pressure to Steam Temperatures

Pressure (lb/in²)	Temperature	
	(°F)	(°C)
0	212	100
5	228	109
10	240	115.5
15	251	121.5
20	260	126.5
40	287	141.5

Factors that may influence the sterilization of canned goods include the following:

- Number and kind of organisms present before sterilization
- Size of container used
- Reaction or pH of the food (e.g., fruits are more easily sterilized than corn or peas)
- Amount of moisture initially present
- Ease of heat conduction and convection
- Agitation during sterilization
- Volume
- Composition

The data in Table 6.4 illustrate the effect of the pH associated with different foods on the time required to destroy spores of *Clostridium botulinum*. In general, the more acid the food, the less time at a given temperature is needed to kill spores. The difference in the heat susceptibility of sporeforming bacteria (e.g., *C. botulinum* and *Bacillus anthracis*) and non-sporeforming bacteria (e.g., *E. coli*) is indicated by the data in Table 6.5. Clearly, *C. botulinum* is very resistant to heat, while

Table 6.4. Time Required to Destroy Viable Spores of *Clostridium botulinum* in Different Foods

Kind of food	pH of food	Minutes at temperature of				
		90°C	95°C	100°C	110°C	115°C
Hominy	6.95	600	495	345	34	10
Corn	6.45	555	465	255	30	15
Spinach	5.10	510	345	225	20	10
String beans	5.10	510	345	225	20	10
Pumpkin	4.21	195	120	45	15	10
Pears	3.75	135	75	30	10	5
Prunes	3.60	60	20	—	—	—

Table 6.5. Time and Temperature Required to Destroy Sporeforming and Non-sporeforming Bacteria

Organism	Time (min)	Temperature (°F)	(°C)
Sporeformers:			
Bacillus anthracis	8	221	105
	10	212	100
	25	203	95
	45	194	90
Clostridium botulinum	0.78	260	127
	1.45	255	124
	2.78	250	121
	5.27	245	118
	10	240	116
	36	230	110
	150	220	104
	330	212	100
Non-sporeformers:			
Escherichia coli	2	212	100
	4	170	77
	10	125	52
	30	115	46

B. anthracis is much less resistant. The non-sporeforming *E. coli* is quite susceptible to heat and can be destroyed at 212°F (100°C) for 2 min.

METHODS OF PRESERVING FOOD BY HEAT

Cooking

The cooking of food generally makes it more palatable and perhaps more easily digested. However, the various cooking methods differ in their ability to destroy bacteria and parasitic organisms.

Boiling

Boiling is one of the most common methods of cooking foods. However, boiling food at atmospheric pressure, which rarely produces a temperature as high as 212°F (100°C), does not insure complete destruction of all organisms. The addition of sugar to a food will raise the boiling temperature, and this procedure was one of the first methods used to preserve food. The food is usually boiled in an open kettle and the contents poured into jars and then securely sealed. This method does not insure sterile food because many of the organisms are not destroyed during the cooking process, especially sporeforming microorganisms and many kinds of thermophilic bacteria.

Boiling does destroy many pathogenic organisms that may be present in foods. Organisms that cause diseases such as tuberculosis, typhoid, paratyphoid, brucellosis, dysentery, tularemia, sore throat, scarlet fever, and diphtheria are easily destroyed by boiling, providing sufficient temperature and time of exposure are maintained.

The toxin produced by *Staphylococcus aureus* is heat labile, but not at the temperatures and times ordinarily used for cooking. Food recently contaminated by these bacteria and then eaten may not cause any ill effects. It is the toxin produced by the organisms as a result of their ability to grow in the food when conditions are favorable that causes the ill effects. Moreover, the mechanism of food poisoning by various *Salmonella* species is not well understood. Illness may result from ingesting the organisms in the food. It appears in this case that intoxication takes place by the liberation of a toxin after the bacteria have entered the intestinal tract.

In general, any sporulating bacteria present in food at the time it is eaten will cause no ill effects in man. For example, *C. botulinum* spores are not usually pathogenic to man when eaten. However, when conditions are favorable for growth in food products, this organism will produce an extremely potent toxin, which is often fatal. Fortunately, heating to a boiling temperature usually inactivates or destroys this toxin. The spores of *C. botulinum* are very resistant to heat and can withstand boiling temperatures for several hours (*see* Table 6.5).

Baking and Frying

During the baking process, it is the temperature that the food attains, not the oven temperature, that is important in terms of destroying microorganisms. As long as a food remains moist during baking, the maximum temperature reached is rarely more than 212°F (100°C), even when the oven temperature is 350°–400°F (175°–205°C). Baked goods are likely to have low heat conductivity. Lack of convection currents prevents heat penetration, and moisture loss also has a cooling effect. Thus, baking is less effective than boiling as a means of destroying microorganisms.

Frying kills most microorganisms but not all. Sporeforming, thermophilic, and heat-resistant strains of bacteria are not killed by frying, especially those present in the interior of the food.

Trichinosis, caused by a parasite infecting pork, can cause serious illness in man. *Trichinella spiralis* larvae are usually killed by heating meat to 137°F (59°C) or storing the product at a temperature of 5°F (−15°C) for 30 days. Adequate cooking of the pork is absolutely necessary to destroy the larvae because meat inspection fails to detect their presence. Cooking garbage fed to swine is one way to control the spread of trichinosis. It has been estimated that at one time 5% of contaminated swine are infected by eating raw garbage and 10% by eating raw offal from slaughterhouses.

Cold Pack

Although the cold-pack method was widely used in the past for preserving many kinds of vegetables and fruits, it is not generally used at the present time. In this method, raw food is packed and then a brine or syrup is added. The containers are closed loosely and heated in a water bath at a temperature well under the boiling point of water. Sometimes the food is blanched before packing, which reduces the microflora on the raw product. Cold-pack foods are quite susceptible to spoilage and have been responsible for many outbreaks of botulism since the method was introduced in 1917. It is a very poor method for preserving vegetables and meats.

Hot Pack

In the hot-pack method, a modification of the cold pack, food is given a short precooking before packing. The temperatures obtained in the hot-pack method are comparable to those obtained in boiling or cooking food in an open kettle. The fundamental factors are temperature and time.

Oven Process

Another method of heat preservation is the oven process, which was recommended at one time by appliance manufacturers. This method is not only expensive but generally unsafe. Food is placed in presumably sterile containers and remains unsealed during the heating period in the oven. The temperature rarely reaches 212°F (100°C), and the heat penetration into the center is very slow.

Studies were conducted at Purdue University on heat penetration in the oven process. A quart of water at 68°F (20°C) was placed in a preheated oven at 275°F (135°C). It required 95-110 min for the water to reach the boiling point.

Pasteurization

One of the most common methods for heat processing foods is pasteurization. In this process, the food is heated to a high temperature and then held at that temperature for a period of time, usually 30 min or longer. The usual method for pasteurizing milk, for example, involves heating it to 142°-145°F (61°-63°C) and holding it at that temperature for 30 min.

Flash (short-time) pasteurization is gradually replacing the holding method in the pasteurization of raw milk. The milk is exposed in a thin film to a temperature of 160°-171°F (71°-77°C) for 15-19 sec. This method is more economical and has several advantages over the traditional method of pasteurization.

Many foods can be preserved, at least for short periods, by the pasteurization process. However, pasteurization should not be confused with complete sterilization, which occurs during canning of many foods and refers to the destruction of all life. Although pasteurization does not destroy all microorganisms present in a food, it can significantly reduce their numbers. For example, pasteurization of figs, prunes, and raisins at 160°-185°F (71°-85°C) with holding times of 30-90 min reduces the microbial content by about 90%.

At the present time there is a tendency to pasteurize all foods that may be consumed in the raw state. Some states have recommended, or have enacted laws requiring, the pasteurization of all milk before it is processed into cheese or ice cream. However, many kinds of food cannot be preserved by pasteurization because the heat treatment damages the physical properties, texture, or flavor of the food. The required temperature and time of pasteurization will vary with the method used and the food product.

Dried Fruits

Pasteurization of dates and figs has been recommended because of public health concerns. Some French packaged dates were examined several years ago and the investigators found large numbers of *E. coli* present. The survival of several species of intestinal organisms is a reflection of the general sanitary practices employed in handling and processing dried fruits. If dried dates are immersed in a viable culture of *Salmonella typhi,* live test organisms can be recovered 72 days after inoculation. Obviously, dried fruits can be an important factor in the spread of certain enteric organisms that are important to public health. Coliforms and *Salmonella typhi* on dried fruits are killed in 30 min at 160°F (71°C) when the humidity is 75° or higher.

Few organisms are found on sulfured fruits or on very acid fruits because of the sulfurous or natural acid present in these products. Hence, these foods are much easier to sterilize than those of lower acidity.

Seafood

Crab meat pasteurized at 145°F (63°C) for 30 min, at 150°F (66°C) for 20 min, or at 170°F (77°C) for 1 min may be kept as long as 5 weeks when stored at 41°–43°F (5°–6°C). Such meat has been shown to be free of *E. coli* and to undergo no appreciable change in color, odor, or flavor during storage.

Dairy Products

The effect of pasteurization on the microflora in milk and other dairy products is well known. The process is so firmly established in the dairy industry that it is considered a routine practice, and the results of pasteurization are unchallenged by public health officials.

Public health reports show a steady decline in brucellosis, streptococcus infections, and tuberculosis when pasteurization of milk is carried on routinely in the dairy plant. Probably no other food product has as good a public health record as dairy products. Obviously, an infected animal must be eliminated from the herd to insure the safety of the product. Good herd management together with proper pasteurization has made milk, although one of the most perishable foods, one of the safest.

Since ice cream is widely consumed, this product can be an important vehicle in the spread of pathogenic bacteria. The low temperature at which this product is kept is not sufficient to insure its safety for the public. Investigators found that samples of ice cream mix before and after pasteurization at 142°–151°F (61°–66°C) for 20 minutes had microbial counts ranging from 56,000 to 15.3 million/ml before pasteurization, and from 170 to 66,000/ml after pasteurization. Pasteurization of ice cream mix at 151°F (66°C) for 30 min reduced the average bacterial count by 98.7%. Paley and Isaacs (1941) reported that *E. coli* survived twice as long in ice cream mix as in milk at 144°F (62°C). They attributed this to the protective action of stabilizers present in ice cream mix. Some of their results are shown in Table 6.6.

Table 6.6. Survival of *E. coli* in Milk and Ice Cream at 143°F

Experiment no.	Number of organisms/ml surviving after[a]				
	10 min	20 min	25 min	30 min	35 min
Milk					
1	0	0			
2	3.7	0			
3	450	0			
4	0	0			
5	0	0			
6	29	0			
7	0	0			
8	0	0			
Average	0	0			
Ice Cream Mix					
1	6,400	1,500	170	18	35
2	7,700	0.4	0	0.4	0.4
3	120,000	1,600	36	0	0
4	890	0.6	1.2	2.4	1.2
5	5,100	180	5	0.3	0
6	2,100	14	2	1	1
7	1,100	33	0	0	0
8	44	2	0	0	0
Average	18,000	416	27	2.8	0.2

Source: Paley and Isaacs (1941).
[a] The data have been calculated on the basis of an initial 1,000,000 organisms.

THERMOPHILIC AND THERMODURIC BACTERIA

Heat-resistant bacteria, or *thermophiles,* constitute an important group of organisms in heat-treated foods. Their ability to grow at temperatures 15°–25°F (8°–14°C) above the pasteurization temperature of milk is well known. Such a heat treatment will coagulate blood serum and egg albumin and produce painful burns on the skin of human beings.

Another group of bacteria closely related to the thermophiles are called *thermodurics.* Thermoduric organisms, as the name indicates, are heat tolerant, whereas thermophilic microorganisms are considered heat lovers or heat resistant. Although thermodurics can survive pasteurization of milk for short periods of time, they cannot grow and multiply at pasteurization temperatures. In contrast, thermophiles can grow and multiply at high temperatures. Lowering the temperature to the minimum range for thermophilic bacteria greatly favors the growth of thermoduric organisms.

How much heat an organism can withstand before it is classified as thermoduric or thermophilic is not yet accurately defined. According to one definition, all organisms that are not killed by pasteurization at 143°F (62°C) for 30 min are classified as thermodurics and those that are metabolically active above 113°F (45°C) as thermophilic. An organism that grows at ordinary temperatures as well as at high temperatures is called a facultative thermophile. Obviously, most thermophilic bacteria are also thermoduric.

Heat-loving bacteria are widely distributed in nature. They are abundant in decaying organic matter, surface soil, water, sewage, hay, ce-

Table 6.7. Thermophiles of Importance to the Food Industries

| Name | Economic importance | Heat-resistant spores | Growth temperatures | | Oxygen require-ment |
			Optimum (°F)	Range (°F)	
Streptococcus thermophilus	Grow during pasteurization of milk. Ripening agent in Swiss cheese	None	120	77–140	Facultative
Lactobacillus bulgaricus	Bulgaricus milk and used in making lactic acid	None	120	77–140	Facultative
Lactobacillus thermophilus	Grow during pasteurization of milk	None	131	86–150	Facultative
Lactobacillus delbruckii	Acidification of brewery mash. Lactic acid manufacture	None	113	?	Facultative
Bacillus calidolactis	Coagulates milk at high temperatures	Yes	131–149	113–167	Facultative
Bacillus thermoacidurans	Flat-sour spoilage of tomato juice	Yes	131	?	Facultative
Bacillus stearothermophilus	Flat-sour spoilage of canned foods	Yes	122	113–169	Facultative
Clostridium thermosaccharolyticum	Hard swells of canned food	Yes	131–143	110–160	Anaerobic
Clostridium nigrificans	Sulfide-stinkers of canned foods	Yes	131	?–158	Anaerobic

reals, milk, dust, dirt, and filth, and are commonly associated with dirty dairy utensils. In fact, their frequency on milk equipment can be used as an index of the quality of sanitary practices that are followed on a dairy farm and in a dairy plant.

Root crops and grains that are grown on heavily manured soils contain large numbers of thermophilic organisms. Thermophilic bacteria do not have any public health significance. No toxins, such as those associated with food-poisoning organisms, are produced by thermophiles.

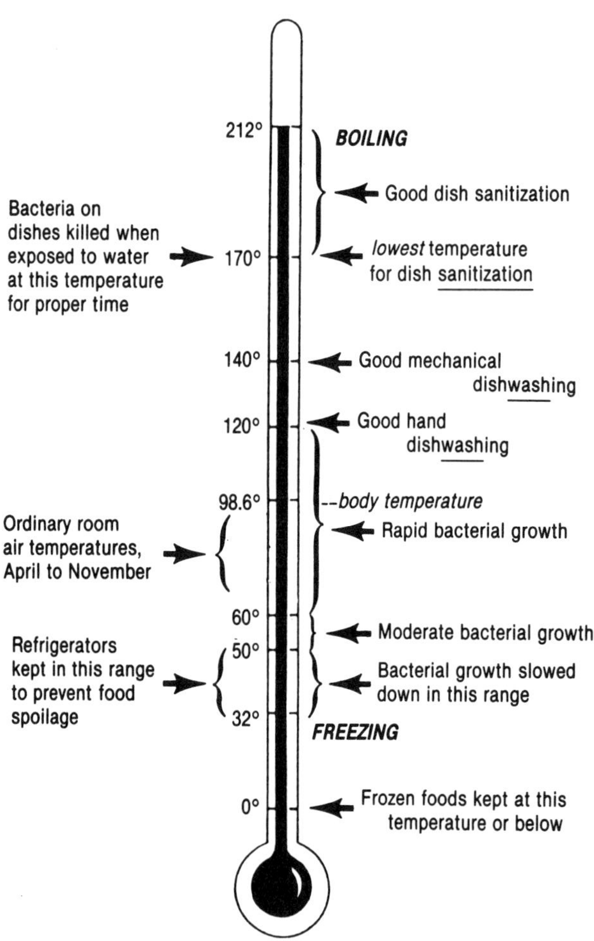

Fig. 6.2. Relationship of temperature to food sanitation. Courtesy Institutions Magazine.

Food products preserved in sealed containers should never be stored near steam pipes, boilers, hot air registers, or direct sunshine because the elevated temperatures may be sufficient to permit thermophilic organisms to grow. During the canning of foods, thermophilic bacteria may survive the heating process. If the container is airtight, anaerobic sporeformers should be suspected. Usually their metabolic activities are marked by putrefactive odors with gas formation. The majority of thermophilic organisms found in foods are aerobic or are facultative with respect to oxygen. Unfortunately, these microorganisms can survive and grow under a wide variety of conditions. Many of these organisms are sporeformers and very resistant to heat.

A summary of the properties of some important thermophiles is presented in Table 6.7. Figure 6.1 illustrates the relation between temperature and food sanitation.

SELECTED REFERENCES

Ayres, J. C., Mundt, J. O., and Sandine, W. E. 1980. "Microbiology of Foods." W. H. Freeman and Co., San Francisco, California.

Buchanan, R. E., and Buchanan, E. D. 1951. "Bacteriology," 5th ed. Macmillan, New York.

Hammer, B. W., and Sanders, L. W. 1919. A Bacteriological Study of the Method of Pasteurizing and Homogenizing the Ice Cream Mix. Iowa Agr. Exp. Stn. Bull. 186.

Jackson, J. M., and Shinn, B. M. 1979. "Fundamentals of Food Canning Technology." AVI Publishing Co., Westport, Connecticut.

Oginsky, E. L., and Umbrett, W. W. 1959. "An Introduction to Bacterial Physiology," 2nd ed. W. H. Freeman Co., San Francisco, California.

Paley, C., and Isaacs, M. L. 1941. The effect of pasteurization on *Escherichia coli* in milk and ice cream mix. *J. Dairy Sci.* 24, 421–427.

Stumbo, C. R. 1965. "Thermobacteriology in Food Processing." Academic Press, New York.

7
Low-Temperature Food Preservation

Effect of Low Temperatures and Freezing on Microorganisms   100
Physical Principles of the Freezing Process   104
   Mechanism of Freezing   105
Survival of Microorganisms on Frozen and Refrigerated Foods   106
   Fruits and Vegetables   106
   Egg Products   108
   Meat Products   108
Effect of Freezing on Certain Food Products   109
   Fish   109
   Poultry   110
   Eggs   110
   Milk Products   111
Quality of Frozen Foods   111
   Blanching before Freezing   111
   Enzyme Activity   112
   Preservation of Fruit Color   112
   Off-Flavors   113
   Vitamin Content   114
   Effect of Thawing and Refreezing   115
Use of Microwave Heat in Cooking and Defrosting Foods   116
Selected References   117

Cold has been used since prehistoric times to preserve food. The development of rapid freezing and low-temperature storage makes it possible to preserve food for a long time, to maintain the food in a good physical state, and to retain its nutritional value. The consumption of frozen foods has increased manyfold in the last 25 years and no doubt will continue to increase in the future.

Freezing is not an effective means of destroying microorganisms in foods. Although the total number of organisms usually is reduced after freezing, there may be as many species present at the end of the storage period as in the original unfrozen food.

Since cold is not an effective means of destroying microorganisms, refrigerated meats pose potential dangers of spreading certain infections of significance in public health. The danger is increased further by lack of antemortem and postmortem inspection of meat placed in home-sized lockers. The handling of such meat under unsanitary conditions and the use of infected animals makes the problem all the more important from the public health standpoint. No food product is rendered safe by refrigeration.

The importance of psychrophilic microorganisms (cold lovers) has not been fully appreciated. As a rule, this group of organisms is not pathogenic, but they can bring about undesirable biochemical changes in many kinds of foods refrigerated for a long time. These organisms have been shown to not only increase in numbers but to cause a pronounced off-flavor in market milk supplies. It is reasonable to conclude that their presence in other foods besides milk would have similar effects.

Most psychrophilic bacteria are easily destroyed by heat, since the majority of the group are non-sporeforming organisms. A few members of this group may acquire a heat-resistant tendency and are able to withstand the conventional method of pasteurization of milk. This may be one explanation for the presence of these bacteria in market milk.

Many species of bacteria are able to carry on their metabolic activities slightly above the freezing temperature of water, 32°F (0°C). Low temperature should not be relied upon to destroy microorganisms; it merely slows metabolic activities. *Staphylococcus aureus*, a food-poisoning species, has survived at 41°F (5°C) for 2 years in ice cream, although this temperature is not the optimum for this organism to multiply and produce an enterotoxin.

The basic objective in low-temperature preservation of perishable foods is to retard the metabolic activity of organisms so that their growth and multiplication are severely retarded.

EFFECT OF LOW TEMPERATURES AND FREEZING ON MICROORGANISMS

At low temperatures, cells metabolize very slowly because their enzymes function slowly. However, low temperatures *per se* do not usually cause a destruction of cell proteins. At or below the freezing point, ice crystals form. The size of the crystals depends upon the rate of freezing. If small crystals are formed, the turgidity of the cell is maintained. The mechanism of death of bacteria at low temperatures is not well understood and leaves much speculation in the minds of bacteriologists.

It is known that low temperature does affect the colloidal state of protoplasm, which in turn would tend to destroy the cell. The crystallization of water by freezing no doubt disturbs the colloidal state, or may result in the mechanical crushing of the bacterial cell. Cell death also may be due to "salt death," that is, the excessive concentration of crystalloids as the pure solvent separates from a solution. Repeated freezing and thawing may be even more destructive to the colloidal state of the protoplasm in the cell. However, bacteria that survive freezing and thawing several times usually can be kept at low temperatures for a long time.

A few bacterial species depend upon animals for their existence. When these organisms are removed from their host, they are able to survive only a very short time even at slightly reduced temperatures. Meningococci and gonococci are very quickly destroyed at freezing temperature. Likewise, *Lactobacillus acidophilus* survives much better at room temperature than at refrigerator temperatures. In one study, the rough strain[1] of *L. acidophilus* was much less viable when held at 41°–49°F (5°–9°C) than when stored at 60°–86°F (20°–30°C). When the initial titratable acidity ranged from 0.6 to 1% and the bacteria were held in a refrigerator, the original count of *L. acidophilus* decreased by approximately 90%. When the culture was stored at room temperature, only 20–75% of the test organisms died.

On the other hand, molds such as *Cladosporium, Sporotrichum, Penicillium,* and *Monilia* will grow slowly at temperatures ranging from 25° to 20°F (−4° to −7°C). Some yeasts and bacteria also will grow at these temperatures. Other bacteria, such as *Achromobacter, Alcaligenes, Flavobacterium,* and *Pseudomonas* species can grow only at temperatures above freezing.

Because of the wide use of refrigeration and freezing units to preserve food, the growth and metabolic activity of organisms at low temperatures has important consequences in terms of the flavor, texture, and safety of foods. A marked reduction in the number of organisms may take place in frozen foods. Any lethal effect probably is caused primarily by some form of denaturation and flocculation of cell proteins, or by the holding of organisms in a bacteriostatic state, comparable to the hibernation of certain animals during the cold months.

Several factors influence the destruction of microorganisms in frozen foods: (1) kind of organism, and whether or not it sporulates (mold spores are very resistant to freezing); (2) fast versus slow freezing (fast

---

[1]The term *rough* refers to the typical appearance of a colony of this strain as distinguished from the smooth appearance of another strain when both are grown on plates.

freezing seems to cause less mortality than slow freezing); and (3) kind of substrate. Sugar, salt, proteins, colloids, and fat may serve as protective agents, whereas high moisture and high acidity may speed destruction of organisms. The viability of organisms in frozen foods tends to decrease as time goes on. If death does occur, it may be caused by starvation, as the cell may be unable to use the available nutrients.

Freeze-drying, or lyophilization, is a well-known method for preserving bacterial cultures. In this process, an aqueous solution containing the organisms is placed in small containers and rapidly frozen by immersion in dry ice ($-78°C$). The frozen ampules are attached to a high vacuum and dehydrated. Lyopholized cultures may remain viable for years. A modification of this technique has been used to preserve certain products such as fruit juices. However, high cost and other limitations restrict the use of this method in the food industry.

Although cold temperatures slow the metabolic activity of most bacteria, many studies have demonstrated that various microorganisms can remain viable for long periods under extreme cold conditions. In general, sudden changes in environmental conditions, such as cold shock, slow the metabolism of organisms but do not necessarily kill them. For example, when *E. coli* were grown at $113°F$ ($45°C$) and then transferred to a culture medium at $50°F$ ($10°C$), a 90% reduction in cell number was observed.

The freezing of a liquid food product subjects the microorganisms present to certain physical and chemical changes related to the viscosity of the product and rate of crystallization. These factors tend to influence dissolved materials in the food during freezing. Obviously, if the frozen mass is not uniform throughout, there may be some areas more favorable for microbial growth and some unfavorable areas in which growth ceases and organisms are destroyed.

Staphylococcus aureus has remained viable after exposure to $-252°C$ for 2 hr. (Presumably absolute zero, $-273°C$, is the temperature at which molecular activity ceases.) Many investigators maintain that staphylococci show a greater resistance to freezing than other bacteria occurring in frozen vegetables. Staphylococci isolated from frozen vegetables have been shown to be capable of producing enterotoxin, the toxin usually associated with staphylococcus food poisoning. Many strains of staphylococci that can produce food poisoning grow very rapidly in certain kinds of vegetables held at room temperature.

Refrigeration can kill the larvae of *Trichinella spiralis* imbedded in the meat. In one experiment, 30 days at $-5°F$ ($-20°C$) were required to kill the larvae; 20 days at $-10°F$ ($-23°C$); and 12 days at $-20°F$ ($-29°C$).

Clostridium botulinum spores and toxin are very resistant to freezing. If the toxin is present in a food before freezing, the food is dangerous for consumption, but the spores may be eaten along with the food with no particular ill effects. Botulinum toxin is readily destroyed by boiling for 10 min. Growth occurs and toxin is produced at 68°F (20°C) when detoxified spores are inoculated. At 59°F (15°C), germination and toxin production from heated spores does not occur very often. Growth at 50°F (10°C) occurs only after 27 days, and germination of spores does not take place in 47 days at this temperature. Hence, frozen foods held at room temperature for a few days may be dangerous, but when kept below 50°F (10°C), the food may remain safe for a long time.

The survival and virulence of *Mycobacterium tuberculosis* (bovis) in milk from an infected cow has been studied. The milk was frozen and at periodic intervals samples were injected into guinea pigs. The infected milk produced active tuberculosis in the experimental animals after the milk was kept at 32°F (0°C) for 32 months. Moreover, large amounts of *M. tuberculosis* (bovis) inoculated into fresh normal raw milk and held at 32°F (0°C) survived up to 12 months. This observation again emphasizes the survival potential of pathogenic organisms in milk during the normal souring process. Apparently, organisms are able to adjust their enzyme systems to such an environment. However, if the pH of the milk is adjusted to the point comparable to sour milk and properly buffered, most organisms will survive only a very short time. Likewise, the natural acidity of fruit juice destroys many kinds of microorganisms very quickly.

Alternate freezing and thawing also have not proved successful in attenuating many organisms. For instance, *Bacillus anthracis* was frozen with carbon dioxide at −90°F (−68°C) and thawed in water 21 times in succession without losing any of its virulence. *Staphylococcus aureus* and *Salmonella typhi* were frozen in liquid air at −182° to −190°C for 20 hr without destroying the viability of these organisms. Likewise, staphylococci have been dried and exposed to liquid air for several months and remained viable. This treatment also failed to destroy *Brucella abortus* (suis) in the tissues of a swine carcass exposed for 30 days at −10°F (−23°C).

In another experiment, a standard ice cream mix was inoculated with *Salmonella typhi* and then frozen. The mix was placed in a hardening room where the temperature was maintained at about −40°F (−4°C). The results, shown in Table 7.1, demonstrate how difficult it is to totally destroy bacteria by cold treatment.

Many kinds of viruses and pathogenic protozoa are likewise very re-

Table 7.1. Survival of Typhoid Bacilli in Frozen Ice Cream Mix

Storage time	Typhoid bacilli/ml
5 days old	51,000,000
20 days old	10,000,000
70 days old	2,200,000
342 days old	660,000
430 days old	51,000
648 days old	30,000
2 years old	6,300
2 years, 4 months old	Viable typhoid bacilli

sistant to low temperatures and may be a threat to public health. Viruses remain viable over very long periods of time under conditions difficult for bacteria.

PHYSICAL PRINCIPLES OF THE FREEZING PROCESS

The freezing of any foodstuff involves the removal of heat from the product until its freezing point is reached, at which point a physical change takes place from the unfrozen to the frozen state. In most cases, the product is further cooled to 0°F (-17.8°C) for storage.

The initial freezing process involves cooling the material to its freezing point. Foods of low moisture content require the removal of fewer Btu's for freezing than those of high moisture content. This step in the cooling process takes place very rapidly because of the wide differences in temperature between the warm food and the cold refrigerating medium. As the food reaches the freezing temperature, the rate of heat transfer slows down. At this point the actual freezing of the food begins.

During the actual freezing process, the temperature of the food changes very little while the latent heat of fusion is removed. The latent heat of fusion ranges from 22 Btu/lb for dried beef to 134 Btu/lb for fruits and vegetables with a high water content. Table 7.2 shows the specific heat of various foods before and after freezing and the corresponding latent heat of fusion.

If the initial temperature, the freezing point, and the storage temperatures are known, it is possible to calculate how much heat must be removed per pound of food in order to freeze it from the following equation:

$$Q = W[C_1 (t_1 - t_f) + h_f + C_2(t_f - t_3)]$$

where Q = total heat removed, w = pounds of food, C_1 and C_2 = specific heats before and after freezing, t_1 = temperature before cooling

Table 7.2. Specific and Latent Heats of Foods

Food product	Specific heat (Btu/lb)		Latent heat of fusion (Btu/lb)
	Unfrozen	Frozen	
Asparagus	0.94	0.48	134
Bacon	0.50	0.30	29
Raspberries	0.85	0.45	122
Strawberries	0.92	0.47	129
Beef (lean)	0.77	0.40	100
Beef (fat)	0.60	0.35	79
Beef (dried)	0.22–0.34	0.19–0.26	7–22
Cabbage	0.94	0.47	132
Carrots	0.90	0.46	126
Eggs (crated)	0.76	0.40	90
Fish (frozen)	0.76	0.41	101
Fish (dried)	0.56	0.34	65
Green peas	0.79	0.42	106
Milk	0.93	0.49	124
Lamb	0.67	0.30	83.5
Oysters (shell)	0.83	0.44	116
Poultry (frozen)	0.79	0.37	106
Pork (fresh)	0.68	0.38	86.5
Beans (string)	0.91	0.47	128
Veal	0.71	0.39	91

Source: Anon. (1967).

($°F$), t_3 = temperature at end ($°F$), t_f = temperature at freezing point ($°F$), and h_f = latent heat of fusion.

For example, suppose you have 2 lb of shelled oysters at 55°F and wish to freeze them to 0°F. The freezing point is 30°F and the specific heat of the unfrozen product is 0.83 Btu; that is, it takes 0.83 Btu/degree/lb to lower the temperature from 55° to 30°F. Thus, the total heat loss would be $2 \times 0.83 \times (55 - 30) = 41.5$ Btu. In addition, since the latent heat of fusion is 116 Btu/lb, then a total of 232 Btu must be removed during the actual freezing process. Finally, the temperature of the oysters is brought to 0°F from 30°F by removing 0.46 Btu/degree/lb, or $2 \times 0.46 \times (30 - 0) = 27.6$ Btu. The total heat removed is the sum of these three quantities: $41.5 + 232 + 27.6 = 301.1$ Btu for 2 lb. This is approximately equivalent to freezing 2 lb of water from a temperature of 55°F and then lowering it to 0°F.

Mechanism of Freezing

Ice crystals form when food products are frozen because of the presence of free water in or on the food. In food packages, air is trapped between the wrapping material and the food or between the portions of food. During the freezing process, moisture from warm food forms as vapor on the surface of the wrapper and condenses and freezes faster than frost forms in the food. If quick freezing is employed, the moisture will freeze in place and form ice crystals.

The presence of salt or sugar in a food lowers its freezing point below that of pure water. In weak salt or sugar solutions, the freezing process results in the formation of pure ice crystals. Obviously, the remaining solution tends to become more concentrated, and the freezing point is lowered to the eutectic point where all the liquid that remains unfrozen freezes solid. Sometimes the eutectic point may be below the freezing point of water, so that some of the liquid may remain unfrozen, as in a heavy syrup.

In freezing some foods, ice crystals may form either within or between the cells. The size of crystals depends upon the rate of freezing. In slow freezing, the ice crystals tend to become larger. They actually rupture the cell walls or membranes causing a partial breakdown of the physical structure of the cell. Such a condition usually contributes to poor product quality. Fruits and vegetables that are fresh, crisp, and firm before freezing may become soft and mushy after slow freezing.

Frozen foods tend to dehydrate very rapidly while in storage unless some precautions are taken. Dehydration is caused by a difference in vapor pressure. The water vapor tends to migrate to the colder area of the cooling coils where the vapor pressure is lower. Here the vapor condenses and forms frost. Unwrapped foods are very susceptible to drying out on the surface. This is sometimes called "freezer burn" as in the case of poultry. Careful wrapping will reduce the air space and thus avoid serious drying.

SURVIVAL OF MICROORGANISMS ON FROZEN AND REFRIGERATED FOODS

Fruits and Vegetables

The fact that many fruits and vegetables are consumed in the raw state makes such products a potential source of danger, especially from the enteric organisms that cause dysentery, typhoid, and paratyphoid infections. It has been shown that certain intestinal bacteria cannot survive more than 4 weeks in cherry juice sealed in cans without heat treatment, but are viable from 2 to 3 months in frozen whole cherries. One should always bear in mind that low temperatures do not sterilize a product but only retard extensive microbial growth.

In one experiment, the reduction in microbial content of strawberries held in sealed tins and other containers for a year at 15°F (−9°C) averaged 99.3%. Fresh strawberries have been found to have 600–2000 bacteria and 80–800 molds and yeasts per gram. The microbial content of shelled beans and peas is appreciably higher than strawberries even following blanching.

Approximately 50 isolations of *Staphylococcus aureus* were studied using frozen-pack vegetables. Twelve cultures were capable of producing a potent enterotoxin. When the vegetables were maintained at a temperature of 41°F (5°C), an increase in the microflora was observed. Bacterial filtrates of the enterotoxin were potent after being held in a refrigerator for 2 months at 40°F.

According to one investigator, fresh washed mushrooms contained 10,000–300,000 microorganisms per gram. However, blanching this product for 5 min in steam reduced the viable microbial content from 275,000 to 720 per gram. Approximately two-thirds of the surviving bacteria died during 6 months of frozen storage. The coliform organisms survived blanching but disappeared after 6 months' storage at 15°F (−9°C). As expected, all the sporeforming bacteria, molds, and yeasts survived.

Studies on shelled raw peas showed 1,000,000 organisms per gram, and the microbial content increased rapidly after 6 hr at 70°F (21°C). Blanching the peas at 210°F (99°C) for 60 sec destroyed 99% of the microflora; the surviving organisms grew rapidly and the peas spoiled almost as readily as the raw peas. The growth of organisms was largely prevented in both raw and blanched peas at 32°F (0°C) for 48 hr. The usual storage temperature of frozen-pack peas is 0°F (−18°C). At this temperature, the bacterial population is appreciably reduced in 2–3 months.

A comprehensive study of numbers and types of microorganisms in asparagus, peas, beans, and corn showed a marked decrease in numbers of organisms during the first 2 weeks of storage. Then the microbial population declined slowly or remained stationary. The method of packaging such as water, brine, or dry-pack did not alter the survival of the organisms. Frozen vegetables and fruits often contain appreciable numbers of organisms even after 9 months' storage at 0°F (−18°C). Obviously when the product is defrosted, these organisms can be responsible for spoilage.

Temperatures of 41°–50°F (5°–10°C) are favorable for the growth of many psychrophilic bacteria, especially those found in vegetables.

Staphylococci and several *Flavobacterium* spp. can withstand freezing much better than other types of bacteria occurring in fresh-pack vegetables. After 9 months of freezer storage, the former types of organisms predominate.

Egg Products

In studies of eggs subjected to fast freezing by local packers, the bacterial count was found to range from 800 to 200,000 bacteria per gram of egg white, with equally high counts in egg yolks. After 9 months' cold storage of the egg products, the decrease in the microbial content ranged from 95 to 99%.

Studies on the coliform counts of fresh eggs have shown that coliform organisms constitute about 10% of the total microflora of fresh eggs. However, after 9 months of cold storage, egg products are free of coliform bacteria.

Coliforms and gram-positive cocci were the principal bacteria found in high standard plate count samples of liquid egg. Freezing at 2°–10°F ($-17°$ to $-12°C$) and storage at the same temperature for 2–3 weeks reduced the number of bacteria but did not eliminate any of the groups of bacteria studied. Pasteurization of liquid whole egg at 143°–144°F (61°–62°C) for 3.7–4.0 min reduced the standard plate count of bacteria more than 99%. This included nearly all the coliforms, gram-positive cocci, and many pathogenic gram-negative enteric bacteria.

Alcaligenes, Flavobacterium, Proteus, and *Pseudomonas* were the principal genera of bacteria found in another study of liquid and frozen egg products. All of these organisms produced spoilage when inoculated into sterile liquid whole egg.

Meat Products

In work reported by Fitzgerald (1947), pork was variously packaged and stored for 12 weeks at 0°F ($-17.8°C$) and 25°F ($-4°C$). Aerobic bacterial counts at 98.5°F (37°C) decreased under all conditions. Lipase-forming microorganisms (counted at 68°F) increased in all storage conditions. Coliform organism counts (determined by most probable number) showed a reduction approaching extinction during storage at 25°F. The count of coliforms was only slightly reduced during storage of protected samples at 0°F. The work of Fitzgerald (1947) clearly indicates that many bacteria can survive and multiply in meat at freezer temperatures. Many coliform organisms survived when they were protected in the meat samples stored at 0°F. Strict sanitation is absolutely necessary for handling meat intended for freezer storage. A low initial bacterial load on the fresh meat is necessary at the time of

Table 7.3. Approximate Number of Months of High-Quality Storage Life

Product	Storage temperature		
	0°F	10°F	20°F
Orange juice (heated)	27	10	4
Peaches	12	<2	6 days
Strawberries	12	2.4	10 days
Cauliflower	12	2.4	10 days
Green beans	11–12	3	1
Green peas	11–12	3	1
Spinach	6–7	<3	¾
Raw chicken (adequately packaged)	27	15½	<8
Fried chicken	<3	<30 days	<18 days
Turkey pies or dinners	>30	9½	2½
Beef (raw)	13–14	5	<2
Pork (raw)	10	<4	<1.5
Lean fish (raw)	3	<2¼	<1.5
Fat fish (raw)	2	1½	0.8

Source: USDA (1960); Van Arsdel (1961).

freezing to insure a low microbial count on the product when it is removed from the freezer.

The literature on meat preservation is not as extensive as that on preservation of other food products. The spoilage of meat and meat products depends on the presence and survival of saprophytic organisms including the pathogenic and toxigenic bacteria. The growth of *Pseudomonas* at subfreezing temperatures plays an important role in meat spoilage. Table 7.3 shows the storage limits for various foods at several low temperatures.

EFFECT OF FREEZING ON CERTAIN FOOD PRODUCTS

Fish

Fish may be prepared for freezing by different methods but the fillet is perhaps the most familiar product. Because of the relatively high fat content of fish, precautions must be taken to maintain good quality. The fat undergoes oxidation and is readily recognized as rancid. Any preparation method that reduces the exposed surface will tend to reduce the oxidation of the fat. In the brine-immersion process, a "protein skin" forms on the fish. Since salt tends to accelerate the oxidation of the fat, the brine treatment of fish should be followed with caution.

The darkening of the surface of herring, mackerel, and salmon is caused by oxidation of the fat.

During the cold storage of fish, the proteins are denatured, lowering their ability to absorb and hold cellular juices. When these juices are free from the cells, autolytic enzymes and certain bacteria in the juices bring about undesirable changes.

Poultry

Most processed poultry has part or all of the outer layer of skin removed during the dressing process. Scalding carcasses at temperatures above 128°F (53°C) results in removal of the skin on at least some parts of the carcass. If the carcasses are scalded at 140°F (60°C) for about 45 sec (subscalding), complete removal of the outer skin takes place upon picking. A common practice is to scald broilers for about 1 min at 125°F (52°C) (semiscald) and then scald the necks and thighs at a much higher temperature, which causes the skin to peel in these areas.

To prevent the surface from becoming darkened by dehydration, carcasses are held in slush ice water or crushed ice until they are consumed or packaged for freezing. It is important to package poultry carcasses while wet in a moisture-vapor-proof material before freezing. Since poultry fat, relative to other animal fats, is unstable, the air should also be removed from the package.

Eggs

Frozen whole eggs, whites, and yolks are the forms available to the consuming public. Freezing during the surplus season is a means of storing eggs to be used later for dried whole eggs. It is possible to keep an egg-drying plant in continuous operation by using the frozen eggs during the offseason. The baking industry uses the whole-egg product in ever increasing amounts.

Frozen egg whites are more desirable than fresh egg whites because the former have more desirable whipping properties. Consequently, this product is used as an ingredient in white cakes, biscuits, cookies, candies, ice cream, and pies.

Frozen egg yolks are very unstable. Since the yolk is rich in protein, it tends to undergo denaturation, and the colloidal structure of the fat emulsion is changed. During the freezing of egg yolk, the lecithin tends to separate and coagulate causing an undesirable lumpy condition. The coagulation of the yolk can be overcome by lowering the freezing point by adding glycerine, salt, or sugar to the yolk prior to freezing.

Plain, glycerine, or sugar yolks can be used in making biscuits, cookies, dark cake, and doughnuts. Plain and sugared yolks are used in custards and ice cream. Plain yolks are preferred in the making of noodles because processed yolks impart an undesirable flavor to the finished product.

The sanitary practices used in plants during the processing and handling of whole eggs, egg whites, and egg yolks before and during the freezing period significantly influence the quality of the final product. Cracked and overripe eggs obviously contribute to a high microbial content of the egg product.

Milk Products

The peculiar chemical and physical composition of cream and milk introduces some significant problems when these products are frozen. Freezing changes the colloidal particles so that when the frozen product is thawed, the insoluble fractions separate. This is known as "oiling off." The butterfat separates and rises to the top as a floc. Likewise, the proteins tend to precipitate when the frozen product is stored.

It is well known that high-fat cream will oil off more readily than a lower-fat product. If cream is homogenized, which is a physical process, the fat globules are greatly reduced in size and they become uniformly distributed throughout the product. Consequently, the emulsified fat is partially protected during the freezing and thawing cycle. The addition of sugar will prevent the separation of fat from frozen cream by the formation of small ice crystals. This alteration of the cream lessens the tendency of the butterfat particles to coalesce.

QUALITY OF FROZEN FOODS

Freezing and subsequent cold storage may affect the flavor, color, and nutritive value of foods. Some of the quality changes that may occur and practices that can enhance the quality of frozen foods are discussed in this section.

Blanching before Freezing

A heat treatment, or blanching, before freezing improves the keeping quality of many frozen products. Blanching accomplishes the following:

- Destruction of large numbers of surface microorganisms, which decreases the microflora during processing
- Inactivation of enzymes that could contribute to deterioration of the product during storage
- Removal of mucilage-like substance, which can contribute to off-flavors
- Color set, which helps preserve the natural color of the product
- Removal of extraneous substances that can contribute to undesirable flavors
- Shrinking, wilting, or softening of tissues, which makes the product easier to pack
- Removal of occluded or entrapped air from tissues

Enzyme Activity

Enzymes differ in their ability to survive at low temperatures. A partial inactivation of some of the more labile enzymes, such as dehydrogenases, may alter the type and rate of biochemical changes produced.

It has been reported that many enzymes survive temperatures varying from $0°$ to $-312°F$ ($-18°$ to $-191°C$) either in tissues or in solution. It is believed that ice formation in the cell has a retarding effect on enzymes in foods in the frozen state. No doubt the physical and colloidal properties of a system, when it passes from a liquid to a solid state, have an effect upon the enzymes.

Endogenous enzymes in foods continue to function slowly in cold storage. Any procedure that stops or retards enzyme activity in a food is highly desirable. Blanching before freezing or the use of chemicals such as vitamin C or citric acid is recommended for fruits or vegetables. Moreover, the addition of syrup to fruit will partially prevent oxidative enzyme reactions such as browning.

Preservation of Fruit Color

The retention of the original color in fruits during storage is very important. There are two causes of browning or discoloration; (1) chemical reaction between the tissue and atmospheric oxygen and (2) activation of enzymes. As mentioned already, blanching and/or the use of certain antioxidants are valuable in minimizing discoloration.

Blanching tends to inactivate endogenous enzymes and thus prevent biochemical changes in the product that may contribute to discoloration and off-flavor development. This method is effective and recommended for all vegetables to be frozen; however, it has a tendency to destroy some of the delicate volatile flavors in fruits.

The use of antioxidants is gaining popularity because the practice tends to avoid some of the disadvantages of blanching. Citric acid, which is an inherent and flavorful acid of citrus fruits, and other organic acids have been used with success in preventing the darkening of peaches, apples, apricots, and pears. Sulfuring of fruits, especially fruits for dehydration, has been used for a long time. Sulfites and sulfur dioxide are the common forms of sulfur used. However, certain precautions as to concentration and time of exposure must be followed in order to make a good product. The use of vitamin C (ascorbic acid), which is a good reducing agent, to prevent browning of cut fruits for freezing has the added advantage of improving the vitamin content of the fruit. More recently, combinations of antioxidants and organic acids have been used to preserve fruit color.

Off-Flavors

The changes in flavor of many vegetables during frozen storage is caused by enzymes that are not appreciably affected by the low temperature of ice formation. A temperature of $-313.6\,^\circ\text{F}$ ($-192\,^\circ\text{C}$) does not completely inhibit the action of autolytic enzymes in peas. The off-flavors of raw or underblanched vegetables can also be attributed to unknown volatile compounds.

The production of off-flavors decreases as the temperature is decreased. In one study, for example, unblanched vegetables showed marked enzymatic activity in a few weeks at $0\,^\circ\text{F}$ ($-18\,^\circ\text{C}$), but when the temperature was lowered to $-50\,^\circ\text{F}$ ($-46\,^\circ\text{C}$) no off-flavors were noticeable in many unblanched vegetables after 6 months.

A good cold-storage temperature for many products is $-10\,^\circ\text{F}$ ($-23\,^\circ\text{C}$). It is necessary that all vegetables be treated in order to stop enzyme action that may be involved. It is believed that one or more respiratory enzymes are involved in the production of off-flavors and that these off-flavors are similar to those produced in plant tissues in which anaerobic respiration[2] is occurring.

The storage periods at various temperatures during which some vegetables change perceptibly in flavor or color are shown in Table 7.4.

According to some studies, acetaldehyde is produced in the tissue of frozen peas by enzymes. This compound serves as an index of quality

[2]Incomplete oxidation with limited amounts of oxygen from sources other than air.

Table 7.4. Months to Reach a Perceptible Flavor or Color Difference

Product	Temperature		
	0°F (−18°C)	10°F (−12°C)	20°F (−7°C)
Flavor			
Green beans	10	3	1
Peas	10	3	1
Spinach	6	2	0.7
Cauliflower	10	2	0.5
Color			
Green beans	3	1	0.2
Peas	7	1.5	0.3
Spinach	—	—	—
Cauliflower	2	0.5	0.2

Source: Tressler *et al.* (1968).

since the quantity decreases with a decrease in catalase activity. Acetaldehyde was highest in off-flavored samples, and smaller amounts were present in normal flavored peas. Other vegetables such as artichoke hearts, green beans, asparagus, lima beans, brussels sprouts, and squash showed considerable variation in acetaldehyde content. Consequently more work must be done to correlate the amount of this compound present with the flavor and keeping quality of these products.

Vitamin Content

Low temperatures do not destroy vitamins in foods. During the storage, however, certain microorganisms and enzymes may alter the physical characteristics of the food to make conditions unfavorable for the retention of certain vitamins. Frozen foods tend to dry out, if not properly packaged, at temperatures ranging from 0°F to 20°F (−18° to −7°C). Ascorbic acid content is rapidly lost as the temperature increases. Thiamine in pork is not particularly affected by temperature fluctuations or period of storage in the freezer, although rancidity usually increases at 0°F (−18°C) or above.

Ascorbic Acid Changes

One way to assess changes in the nutritive value of a food product during cold storage is to study the ascorbic acid (vitamin C) content. Since this vitamin is sensitive to heat and oxidation and is soluble in water, it serves as a useful indicator for detecting any changes in the nutritive value of foods.

Low storage temperature tends to slow the disappearance of ascorbic acid normally present in milk. When free oxygen is removed by spraying the warm milk into a vacuum chamber, oxidation is retarded. For this reason the sooner fresh milk is deaerated, cooled, pasteurized, homogenized, or ascorbic acid added, the better milk will keep.

The blanching process destroys some of the vitamin C in foods. Foods naturally high in this vitamin lose a substantial amount during frozen storage. Cauliflower has a high vitamin C content, and may retain its value through several months of storage but loses about one-half of its vitamin C by the end of a year. Peas, snap beans, and asparagus are low in vitamin C. Their loss of vitamin C is usually small in spite of blanching and frozen storage. Frozen cranberries retain their color and flavor because their skin is tough and serves as a protection against biological changes.

Effect of Thawing and Refreezing

Many foods are affected by thawing. In fact, to prevent loss of quality, some foods should not be thawed prior to cooking. The rate of thawing and the medium used play an important role in the final quality of food.

Generally, the faster a food is thawed, the more likely will be the maximum retention of quality. In foods where there is a leakage of fluid, a slow rate of defrosting is desirable because this procedure will allow the solids to reabsorb some of the fluid, which may exist as ice crystals. Microwave heating has many advantages over the conventional methods of defrosting foods because heat is generated in the interior of the food.

The method of thawing also influences the microbial count of the thawed products, as is illustrated by the data in Table 7.5 on frozen eggs.

The practice of refreezing foods should be discouraged. Although the microflora of frozen foods is largely made up of soil organisms, which may not be significant from the public health standpoint, they are capable of causing spoilage. In addition to the microflora of the frozen product, enzymes are present that may induce undesirable changes. Enzymes, like microorganisms, are active at certain temperatures especially when the food is in the process of thawing. The microbial content of different foods after thawing for 24 hr is shown in Table 7.6.

Table 7.5. Increase in Microbial Count of Frozen Whole Egg Meats during Thawing

Thawing method	Hours required	Increase in microbial count (%)
In air at 80°F (27°C)	23	1,000
In air at 70°F (21°C)	36	750
In air at 45°F (7°C)	63	225
In running water 60°F (16°C)	15	250
In running water 70°F (21°C)	12	300
Agitated water 60°F (16°C)	9	40
Dielectric heat	15 min.	Negligible

USE OF MICROWAVE HEAT IN COOKING AND DEFROSTING FOODS

In microwave heating alternating current of a few megahertz frequency is applied through condenser plates or electrodes between which the food is placed in insulating containers such as glass. The food becomes the dielectric of the capacitor. Its dielectric properties are such that strong currents are induced in all parts of the food. This raises the temperature of the food much more quickly than conventional methods, which depend on leading heat in from the outside. Because little or no heat is generated in anything but the food, the method is efficient and economical in the use of power.

Experiments have been conducted using microwave heat. *Aspergil-*

Table 7.6. Bacterial Counts in Frozen Foods after 12 Months' Frozen Storage and after Thawing 24 Hours at 70°F

Product	Bacteria per gram	
	Frozen	After thawing
Beef stew	390	1,400,000
Beef steak	390	1,400,000
Carrots scalded	3,000	5,800,000
Eggs (in tin)	190,000	70,000,000
Green beans scalded	1,000	40,000,000
Haddock	38,000	770,000
Oysters	22,000	320,000,000
Peaches, with sugar 3:1	60	700
Peas scalded	1,000	24,000,000
Pork chops	1,300	8,700,000
Raspberries, with sugar 3:1	3,000	8,000
Sour cherries, with sugar 3:1	0	20
Strawberries, with sugar 2:1	200	2,000
Sweet corn scalded	1,500	60,000,000

lus, Rhizopus, and *Penicillium* spp. on unwrapped bread were reduced after exposure to microwave for 2 min. Other work indicates that the use of microwaves for freeze-drying deboned poultry meat results in more rapid and uniform removal of water vapor than with conventional units. Litton Food Industries reported the use of a conveyor-type microwave oven for precooking chicken parts.

Microwave cooking does not eliminate strains of bacilli, staphylococci, and clostridia. Microwave exposure for 3 min was an unreliable method of reducing bacterial counts in mashed potatoes, and salmonellae survived in chicken parts cooked by microwaves.

SELECTED REFERENCES

Anon. 1961. *Proc. Low Temperature Microbiol. Symp.* Campbell Soup Co., Camden, New Jersey.

Anon 1967. Handbook of Fundamentals, Heating, Refrigerating, Ventilating and Air Conditioning. Am. Soc. Heating Refrigeration and Air Conditioning Engineers, New York.

Anon. 1982. Refrigeration Data Book. American Society of Heating and Refrigeration Engineers, Atlanta, Georgia.

De Figueirdo, M. P., and Splittoesser, D. F. 1976. "Food Microbiology: Public Health and Spoilage." AVI Publishing Co., Westport, Connecticut.

Derosier, N., and Tressler, D. K. 1977. "Fundamentals of Food Freezing." AVI Publishing Co., Westport, Connecticut.

Fitzgerald, G. A. 1947. Are frozen foods a public health problem? *Am. J. Public Health* **37**, 675–701.

Gunderson, M. F., and Rose, K. D. 1947. Survival of bacteria in a precooked frozen food. *Food Res.* **13**, 254–263.

Haines, R. B. 1938. The effect of freezing on bacteria. *Proc. R. Soc. London* **124**, 451–463.

Paul, C., Wiant, D. E., and Robertson, W. F. 1949. Freezing Temperature and Length of Frozen Storage for Foods Frozen in Household Freezers. Mich. State Coll. Agr. Exp. Stn. Bull. *213.*

Tressler, D. K., Van Arsdel, W. B., and Copley, M. J. (Editors). 1968. "The Freezing Preservation of Foods, Vol. 2. Factors Affecting Quality in Frozen Foods." AVI Publishing Co., Westport, Connecticut.

USDA 1960. Conference on Freezer Food Quality. U.S. Dept. of Agric. Agric. Res. Ser. 74–21.

Van Ardsdel, W. B. 1961. Alignment chart speeds computation of quality changes in frozen foods. *Food Process.* Dec. 1961.

8
Acids and Alkalies in Food Preservation

Acid-Producing Organisms   120
Acids as Germicidal Agents   120
  Acetic Acid   121
  Benzoic Acid   123
  Boric Acid   123
  Citric Acid   123
  Formic Acid   124
  Lactic Acid   124
  Salicylic Acid   124
  Hydrochloric Acid   125
  Pyroligneous Acid   125
  Sulfurous Acid   125
  Monochloracetic Acid   126
  Propionic Acid   126
  Sorbic Acid   127
  Vanillic Acid   127
Alkali-Producing Organisms   127
Alkalies as Germicidal Agents   128
Selected References   130

The food industry has an acute need for nontoxic materials that are effective in preserving perishable food products not stored under refrigeration. Although the addition of certain chemicals to foods can delay the action of spoilage organisms, the use of chemical preservatives must be carefully regulated. Otherwise, the practice of adding preservatives to unwholesome foods can lead to unscrupulous practices by the food industry. In the past, for example, food dealers added formaldehyde to poor-quality milk to prolong sweetness. In some instances, catsup canneries have used sodium benzoate as a cheap substitute for cleanliness and good-quality tomatoes. Nowadays, the use of preservatives in commercial food processing operations is governed by federal regulations.

In general, good clean food that is produced, processed, and handled in a sanitary manner does not require addition of preservatives. Nonetheless, the use of preservatives with certain products is a common practice to extend their shelf life and preserve their quality. The following preservatives are permitted under selected conditions.

- Acetic acid used as vinegar in pickling
- Benzoic acid in the form of the acid or as benzoates
- Sulfur dioxide (SO_2) in which the foods are exposed to the sulfur fumes, forming sulfurous acid
- Carbonic acid in the form of CO_2 maintained under pressure
- Propionic acid in the form of calcium and sodium salts to delay mold growth
- Sorbic acid as a fungistat in bread and cheese wrappers
- Antioxidants to delay oxidation, thus preventing off-flavors and rancidity

As this list indicates, acids or their salts are among the most common compounds used as food preservatives.

ACID-PRODUCING ORGANISMS

Many microbial species produce organic acids during normal metabolism. The fermentations that typically occur in a food rich in carbohydrates may produce enough acid to suppress further microbial activity. For example, the souring of milk results from a normal fermentation in which appreciable amounts of lactic acid are formed. This acid helps to preserve the milk from further decomposition, so it remains sour for a long time without spoiling. However, if the acid is neutralized, mold growth is encouraged. Molds have the ability to use at least a certain level of acid as a source of energy, thus increasing the pH nearly to neutral. Under these conditions the milk may then be spoiled by bacteria. Sauerkraut and pickles are examples of other foods preserved by acids produced during fermentation.

ACIDS AS GERMICIDAL AGENTS

The observation that acid-producing fermentations often produce an environment unfavorable to growth of many microorganisms suggests that organic acids might be useful preservatives for a variety of foods. Information on the germicidal properties of various acids is presented in this section.

Acetic Acid

The toxicity of acetic acid and other organic acids seems to be caused by factors other than pH alone. The idea that the undissociated molecule is toxic to many organisms associated with food spoilage seems to be an acceptable explanation.

One study showed that acetic acid adjusted to a certain pH was more toxic to bacteria, molds, and yeasts than were lactic acid or hydrochloric acid at the same pH. The addition of sugar syrups or salt brines did not appreciably alter the toxicity of the acids studied. It appeared that acetic acid was inhibitory to selected microorganisms in direct proportion to the amount of acid used.

When different concentrations of acetic acid were prepared and various test organisms were exposed, the following results were observed: 2% acid destroyed *Pseudomonas aeruginosa* in 15 min; 3% killed *Salmonella typhi;* 4% killed *Escherichia coli;* and 9% killed *Staphylococcus aureus.*

Commercial strength vinegar is usually less than 5% acetic acid. A normal solution of acetic acid is approximately 6% and has a pH of 2.36. Table 8.1 shows the effectiveness of acetic acid in inhibiting and destroying various microorganisms.

It appears that acetic acid has a greater germicidal effect against *Staphylococcus* food-poisoning strains than other acids. Several workers have studied the dissociation of organic acids in relation to food-spoilage organisms. They observed marked variation in the dissociation constants of different acids and in their effect on a variety of microorganisms involved in undesirable biological changes in foods.

They concluded that the germicidal and antiseptic effect of organic acids are not directly related to the hydrogen ion concentration (pH)

Table 8.1. Inhibiting Action of Acetic Acid on Various Microorganisms

Organism	Inhibiting pH[a]	Lethal pH[b]
Salmonella aertryche	4.9	4.5
Micrococcus aureus	5.0	4.9
Phytomonas phaseoli	5.2	5.2
Bacillus cereus	4.9	4.9
Bacillus mesentericus	4.9	4.9
Saccharomyces cerevisiae	3.9	3.9
Aspergillus niger	4.1	3.9

Source: Levine and Fellers (1940).
[a] The pH at which no visible growth occurred yet the organisms remained viable.
[b] The pH at which the organisms were destroyed.

Table 8.2. Antiseptic and Germicidal Effect of Acids in Decreasing Order of Effectiveness

Acid	Germicidal pH	Inhibiting pH
Acetic	4.37	4.59
Citric	3.87	4.06
Hydrochloric	3.80	4.27
Lactic	2.43	2.94
Malic	3.74	3.98
Tartaric	3.65	3.92

produced. The summary in Table 8.2 shows the relative antiseptic and germicidal effects of different acids.

Apparently the action of HCl, a highly dissociated mineral acid, is caused by the hydrogen ion concentration, while the germicidal and antiseptic actions of the organic acids is not in proportion to the pH produced. Perhaps the unionized molecule or the anion or both are additional factors that affect the survival of microorganisms in the presence of organic acids.

Vinegar added to foods prior to heat treatment may increase the effectiveness of the treatment in destroying organisms. As the data in Table 8.3 show, a decrease in pH reduces the lethal temperature for some bacteria.

Table 8.3 Effect of pH on Lethal Temperature for Selected Microorganisms

Organism	pH	Lethal temperature[a]	
		C°	F°
Salmonella aertrycke	6.0	55	131
	5.0	50	122
Mirrococcus aureus	6.6	65	149
	5.5	60	140
Bacillus cereus	6.6	100	212
	5.5	60	140
Bacillus mesentericus	6.6	100	212
	6.6	60	140
Saccharomyces cerevisiae	6.8	60	140
	4.5	60	140
Aspergillus niger	6.8	60	140
	6.0	60	140
	4.5	60	140

Source: Levine and Fellers (1940).
[a] With 10-min exposure.

Benzoic Acid

Benzoic acid, especially in the form of salts, has been used in food preservation for a long time. At a concentration of 0.2%, sodium benzoate retards the growth of *E. coli* in glucose broth but is ineffective when calcium carbonate is added. *Saccharomyces cerevisiae* fails to grow in the presence of 0.5% sodium benzoate and is not viable in a 3% solution. These findings seem to show that sodium benzoate can be safely added to acid fruit juices and other acid foods to retard fermentation without being harmful to humans. In the case of low-acid foods, benzoates will not work, but sorbic acid and sorbates can be used.

A mixture of borax and sodium benzoate has been used for many years with good results in the salting of cod fish. The Canadian Pacific fisheries used a 15-min dip of the fillets in 20% salt brine plus 0.1% benzoic acid as a routine procedure in handling their fish in storage. This treatment definitely prevents the formation of trimethylamine, an objectionable product of bacterial activity. Antioxidants are important in delaying undesirable changes such as oxidative rancidity in mackerel, herring, and salmon. These fish have appreciable amounts of unsaturated fatty acids, which readily oxidize and turn rancid. To prevent these changes, various methods such as storage in carbon dioxide, the addition of ascorbic acid, and finally ice-glaze coating on the fish may be used. It has been suggested that benzoate offers greater effectiveness in controlling the taste and odor of fish fillets than in suppressing the total number of bacteria. Perhaps benzoic acid affects only certain kinds of bacteria.

Boric Acid

Boric acid has been used in the preservation of milk samples for bacteriological examination. However, in most instances 1% boric acid solution fails to prevent most non-sporeforming organisms from growing. The borate anion has been found to be less toxic than the benzoate and more toxic than the salicylate anion.

Citric Acid

Citric acid is a tribasic acid whose normal solution of around 6% has a pH of 1.73. This acid is more effective than either acetic or lactic acid in lowering the pH even though the citrate anion is thought to be less toxic to most bacterial cells.

It has been known for a long time that citrates are an excellent source of carbon for many kinds of bacteria. In fact, nonfecal types of coliform bacteria are differentiated from the fecal forms such as *E. coli* by their ability to use citrates as their sole source of carbon.

Formic Acid

The germicidal action of formic acid is caused by the hydrogen ion concentration. Although not considered to be an efficient germicide, formic acid has been used as a food preservative. In one study, when *Salmonella typhi* was exposed to 0.5% formic acid, the organisms were killed in 25 min, but with 0.1% formic acid, 205 minutes were required to kill the organisms.

Lactic Acid

Lactic acid is reported to be less active as a germicide than acetic or citric acids but much more active than malic or tartaric acids. Lactic acid at pH 5.2 is effective in destroying *Salmonella aetrycke* and at pH 4.27 inhibits the growth of food-poisoning staphylococci. A normal solution of lactic acid is around 9% and has a pH of 1.9.

Food microbiologists are familiar with the production of lactic acid in milk by lactic acid-producing bacteria. Lactic acid exists in milk partly as lactate and partly as a free acid. The buffering action of the milk makes it difficult to assess the true effect of lactic acid upon the survival of microorganisms.

Lactic acid has been observed to inactivate *Pseudomonas aeruginosa* in a 0.3% solution, while *Salmonella typhi* required almost twice this concentration. Escherichia coli required 2.25% lactic acid and *Staphylococcus aureus* 7.5% for inactivation.

Not all species of milk-souring organisms are tolerant to the acids they produce. As a matter of fact the acids formed are toxic to most organisms after sufficient time. However, the buffering action of milk helps to prolong the viability of these bacteria.

Salicylic Acid

Since salicylates have been widely used to inhibit the growth of bacteria, salicylic acid has been suggested as a germicidal agent in the control of food-spoilage organisms. Although salicylic acid has been found to be a very weak germicide, it is effective as a bacteriostatic agent. A few enthusiastic housewives have related experience in canning peaches with aspirin, which contains salicylic acid. The idea spread from one household to another resulting in specific directions as to the use of aspirin. Obviously such a procedure is not economical nor practical, and such practices or fads should be discouraged. This use of aspirin would not be permitted by the U.S. Food and Drug Administration in commercial practice.

Hydrochloric Acid

The effects of hydrochloric acid on the preservation of certain canned vegetables have been studied. In the subsequent preparation of the canned product for the table, the addition of a little baking soda can be used to remove the sour taste. In one study, it was found that 25 ml of 0.5 N hydrochloric acid added to 1 pint of snap beans gave a pH of around 3.6; in mixed vegetables, a pH of 3.5–3.7 was obtained. Obviously, such pH values are unfavorable for growth of bacteria and for most enzyme activity. Cans of snap beans and cans of mixed vegetables were stored for 6 months to 1 year without any appreciable spoilage occurring in the products.

One quarter of a teaspoon of baking soda was found to be adequate to remove the sour taste and yet not completely neutralize the product, which would tend to destroy vitamin C during the heating preliminary to serving. After neutralization by baking soda, the beans had a pH of 6.2, and the vegetable mixture a pH of 4.9–5.0.

Pyroligneous Acid

Pyroligneous acid is a mixture of acetic acid, methanol, wood oils, and tars and is formed and absorbed by meat during the smoking process. Although the underlying principle involved in the smoking of meats is not well understood, it is known that a dehydration process takes place and a hard rind is formed, all of which are unfavorable to microbial growth. In addition, the absorption of formaldehyde by the meat and the formation of other chemical compounds play an important role.

There are certain commercial preparations on the market as substitutes for smoking, such as smoked salt and smoking liquids. Some mixtures contain pyroligneous acid, while others are made from the condensation of hickory wood smoke upon salt.

Sulfurous Acid

Sulfur fumigation has been used in the past following outbreaks of contagious diseases because the burning of sulfur was considered an excellent germicide. Later sulfur dioxide was used to kill insects. It is not effective against microorganisms unless moisture is present to form a film around the cell. The moisture and sulfur dioxide combine to form sulfurous acid, which is a fairly effective germicide. Its effectiveness is thought to be caused primarily by its hydrogen ion concentration.

Sulfurous acid is not widely used as a food preservative except in wineries. Sulfur dioxide is objectionable because it attacks metals very readily and may cause allergic reactions. There are many other more effective germicides that have fewer disadvantages. The sulfuring of certain kinds of dried fruits is a common practice. The formation of sulfurous acid on the sulfured fruits improves the keeping quality of these products.

Monochloracetic Acid

Monochloracetic acid is used as a preservative for foods and beverages, such as fruit juices, soft drinks, beer, and wine, although it possesses certain disadvantages that may limit its acceptance. In concentrations of about 5%, it may cause local irritation of mucous membranes or skin. When used as a beverage stabilizer in concentrations of 0.05%, it poses no significant public health hazard. At the present time, however, the use of monochloracetic acid is prohibited by the Food and Drug Administration.

This acid is comparable to benzoic acid in its preservative action. Both chemicals are more effective against yeasts than against acid-producing bacteria.

Propionic Acid

Sodium or calcium propionates (salts of propionic acid) are compounds, which are inhibitory toward molds and many undesirable bacteria found in food products, and are widely used. For example, butter wrapping paper and butter tubs treated with calcium or sodium propionates help protect butter against mold growth. Smoked fish fillets, smoked meats, and many other foods have been similarly treated with good results.

A concentration of 0.1% propionic acid, with a pH of about 3.5, is sufficient to protect fruit fillings from mold growth for long intervals. Smoked fish fillets dipped in 6% propionic acid also are markedly resistant to mold growth. However, this practice is not permitted by the U.S. Food and Drug Administration.

Propionic acid as a rule is nontoxic, since it is a normal constituent present in the body resulting from the metabolisn of fatty acids. Approximately 2–3 oz of calcium propionate is added to each 100 lb of wheat flour used for breadmaking. The amount is usually double in dark bread.

Sorbic Acid

Sodium, calcium, and potassium salts of sorbic acid are used as fungistats in apple cider, cheese, cakes, dried fruits, macaroni salads, and similar preparations.

Vanillic Acid

Workers at the Institute of Paper Chemistry (Pearl 1945) developed a process for the quantitative transformation of vanillin to vanillic acid and its esters. It was found that these esters have inhibiting properties. Vanillic acid is a derivative of p-hydroxybenzoic acid, and esters formed from the latter acid are known to have the properties of an acceptable preservative. Salt fish, fresh fruit juices, vegetable juices, cheese spreads, and bread have been tested using vanillic acid esters as the preservative.

The toxicity of ethyl vanillate as compared with benzoate was studied in rabbits, guinea pigs, and rats by the Kettering Research Laboratories of Applied Physiology of the University of Cincinnati.[1] When administered in olive oil, the two agents had about the same lethal effect on the test animals. However, in aqueous suspension, the vanillate was less toxic than the benzoate.

ALKALI-PRODUCING ORGANISMS

Alkali-forming bacteria in milk and other food products have been defined as bacteria that produce an alkaline reaction in litmus milk in 14–18 days, with no visible evidence of peptonization. It has been suggested that the alkalinity is caused by the fermentation of citric acid salts in milk to alkali carbonates. Perhaps other salts of organic acids may cause an alkaline reaction if they are present in milk. Apparently, ammonia is not involved in this reaction.

Alkali-forming bacteria are thought to belong to a large group of

[1]Private communication.

organisms normally found in milk. The ability of many kinds of micro-organisms to attack proteins, liberating basic products, must be considered along with the true alkali-forming bacteria. A few species of molds are capable of forming basic products when permitted to grow in certain foods.

The presence of alkali-forming organisms in food is usually undesirable, especially when they gain ascendancy over the acid-forming microorganisms. However, these bacteria may be desirable in the soft cheese industry to break down casein, a desirable process in the making of cheese.

ALKALIES AS GERMICIDAL AGENTS

The germicidal action of strong alkalies upon microorganisms has been recognized for a long time. The use of lye made from wood ashes for scrubbing contaminated objects dates back to antiquity. The germicidal efficiency of an alkali is reduced by the presence of organic matter that has the ability to neutralize part of the alkalinity.

Gram-positive bacteria are very resistant and bacterial spores are extremely resistant to the action of alkalies. Gram-negative nonsporulating organisms are very susceptible to alkalies, as are filtrable viruses. The cocci and non-sporeforming bacteria are more resistant to strong alkalies. The acid-fast group is very resistant. *Mycobacterium tuberculosis* can withstand a 2-hr exposure to 2% sodium chloride or 10% calcium chloride. Protozoan parasites seem to be very resistant to alkali solutions.

In general, a pH of 10–12 is lethal for most organisms with the exception of bacterial spores. McCullough (1945) found household lye to be effective in destroying streptococci, even in the presence of 5% skim milk, as shown in Table 8.4.

A 5.6% solution of potassium hydroxide has been found to kill the spores of *Bacillus anthracis* in 10 hr. This concentration is equivalent to a 4% solution of sodium hydroxide. A 5% solution of potassium hydroxide has been recommended as an effective disinfectant against anthrax organisms.

Household lye has been recommended as a practical and economical disinfectant to use around the farm. A solution of lye containing 13 oz dissolved in 2 gal of water, will kill most bacteria except members of the acid-fast group and sporeforming organisms.

There are many alkalies that have been recommended with varying

Table 8.4   Effect of Lye on *Mastitis Streptococci* at 40°C in the Presence of 5% Skim Milk

		Survival after[a]			
97.5% NaOH	pH	1 min	2 min	5 min	10 min
1:300	11.22	+	−	−	−
1:350	11.21	+	−	−	−
1:400	11.20	+	+	−	−
1:500	11.18	+	−	−	−

[a] + indicates growth in subculture; − indicates no growth in subculture.

degrees of effectiveness as a disinfectant. Trisodium phosphate should be mentioned since it is widely used as a detergent in the dairy industry and for washing dishes and glassware where a bright, shiny surface is desired. This compound will destroy less-resistant pathogens. Time of exposure and concentration are two important factors in determining the potency of trisodium phosphate.

One early report suggests that temperature has little effect on the germicidal action of alkalies. In this study, no increase in the germicidal efficiency of sodium hydroxide occurred as the temperature was increased from 77° to 104°F (25° to 40°C).

The hydroxyl ion concentration, or the degree of alkalinity, is largely responsible for the germicidal action of the strong alkalies. However, other factors such as cell permeability, surface tension, oxidation–reduction potential, and osmotic pressure may modify the germicidal action of an alkali.

Neutral salts tend to reduce the dissociation of sodium hydroxide and may alter the resistance of the bacterial cell to hydroxyl ions, or these salts may influence the permeability of the cell wall. If sodium chloride is added to a sodium hydroxide solution, the germicidal efficiency is not appreciably increased at ordinary temperatures. But at higher temperatures the action is much greater. This may be caused by the increased permeability of the bacterial cell walls.

If a weak alkali is added to a sodium hydroxide solution, the germicidal efficiency is increased, especially in the presence of organic matter. In one study, sodium chloride or sodium carbonate added to sodium hydroxide solutions increased the destruction of bacterial spores at 122° and 140°F (50° and 60°C). The killing time was reduced 28% by the addition of sodium chloride and 30% by the addition of sodium carbonate. This effect was much more pronounced at 104°F (60°C) than at the lower temperature.

SELECTED REFERENCES

Bosund, I. 1962. The action of benzoic and salicylic acids on the metabolism of microorganisms. *Adv. Food Res.* 11, 331–353.

Levine, A. S., and Fellers, C. R. 1940. Action of acetic acid on food spoilage microorganisms. *J. Bacteriol.* 39, 499–515.

Mountney, G. J., and O'Malley, J. 1965. Acids as poultry meat preservatives. *Poultry Sci.* 44, 582–586.

McCullough, E. C. 1945. "Disinfection and Sterilization," 2nd ed. Lea and Febiger, Philadelphia, Pennsylvania.

Myers, R. P. 1929. The germicidal properties of alkaline washing solutions. *J. Agric. Res.* 38, 521–563.

Pearl, I. A. 1945. Vanillic acid esters as preservatives. *Food Ind.* 17, 1173.

Reid, J. D. 1932. The disinfectant action of certain organic acids. *Am. J. Hyg.* 16, 540–566.

9

Sugar and Salt in Food Preservation

Salt Preservation of Foods   132
  Response of Microorganisms to Salt   132
  Effect of Salt on the Keeping Quality of Cream   134
  Preservation of Meat by Salt   134
Halophilic Organisms   135
Use of Sugar in Preserving Foods   136
  Microbial Content of Sugar Products   138
  Cellulose   138
Selected References   139

Preservation of meats and vegetables with salt brine or dry salting is one of the oldest techniques known. No doubt salt was added to the food as a seasoning and was observed to preserve the food as well.

Salt or sugar in solution exerts an osmotic pressure and thus affects the growth of microorganisms. Osmotic pressure depends upon the number of molecules in solution. Equal amounts of salt therefore exert greater osmotic pressure than sugar because of the different mass of the molecules.

As discussed in Chapter 5, the most favorable environment for bacterial growth occurs when the bacterial cell contents have the same concentration as the surrounding medium; that is, the medium is isotonic. If conditions are favorable, organisms grow most rapidly in isotonic media. However, if the solute concentration of the medium differs from that of the cell, growth is retarded and an organism may die.

Plasmoptysis occurs when the medium is hypotonic, i.e., has a lower concentration than the cell contents. Water passes from the medium to the cytoplasm, causing the cell to burst in some cases. This condition has very little application to food preservation. Plasmolysis, which occurs in hypertonic solutions, has a practical application. In this case, the medium or food has a higher concentration of solutes than the cell. Since water passes from the higher water concentration to lower concentration, water passes out of the cell, and the cell shrivels. The use of sugar and salt in preserving foods applies this phenomenon. For example, the bacterial spoilage of butter is inhibited in part by the salt incorporated in the water dispersed throughout the butter.

Salt and sugar, have much the same effect upon microorganisms. If the salt or sugar concentration is sufficiently high, it acts as a preservative by increasing the osmotic pressure. The moisture in the food is withdrawn and the tissues are plasmolyzed, creating a condition where there is insufficient moisture for growth of microorganisms.

SALT PRESERVATION OF FOODS

Salt is added to many foods, such as butter, cheese, cabbage, cucumbers, green tomatoes, eggs, meat, fish, and bread to control microbial activity partially as well as to give a characteristic flavor to the food. The concentration of salt determines what type of organisms will grow, which influences the type of fermentation. In many cases undesirable organisms are held in check, thus favoring a desirable fermentation.

Response of Microorganisms to Salt

Many microorganisms are salt tolerant, and some can tolerate a salt concentration up to 25%. Other bacteria require a high salt concentration for optimum growth; these are called *halophilic* bacteria. Some molds are more resistant than bacteria to high concentrations of salt. Yeasts as a rule are very susceptible to salt, except certain species in the genus *Torula*, which are comparable to bacteria and molds in tolerance. As a rule, the common mold species are inhibited in salt concentrations of 20–25%. The U.S. Dept. of Agriculture suggests a brine strength of 18–25% to check microbial growth. Many bacilli are not able to grow in salt concentrations over 10%, although a few can tolerate concentrations much greater than this. Staphylococci are killed in 20% salt and inhibited by 15%.

Several studies may be summarized to illustrate the effect of salt upon microorganisms. A study of 36 nonpathogenic bacteria showed that none of them grew in concentrations of over 16% salt. Members of the genus *Torula* were the most resistant of all the yeasts studied. *Clostridium botulinum,* an anaerobic bacterium, was inhibited under certain conditions at 10% salt concentration.

The growth of *Clostridium botulinum* in different culture media all containing the same concentration of NaCl differed depending on the medium. In glucose agar, 6.5% salt inhibited the growth of this organism. In one experiment with 6.7% salt in nutrient broth, there were visible signs of growth for one culture and five strains were able to produce toxin. The addition of 7.8% salt to the broth was necessary to completely inhibit growth of *C. botulinum*. It was also observed that 7.3% salt in glucose broth and 7.8% in pork infusion broth allowed growth and toxin production to take place. In glucose broth, 12% NaCl was required to inhibit growth completely. When the pork was cooked and 5% NaCl added, toxin production was checked.

Many species of aerobes, facultative anaerobes, and true anaerobes are able to grow in high concentrations of salt brines containing large pieces of animal tissues. The growth is largely confined to the interfaces of brine and tissue, and does not occur in the brine itself. In 15–20% brines, the growth of salt-tolerant microorganisms was stopped even if whole blood was added. If pieces of meat were substituted for the blood, growth took place at the interfaces.

Chemical and Physical Effects of Salt

Microorganisms may be sensitive to the toxicity of sodium. Sodium actually combines with protoplasmic anions of the cell and thus exerts a toxic effect upon the organism. The chloride ion also is toxic. The chloride ion is firmly bound to the sodium ion and the amount of chlorine liberated is very small unless a great deal of energy is exerted to break the bond between the sodium and chloride ions. However, if free chlorine is liberated, it may combine with the cell protoplasm and thereby cause death of the cell.

It is believed that salt interferes with the enzyme systems, although the mechanism is not well understood. In any case, salt has a marked effect in suppressing the growth of many undesirable organisms in food products.

Salt in low concentrations may accelerate the growth or favor certain physiological activities of certain organisms. Esty and Meyer (1922) reported that 0.5–1% salt increased the thermal resistance of *C. botulinum* spores. If the salt was increased to 2% this effect was lost.

Sodium chloride has a very definite lethal action in the presence of heat against many kinds of micrococci. Likewise, alkaline solutions of salt tend to reduce the thermal death times of spores. It is the opinion of several investigators that NaCl in concentrations used in food preservation is not a bactericide but serves as a bacteriostatic agent against most species of microorganisms.

The principles of food preservation by salting or the use of sugar solutions emphasizes the maximum limit for osmotic pressure for many kinds of microorganisms. Obviously, the effectiveness of these methods depends very largely upon the presence of osmophilic organisms naturally present in food. The molecular weight of salts and their ability to ionize determine the critical osmotic concentration. At certain concentrations of solutes, a toxic reaction may take place that is entirely independent of osmotic pressure.

In the food industry it may be difficult to separate chemical effects and osmotic effects. In the case of sugars, definite osmotic pressures can be observed for many kinds of microorganisms. On the other hand, varying concentrations of salt will support many kinds of organisms even though the compositions of the salts vary widely. For example, Na^+, K^+, Li^+, Mg^{2+}, and Ca^{2+} all supported growth, in a decreasing minimum total ionic concentration, of an obligate anaerobic halophile isolated from salted anchovies.

Effect of Salt on the Keeping Quality of Cream

Thompson and Macy (1940) stated that 7.5–10.0% salt added to cream retarded bacterial growth and acid development. No appreciable off-flavors were observed in the treated cream, whereas the control cream turned cheesy and yeasty.

Castell and Garrard (1939) claimed that cream was very satisfactory after being stored for 8 days at 77°F (25°C) when 7% salt was added. No cheesy or rancid flavors were observed because of the inhibition of certain oxidizing bacteria, which were largely responsible for these defects. Cream stored for 8 days at 60°–77°F (15°–25°C) was found to be satisfactory for the making of good-quality butter.

Preservation of Meat by Salt

The use of sodium nitrate prior to salting of hams tends to increase the permeability of the muscle fibers. The preserving effect of NaCl involves more than its dehydrating action on the meat. For instance, common salt is less effective as a dehydrating agent than many other salts such as magnesium sulfate, which will dehydrate proteins very rapidly.

Four factors contribute to the preserving action of NaCl: (1) direct effect of the chloride ion, (2) removal of oxygen from the medium, by reducing the solubility of the gas, (3) sensitization of the test organism to CO_2, and (4) interference with rapid action of proteolytic enzymes.

Effect of Calcium and Magnesium in Salt

A few investigators believe that calcium and magnesium in salt affect the quality of meat during the curing process. Others are of the opinion that the salt penetrates into the meat more rapidly when these elements are present. Other studies show no appreciable differences in the rapidity of salt penetration into meat in the presence or absence of these impurities. Hess (1942) observed that a culture medium not containing calcium and magnesium ions (prepared from dialyzed drip of fish muscle, washed agar, and pure NaCl) did not support the growth of red halophilic bacteria. When calcium and magnesium were added to the salt, good microbial growth took place.

Sodium nitrate is rubbed over the ham in the curing of southern-style hams. The nitrate tends to increase permeability, so that after salt or salt and sugar are applied to the ham repeatedly over a 7-week period, the smoked ham very rarely shows any sign of spoilage.

The toxic effect of a monovalent salt can be neutralized by a divalent salt in certain concentrations. The work of Winslow and Falk (1923) showed that $0.145\ M$ $CaCl_2$ mixed with $0.290\ M$ NaCl was toxic to *E. coli*, whereas a solution of $0.145\ M$ $CaCl_2$ + $0.680\ M$ NaCl was nontoxic. These workers concluded that sodium ions actually combined with protoplasmic anions of the cell and thus exerted a toxic effect upon the organism.

The salty fishy odor and taste of salted fish is increased by the presence of calcium and magnesium compounds as impurities in salt. The impurities have a depressing effect on the permeability of cell walls, thus slowing the penetration of salt into the muscle tissue. Less spoilage, as measured by an increase in bacterial count and the formation of trimethylamine, occurs in cod press juice containing about 80% pure salt than in untreated juice.

HALOPHILIC ORGANISMS

As noted already, halophilic microorganisms grow best in high concentrations of salt and in brines, while salt-tolerant organisms can grow in high or low concentrations of salt. *Micrococcus, Halobacterium, Pseudomonas, Flavobacterium, Sarcina,* and *Leuconostoc* are a few bacterial genera that may be classified as salt-tolerant.

Halophilic organisms tolerate varying concentrations of salt. The hypothesis has been proposed that halophilism is the result of a physiological faculty by which certain organisms can adapt themselves to artificial saline media resulting from a prepotency latent in the cell in the saltless environment. Hence, one may conclude that halophilic organisms are growth or mutant forms of ordinary soil bacteria adjusting themselves to a new environment.

Another hypothesis is that halophiles can maintain a low intracellular salt concentration in the presence of a high-salt extracellular environment. For example, the enzyme nitratase of *Micrococcus halodenitrificans* was found to be salt sensitive in the cell-free state and appeared to be salt tolerant in the intracellular state. However, when respiratory inhibitors were added to destroy resting cell suspensions at high salt concentrations, the reaction stopped. It has been postulated that interference with metabolism prevents or blocks the energy necessary for maintaining a low intracellular concentration of salt in the presence of high extracellular concentration of salt.

Salt-tolerant and halophilic bacteria, molds, and yeasts have been associated with food spoilage. They are widely distributed in nature and can be cultivated from sources other than salt or brines when incubated for a long time. It has been noted that the red discolorations caused by halophilic microorganisms occur after a lag period on old salted hides salted with uncontaminated salt. Microorganisms isolated from sea water with a salinity of about 5% will not grow on culture media unless a high concentration of salt is added.

Staphylococcus aureus has been found to persist in table-grade salt contaminated from a human source. An individual engaged in the routine preparation of foods finds it convenient to use the hand in applying the salt from the salt container for seasoning purposes. Staphylococci are frequently present on the surfaces of the skin, the throat, and nasal passages of nearly every normal person. Fortunately, a high percentage of staphylococci isolated from humans are nonenterotoxin producers and do not cause staphylococcus food poisoning.

Salt brines as used in the food industry are not free of halophilic bacteria, molds, and yeasts. Moreover many of these microorganisms may be responsible for undesirable changes in the food product. Dry salt also may contain microflora as a result of unsanitary practices used in salt processing, impurities, and the conditions under which the salt is stored.

USE OF SUGAR IN PRESERVING FOODS

Sugar may alter the growth of microorganisms in foods by the process of plasmolysis. It may be added to foods in a dry form or in the form of syrup. Sugar is also added to foods to help maintain the appearance of the product and improve the flavor of the material.

The kind of sugar added and its concentration determine its ability to accelerate or stop the growth of microorganisms. A concentration of 1–10% sugar significantly increases the growth of certain kinds of organisms, whereas strong sugar solutions usually have a bacteriostatic effect. A concentration of 50% stops the growth of most yeasts, while 65% sugar generally inhibits the growth of bacteria and 80% inhibits molds. A 60% sucrose concentration inhibits many types of food-spoilage organisms. There are, however, many types of sugar-tolerant organisms, and some yeasts are very resistant to high sugar concentrations. The spoilage of honey is often caused by sugar-tolerant yeasts.

The relationship between sugar concentration and bacteriostatic effect differs among different sugars. For instance, 35–45% glucose or 50–60% sucrose is bacteriostatic against food-poisoning staphylococci. These organisms are killed in 40–50% glucose or 60–70% sucrose.

The germicidal action of equal weights of various sugars has been investigated. Glucose and fructose are more effective than sucrose or lactose. One explanation for this is that plasmolysis of the microbial cell, which is a principal mechanism for the bacteriostatic effect of sugars, depends upon the number of particles present in solution. Thus, glucose or fructose with a molecular weight of 180 contain more molecules per unit weight than sucrose or lactose with a molecular weight of 342. Generally, the bacteriostatic activity of sugars decreases as their molecular weight increases.

However, the relative germicidal effect of glucose and fructose with respect to bacteria may be explained on a chemical rather than a physical basis. Fructose is a keto sugar, while glucose is an aldehyde sugar. Moreover, fructose is very active, but glucose requires heat to accelerate a chemical reaction as revealed by chemical tests.

A 40% glucose solution inhibits the growth of yeasts in apple, grapefruit, and pineapple syrups and is more effective than 40% sucrose. An equal concentration of glucose is more inhibitory than sucrose to *Saccharomyces cerevisiae* and *Aspergillus niger.* Equal mixtures of glucose and sucrose, inhibited growth equally with either sugar alone at the same concentration.

Glucose solution heated for 15 min at 212°F (100°C) and cooled was markedly more inhibitory to the growth of yeasts than was unheated sugar. The heat treatment of the sugar had no noticeable effect upon mold growth, especially with respect to *Aspergillus niger.*

Sugar occupies an important place in cooking, nutrition, and preservation of foods. Sugar, like salt, will support the growth of many kinds of microorganisms, depending on its concentration. The role of sugar as a sweetener in cooking and in the processing of many foods is well known.

The term *saccharolytic* is widely used to designate those organisms capable of growing in high concentrations of sugar. There are a few bacteria that have a high sugar tolerance and are classified as saccharolytic bacteria. Many molds and some yeasts are capable of surviving in much higher concentrations of sugar than bacteria.

Microbial Content of Sugar Products

Sugar beets, cane, and maple sap are the principal sources of sugar; the latter, being somewhat of a luxury, is not an important source of sugar. Raw sugar from these sources is not free of microorganisms. Purification of sugar for human consumption removes large numbers of organisms, but saccharolytic bacteria are usually present.

Unfortunately, many of these resident bacteria can contribute to spoilage in certain foods, especially if the concentration of sugar ranges from 20 to 30%. *Leuconostoc* spp. seem to be more prevalent in foods with high concentrations of sugar than other bacterial species.

As a rule, granulated sugar has a low microbial content, usually in the neighborhood of a few hundred organisms per gram. As would be expected, the surviving organisms are sporeforming organisms, caused by the method of processing the sugar. The raw product may be contaminated with spores of thermophiles and with bacteria such as *Leuconostoc mesenteroides, L. dextranicum,* and *Bacillus mycoides.* Molds commonly present include species of the genera *Aspergillus, Cladosporium, Penicillium,* and *Monilia. Schizosaccharomyces, Zygosaccharomyces,* and *Asporogenes* are yeast genera usually found in commercial sugar. In syrups, *Clostridium butyricum* and asporogenous yeasts have been involved in spoilage. Osmophilic yeasts have been involved in molasses and syrup because of their heat tolerance.

The microbial content of frozen fruit juices depends upon the contamination of the fruit prior to freezing and unsanitary practices in the processing of the juice. A variety of organisms may be present including yeasts and molds. The bacteria may include *Lactobacillus, Achromobacter,* and coliform species.

Cellulose

Cellulose, a complex carbohydrate, is an important substance in nature, as nearly all cell walls of plants contain this material. It is not digestable by humans, although certain microorganisms possess cellulase, which catalyzes hydrolysis of cellulose. The decomposition of cellulose yields simple sugars, which are a source of energy.

SELECTED REFERENCES

Castell, C. H., and Gerrard, E. H. 1939. Preserving cream with salt. *Can. Dairy Ice Cream J.* 18, 19.

Dunn, J. A. 1947. Salt and its place in the food industry. *Food Technol.* 1, 415–420.

Esty, K. R., and Meyer, K. F. 1922. The heat resistance of the spores of *B. botulinus* and allied anaerobes. *J. Infect. Dis.* 31, 650–653.

Fabian, F. W. 1951. The preservation action of salt. *In* "The Chemistry and Technology of Food and Food Products," Vol. 3. M. B. Jacobs (Editor). Interscience Publications, New York.

Hess, E. 1942. Studies on salt fish. *J. Fish. Res. Board Can.* 6, 1–23.

Hucker, G. J., and Penderson, C. S. 1942. A review of the microbiology of commercial sugar and related sweetening. *Food Res.* 7, 459–480.

Jenson, L. B. 1954. "Microbiology of Meats." Garrard Press, Champaign, Illinois.

Thompson, W. A., and Macy, M. C. 1940. Effect of salt on the microflora and acidity of cream. *Natl. Butter Cheese J.* 31, 12–14.

Winslow, C. E. A., and Falk, I. S. 1923. Studies on salt action. *J. Bacteriol.* 8, 215–236.

10
Radiation and Ultrasound in Food Preservation

Development and Safety of Irradiated Foods 141
Nature and Sources of Radiation 143
Effect of Radiation on Microorganisms 144
Irradiation in the Food Industry 145
  Development of Off-Flavors 146
  Effect on Shelf Life 146
  Packaging 147
  Use of Cathode Rays 147
Ultraviolet Light in the Food Industry 148
  Mechanism of Lethal Effect of UV Light 150
  Uses of UV Light to Control Microorganisms 151
  Undesirable Effects of UV Light on Foods 153
Ultrasound in the Food Industry 153
  Physical Effects 153
  Biological Effects 154
Selected References 155

DEVELOPMENT AND SAFETY OF IRRADIATED FOODS

Although preservation of selected food products by radiation has been successful, there are still questions concerning the safety of this practice. These concerns need to be resolved before such foods are made available to consumers.

The following events highlight the historical development of irradiated foods in the United States:

- 1940s and 1950s—The Department of the Army (DOA), Atomic Energy Commission (AEC), and private industry did exploratory work on irradiated foods.
- 1953—DOA was recommended to be the lead agency in researching food irradiation.
- 1959—DOA suspended plans for a food irradiation pilot plant.

- 1960—DOA program was reduced, delaying completion of long-term animal-feeding studies. AEC began low-dose irradiation studies, while DOA concentrated on high-dose applications.
- 1963—FDA granted clearance on irradiated bacon under vacuum (withdrawn in 1968) and for use of irradiation for insect disinfestation of wheat and wheat products.
- 1964—FDA granted clearance for use of irradiation for sprout inhibitation in white potatoes.
- 1964–1967—Several packaging materials were approved for irradiated foods.
- 1968—A petition for irradiation of ham, submitted in 1966, was withdrawn.

Extensive short- and long-term toxicity studies have been conducted on irradiated foods. In the late 1950s, problems unrelated to toxicity were encountered delaying the program, although no evidence of any toxicity was discovered. The Office of the Surgeon General of the Army sponsored studies on a group of foods considered to be representative of all foods. These studies indicated that foods irradiated under controlled conditions were safe and nutritionally adequate. They recommended broad clearance because of these studies. Some clearances were granted, but doubts about the study techniques, strengthened by unrelated food and drug safety problems, slowed the granting of clearances. Requirements for testing were changed at this time, and the AEC withdrew from the program.

In the fall of 1979, the Director of the FDA Bureau of Foods established an Irradiated Food Committee, consisting of six scientists. The Committee was to review the agency's practices for evaluating the safety of irradiated foods and to recommend standards for safety evaluation according to current knowledge in toxicology, nutrition, and radiation chemistry. The basic objective was to establish criteria that would, with minimal effort and cost, guarantee the safety of irradiated foods to the same degree that is expected for other foods in the American food supply.

The Committee focused on how the safety of irradiated foods can be evaluated by applying scientific principles and not on whether any particular irradiated food had been demonstrated to be safe. The Committee considered the following: (1) qualitative and quantitative estimates of radiolytic products, (2) comparisons of these radiolytic products with common food constituents in the human diet, (3) projected levels of human exposure to radiolytic products in food, and (4) sensitivity of current toxicological tests. The following recommendations were submitted:

1. Food irradiated at doses of 100 krad or less should be considered wholesome and safe for human consumption. This dose was based on estimated yields of unique radiolytic products that would be well below the limits of detectability of a bioassay. The major radiolytic products are food constituents derived from food during normal processing conditions.

2. Food irradiated at doses exceeding 100 krad should be subject to toxicological testing consisting of four short-term mutagenicity tests and two 90-day feeding studies (one rodent and one nonrodent mammalian species).

3. A food class comprising no more than 0.01% of the daily diet and irradiated at doses of 5 krad or less should be considered safe for human consumption without toxicological testing.

The Committee categorized irradiated foods as separate foods or food classes based on formation of a unique set of radiolytic products. This concept of recognizing characteristic patterns of radiolytic products to categorize foods is the basis for the Committee's support of a generic approach. Such a generic scheme will facilitate future evaluations by relying on the accumulated radiation chemistry data from various foods.

NATURE AND SOURCES OF RADIATION

Radioactive substances give off energy and matter, which is invisible to the naked eye but can be detected on a photographic plate. Alpha (α) and beta (β) particles and gamma (γ) rays emitted by radioactive substances.

Alpha particles consist of positively charged particles, which travel at a velocity of about 20,000 miles/sec. Such particles are readily stopped. Beta particles carry a negative charge and travel at a much higher speed; they have a velocity of up to or over 170,000 miles/sec and can penetrate several millimeters of iron. Beta particles are about 1/1800 the mass of a hydrogen atom nucleus. Gamma rays are able to pass through several inches of iron or lead and several feet of water. They carry no electrical charge. These rays have the property of neutralizing an electrified body. They produce a large number of positively and negatively charged ions in a neutral substrate. This latter property is of the utmost importance in the processing of food.

The emission of radiation proceeds at a definite rate expressed as the half-life (the time in which one-half of a substance will have decayed). This value varies for different radioisotopes. Such atomic reactions give off a tremendous amount of energy. Most of the heating effect of an element such as radium is directly caused by α particles, which are emitted with a considerable energy of motion; this energy is transferred into heat when the particles strike other matter. Because α particles are stopped easily by matter, very few escape the radium, which is heated by the unceasing self-bombardment.

As previously mentioned, α particles are positively charged particles and carry most of the energy from the fission process. These high-energy particles can readily destroy bacteria, but they are readily stopped by material as thin as a sheet of paper. For this reason, α particles are not useful in food processing.

Gamma rays are electrically neutral and have a much greater power of penetration than β particles. Gamma rays, actually a type of X ray, are by-products of atomic fission. By bombarding a heavy metal target with cathode rays, X rays can be produced. Only 3–5% of the electron energy is used in this procedure; the remainder of the energy is released as heat on the target. Thus, continuous use of such units would require extensive cooling. In medical X-ray equipment, the actual exposure time is seldom over a second; therefore, the target can be cooled between exposures by a fan-type cooler. Because a No. 2 can of food would require 10- to 20-min exposure to achieve sterilization, the use of X rays in food processing would be practical only if improved equipment were developed.

Gamma rays can be obtained from radioactive fission products more efficiently than X rays can be produced. If an adequately low-cost radioactive source material were available, then γ rays might be practical in the food industry. Cobalt-60 has been the principal source of these rays.

EFFECT OF RADIATION ON MICROORGANISMS

All kinds and types of microorganisms can be destroyed by irradiation, but the degree of sensitivity varies with species. Resistance of an organism to radiation seems to parallel resistance to heat. That is, sporeforming bacteria are more resistant to irradiation than non-sporeforming species. In general, the lower the order in the plant and animal kingdom, the greater the resistance to radiation. Insects in food products such as dried grains can be easily killed at relatively low dosages of radiation but considerably higher dosages are required to destroy bacteria or fungi.

Studies on the radiation resistance of the natural bacterial flora of canned ham show that most strains of bacteria are sensitive to radiation and are destroyed at low levels. Spores isolated from bacteria in the ham were resistant to 0.5 megarad radiation when incubated at 80°F (27°C) for 2 months. No differences were found in the destruction of organisms when gamma rays or an electron beam were the source of radiation.

Destruction of single-celled organisms by radiation has been explained by the direct hit theory. According to this theory, the passage of a single ionizing particle through or near the cell causes ionization and subsequent death of the organism. This theory is supported by the fact that survival curves are exponential. That is, the number of viable organisms decreases in geometric proportion to the radiation dose. The dose rate is independent of the amount of destruction; therefore, the effect of a given dose is the same regardless of the period of time involved. Destruction is independent of temperature, and the percentage of survival is not affected by the concentration of organisms. Such facts support the direct hit theory, but some destruction may be attributed to other effects.

Some lethal action also may be attributed to chemical effects of free radicals from the solvents. Irradiation of water causes the following reactions:

$$H_2O \longrightarrow H_2O^{\cdot\,+} + e^-$$
$$e^- + H_2O \longrightarrow H_2O^{\cdot\,-}$$
$$H_2O^{\cdot\,+} \longrightarrow H^{\cdot\,+} + OH^{\cdot\,-}$$
$$H_2O^{\cdot\,-} \longrightarrow H^{\cdot\,+} + OH^{\cdot\,-}$$

When γ rays bombard water, a positive water free radical is formed and an electron (e^-) is liberated. This electron reacts with another water molecule and thus produces a negative water free radical. Since β particles are negatively charged electrons, they may directly combine with water molecules without this first step. These positive and negative free radicals then dissociate into free hydrogen and hydroxyl radicals. The hydroxyl free radicals are strong oxidizing agents; the hydrogen free radicals are strong reducing agents. If no oxidizable or reducible solute is present, these free radicals may combine to form water peroxides or other radicals. Such reactions are important in understanding certain side effects produced in irradiated products.

IRRADIATION IN THE FOOD INDUSTRY

Heating of food products often brings about undesirable changes in the food. The radiation necessary for complete sterilization of a product usually does not raise the temperature more than 50°F (28°C). Radiation may produce physical or chemical changes in a food, causing alteration of its flavor, color, and sometimes texture. The packaging, storage life, and toxicity of foods treated by radiation involve problems that have not been solved; nevertheless, some foods can be successfully preserved by radiation (Desrosier and Rosenstock 1960).

Radiation has been successfully applied to the processing of several foods such as meat, fish products, vegetables, and spices. Milk and milk products, however, appear to be extremely sensitive to ionizing radiations and undesirable changes in flavor frequently follow irradiation. Research at the University of Michigan has shown that irradiated potatoes can be preserved at 48°F (9°C) for 1 year, whereas untreated potatoes usually sprout of rot after 6 months. A test panel compared irradiated potatoes with untreated potatoes and found a slightly different taste. The flavor of irradiated potatoes was actually preferred by some panel members.

Development of Off-Flavors

The free radicals that form during irradiation of foods may react with flavor compounds producing off-flavors. This can be alleviated by adding a substance with a greater affinity for the radicals involved than the flavor compounds. Many attempts have been made to find such compounds which do not alter the product. Some of the results have been encouraging. For example, vitamin C and its analogues will act as free-radical acceptors in many cases. Ascorbic acid would also help prevent flavor changes and would have no toxic effects. Freezing food before it is irradiated reduces the diffusion rate of the free radicals produced, thus minimizing the production of off-flavors.

The degree of off-flavor produced in foods increases generally as the dosage of the ionizing radiation is increased. Off-flavor development is definitely noticeable at dosages far below those necessary to sterilize many kinds of foods. Some foods, because of their chemical composition, may be able to withstand a complete sterilization dosage with little or no off-flavor development. Because of the chemical composition of different foods, their reaction to ionizing radiation differs. Each product requires individual evaluation.

Effect on Shelf Life

Somewhat related to the development of off-flavor is the effect radiation may have on the storage life of a product. Generally, the storage life of foods sterilized by irradiation is markedly different from that of foods sterilized by heat.

Enzymes catalyze chemical transformations in foods, which may result in changes in taste, appearance, odor, or texture. Ten times more energy is necessary to inactivate enzymes completely than to destroy food-spoiling organisms. In many cases, the high doses necessary to inactivate enzymes make irradiation processing too expensive and may cause marked side effects in the product. Since most enzymes can be readily inactivated by low heat, a blanching process might be combined with radiation treatment. Blanching consists of a very rapid exposure to a moderate temperature and, therefore, does not have the harmful effects on food quality that may occur with heat sterilization.

Packaging

Packaging is an important consideration in irradiation processing. Steel and glass containers reduce the effect of a given dose of radiation because of their thickness and density. Aluminum or plastic containers readily transmit electrons or X rays. Recent developments make various plastics very desirable as packaging materials. Vacuum packaging and inert-gas packaging would eliminate oxygen from foods, thus reducing the side effects of irradiation. However, because bacteria are more radiosensitive in the presence of oxygen, the sterilization dose may be considerably higher in the absence of oxygen.

Investigations carried out at Massachusetts Institute of Technology and elsewhere have shown that all kinds of microorganisms can be killed by radiation in any type of container, provided that it is not too large, and in any solvent. The basic requirement is that the dimensions of the container must be within practical limits for penetration of the particular type of radiation used.

Use of Cathode Rays

Cathode rays, which consist of high-speed electrons, are similar to the β particles emitted from radioactive materials. Cathode rays have served as a quick and practical way of sterilizing dairy and other food products. Already investigations have shown this is very effective in killing insects in grain, reducing a big waste in this raw food product.

Cathode rays are relatively efficient, since approximately 75% of the energy emitted in the electron beam can be used. Because of this efficiency, cathode rays sterilize in a matter of seconds, an important consideration in production line processing. The range of penetration of these rays warrants their consideration if a food product can be processed in fairly thin containers. However, cathode rays could not be readily used with larger bulk packaging without increasing the cost of packaging. Cathode ray irradiation equipment uses one tube of a single generator and requires elaborate facilities in a production setup. At least two such processing units have been devised and tested on a large scale.

X rays require from 15 to 30 min to destroy bacteria in food, whereas cathode rays work in a few seconds. However, cathode rays will not penetrate much beyond 1 in., which may limit their application in some cases. One advantage of cathode-ray sterilization is that it does not increase the temperature of the product. Also the nutritive value, the vitamin content, and the flavor of food are not altered as much as by cathode rays as by other types of radiation.

One of the disadvantages in the use of cathode rays is that the composition of the food is slightly changed. Fats and oils have a tendency to turn rancid, and carbohydrates are slightly modified. Proteins, vitamins, and enzymes are not affected to any great extent. Vitamin A, carotene, vitamin C, and several components of the vitamin B complex are not appreciably affected by cathode rays; if these vitamins are irradiated in solution instead of in the food, they suffer severe damage.

Vitamin C is relatively sensitive and niacin relatively resistant to cathode rays. When the two vitamins are mixed and irradiated, the niacin protects the ascorbic acid; hence, there appears to be a greater destruction of niacin and a sparing of ascorbic acid. Other vitamin combinations seem to behave very much in the same way when exposed to cathode rays.

ULTRAVIOLET LIGHT IN THE FOOD INDUSTRY

Radiation includes not only the energy emitted by radioactive substances but all forms of radiant energy. Indeed, the most familiar form of radiation—sunlight—surrounds us. Sunlight is part of the electromagnetic spectrum of energy, which includes radio waves, infrared, visible light, ultraviolet (UV), X rays, and γ rays. Each portion of the spectrum (i.e., each type of radiation) has characteristic wavelengths and frequencies, as depicted in Fig. 10.1. The wavelength of electromagnetic radiation is commonly expressed in nanometers (nm) and the frequency in cycles per sec or Hertz (Hz). One nm = 1×10^{-9} meters.

The wavelengths of visible light range from 390 to 800 nm. Within this range, the human eye can detect various colors, which correspond to different wavelengths, ranging from violet (shorter wavelength) through blue, green, orange, and red (longer wavelength). If sunlight, which consists of radiation of many wavelengths, is passed through a

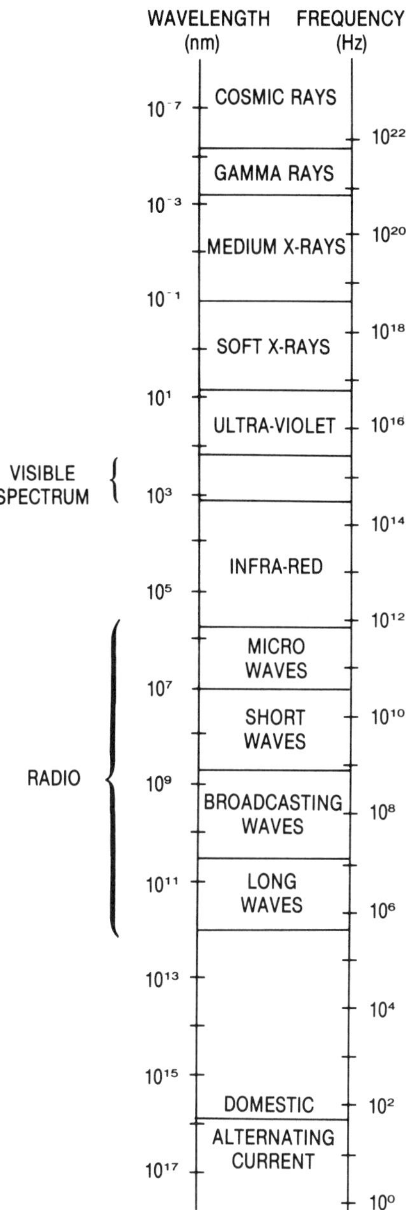

Fig. 10.1. The electromagnetic spectrum.

prism, it is broken up into its component wavelengths and the individual colors can be distinguished. Radiation characterized by wavelengths longer or shorter than visible light is not detected by the human eye and thus is invisible to us.

In general, visible light has very little germicidal power. However, ultraviolet radiation, especially in the range of 260–280 nm, has strong germicidal properties. Sunlight itself is germicidal, largely because it contains UV radiation. The germicidal effect of sunlight varies with climate, season, time of day, wavelength, intensity, angle of incidence, and length of exposure. An incandescent light bulb has about the same germicidal properties as sunlight passed through an ordinary window glass. Table 10.1 summarizes the germicidal and other effects of different wavelengths of radiant energy.

Although UV radiation at 200–280 nm has a strong lethal effect on microorganisms, it has very low penetrating power. Indeed, ordinary window glass or even a thin coverslip will block UV radiation. Certain proteins act very much like glass in this respect; that is, they absorb UV light.

Mechanism of Lethal Effect of UV Light

There is some evidence that UV radiation produces a chemical change in the medium or the cell protoplasm itself. For instance, if a

Table 10.1. Effects of Different Wavelengths of Radiant Energy

Classification	Wavelength (nm)	Effects
Invisible long		
Radio	Very long	None
Infrared heat	800 and larger	Temperature may be raised
Visible	400–800	Little or none
Red, orange, yellow, green, blue, violet		
Invisible short		
Ultraviolet	13.6–400	Total range; various effects
	320–400	Photographic and fluorescent range
	280–320	Human skin tanning; activation of vitamin D precursor
	200–280	Maximum germicidal power
	150–200	Shuman region
	100–150	Ozone forming; germicidal in proper concentration
Alpha, beta, and gamma rays	Less than 100	Germicidal
Cosmic rays	Very short	Probably germicidal

nutrient agar medium is irradiated with UV rays and then inoculated with bacteria, the resulting culture grows much less rapidly than it does on a nonirradiated medium. This result suggests that UV irradiation resulted in the formation of some toxic substance(s) in the medium.

Highly lethal UV rays seem to have enough energy to displace some electrons in the cell protoplasm. Thus, when microorganisms are irradiated directly, displacement of certain molecular groupings in the cell with a high absorption for UV may occur. However, for any radiation to have an effect, it must be absorbed. Although most components of microbial cells do not absorb UV light, the amino acids alanine, tyrosine, and tryptophan do. Thus, any protein that contains these amino acids will absorb UV light. If a dye such as eosin is added to a protein, its ability to absorb UV light is increased.

One explanation for the lethal effect of UV radiation on cells is that it induces some chemical change in the superficial membrane of cells, making them more permeable to hydroxyl ions. As a result of this, hydroxyl ions from the medium can enter the cells and eventually coagulate the cell proteins. The increased intracellular alkalinity of cells exposed to UV radiation has been demonstrated with the eggs of sea urchins. In this experiment, the eggs were stained with neutral red and exposed to UV light; within a few hours, a deep yellow color developed, indicating the presence of alkali within the egg cells.

Ultraviolet light also may produce some oxidizing substances (e.g., ozone) that have a marked germicidal effect. Recently, it has been shown that UV rays are destructive in the absence of atmospheric oxygen. It is known that a consumption of oxygen takes place when tetanus and cholera cultures are irradiated in the presence of air.

A few compounds such as formaldehyde are capable of bringing about a coagulation of the cell protoplasm, which is largely protein. A similar phenomenon is widely accepted as being responsible for the destruction of many kinds of microorganisms. For example, when various albumins and globulins are exposed in thin layers to UV rays, these proteins coagulate and are no longer soluble in weak acids or alkalies.

Uses of UV Light to Control Microorganisms

Although a number of attempts have been made to control food-spoilage microorganisms by the use of ultraviolet light, the most common application has been to reduce the number of organisms in the air. The chief limitation of UV sterilization has been the inability of the rays to penetrate liquids and solids.

Ultraviolet light is used in meat coolers to reduce the numbers of microorganisms in the air. Because of the reduced microbial population, it is possible to hold meat at higher temperatures and speed the aging and tenderizing process considerably.

Ultraviolet light also is used to maintain a relatively microbial free air in food-handling establishments. It has been used with a high degree of success in bakeries in controlling the spores of *Bacillus mesentericus,* which are largely responsible for ropy bread. In one study, ultraviolet light ranging from 200 to 295 nm was effective in reducing the numbers of *Staphylococcus aureus* and *Salmonella enteriditis* when they were sprayed into the air and inoculated onto smooth hard surfaces. However, when these organisms were added to custard-like foods, they were not effectively destroyed by irradiation.

Bacterial spores are five to ten times more resistant than vegetative cells to UV radiation. Bacteria and yeasts are comparable in terms of their resistance, while molds are much more resistant, especially if they are pigmented.

During Tenderization of Meat

The aging of meat is caused primarily by the action of certain enzymes present in the meat. Cold storage retards the activity of these enzymes. If the temperature of storage is raised, enzyme activity increases and many undesirable contaminating bacteria become active, bringing about undesirable changes. Ultraviolet light can be effective in controlling the bacteria on the surface of the meat but not those organisms beneath the surface. Meat that has been exposed directly to UV light sometimes develops certain off-flavors such as a cooked or rancid taste. This is largely caused by oxidation of fats in the meat by ozone formed during irradiation.

Pasteurization of Milk

A 90% reduction in bacterial counts on market milk after irradiation has been reported. Radiation at 2547 nm was more effective than shorter wavelengths between 220 and 230 nm. In general, irradiation of milk to activate the precursor of vitamin D has little bactericidal effect because short exposure times and wavelengths of 280–320 nm are used for this purpose.

Sanitization of Glassware

Clean water or drinking glasses can be effectively sterilized by UV light. Mechanical difficulties in applying UV radiation have restricted the use of this method of sanitization in restaurants and foodservice establishments.

Undesirable Effects of UV Light on Foods

As indicated already, UV irradiation may cause undesirable biochemical changes in some foods that influence their color, odor, or flavor. Some of the changes that have been observed include the following:

- Rise in pH of meats
- Destruction of glutathione
- Increase in hydrogen sulfide
- Destruction of natural antioxidants in fats
- Destruction of thiamin, ascorbic acid, riboflavin, niacin, and other vitamins in addition to many endogenous enzymes

Despite these undesirable effects and other problems associated with UV sterilization, this method has been partially successful with fresh pork and pork products in commercial operations. Ultraviolet sterilization of certain vegetables, milk, and eggs also has been encouraging.

ULTRASOUND IN THE FOOD INDUSTRY

Sound waves in the ultrasonic range, although different in nature from electromagnetic radiation, also have germicidal and other effects of potential interest to the food industry.

Ultrasonic (or supersonic) waves are unaudible to humans. The threshold of audibility for the human ear depends on the pressure and the frequency. The lowest frequency audible to the human ear is about 31 Hz at a pressure amplitude of 1 dyne. The highest audible frequency is around 14,000 Hz at a pressure of 0.05 dyne. The human ear distinguishes variations in frequency most readily when the frequency of the sound lies between 1000 and 4000 Hz.

It has been claimed that ultrasonic waves may be beamed from an oscillator and that the beam may be confined within a relatively small angle and so can be made directional. Thus, ultrasonic waves can be controlled much better than ordinary sound waves.

Physical Effects

Immiscible liquids, such as oil and water or water and mercury, are transformed into remarkably stable emulsions by ultrasonic vibrations. Long-chain molecules have been depolymerized and dispersions in water, alcohol, and oil have been made with many of the metallic elements, as well as with fusible alloys, by the use of ultrasound.

Various ultrasonic frequencies cause accumulation and coagulation of precipitates in liquids, suggesting application to the filtration field. Other frequencies cause smoke or dust particles to agglomerate, suggesting applications in the smoke prevention field. Some frequencies produce colloidal suspensions of solids in liquids; others break up suspensions, yielding precipitated lumps.

An attempt to produce sonic soft-curd vitamin D milk by using subsonic waves has met with partial success. The milk is pumped in a thin film between two stainless steel circular plates. Behind one of the stainless steel plates is an oscillating electrical magnet of high voltage that sets up a vibration in the diaphragms, resulting in intense waves of energy at a frequency of 360 Hz. This breaks up and disperses the fat globules to form a product similar to homogenized milk.

Biological Effects

The biological effects of ultrasound are one of the most fascinating and unexplored areas of research. Protozoa, bacteria, and small animals are all affected by inaudible soundwaves. Some effects are adverse and some beneficial. For example, yeast cells lose their reproductive ability when exposed to ultrasound and luminous bacteria lose their luminosity. In contrast, some bacteria are stimulated to increased activity and virulence.

The lethal effect of ultrasonic vibrations depends on the volume of cells being vibrated, the concentration of organisms, and the shape and the size of the microorganisms.

The effect of cell concentration is easily demonstrated. In one experiment, for example, a suspension of *E. coli* containing 3,000,000 organisms/ml was sterilized after 20-min exposure to ultrasound, whereas a suspension of 45,000,000/ml required 38 min of treatment.

In general, bacilli are more easily killed by ultrasonic vibrations than cocci, and large bacteria are more quickly destroyed than small. Large microorganisms also are more susceptible to lower frequencies. In one study, *Saccharomyces ellipsoideus* (a wine yeast) in a grape juice suspension was quickly killed by ultrasonic treatment, but a bacteriphage present was not affected.

The presence of protein in solution reduces the germicidal effect of ultrasonic treatment. Thus, the ultrasonic treatment of milk is less efficient bactericidally when small amounts of protein are added. The data

Table 10.2. Germicidal Efficency of Ultrasonic Vibration on Bacteria in Milk

Expt. No.	Treatment time (min)	No. of organisms/ml		Efficiency (%)
		Before	After	
1	5.0	1,390,000	26,000	98.9
2	15.0	26,200,000	21,000,000	19.8
3	8.5	2,200	50	97.7
4	10.0	60,000	7,000	88.3
5	10.0	3,500,000	1,040,000	70.3
6	6.0	14,500	5,800	60.0

Source: Beckwith and Weaver (1936)

in Table 10.2 illustrate the effect of bacterial concentration on the germicidal efficiency of ultrasonic treatment of milk.

SELECTED REFERENCES

Anon. 1978. Food Irradiation in the United States. Report Prepared by the Interdepartmental Committee on Radiation Preservation of Food. U.S. Army Natick Research and Development Command, Natick, Massachusetts.

Beckwith, T. D., and Weaver, C. E. 1936. Sonic energy as a lethal agent for yeast and bacteria. *J. Bacteriol.* 32, 361–373.

Josephson, E. S., and Peterson, E. S. 1982. "Preservation of Food by Ionizing Radiation," Vol. I. CRC Press, Boca Raton, Florida.

Larson, B. L. 1960. Significance of strontium in milk. *J. Dairy Sci.* 43, 1–21.

Proctor, B. E. 1952, Cathode rays and the food industry. *Cert. Milk* 27, 15–16.

Proctor, B. E., and Goldblith, S. A. 1951. Electromagnetic radiation fundamentals and their application in food technology. *Adv. Food Res.* 3, 157–167.

Maxcy, R. S. 1981. Irradiation of food for public health protection. *J. Food Protect.* 45(4), 363–366.

11
Red Meat and Poultry

Microorganisms Commonly Present in Meat   158
Microbial Counts of Meat Products   158
   Frozen Meats   159
Spoilage of Meats   161
   Ham Sours   161
   Spoilage of Sausage   162
   Spoilage of Cured Meats   163
   Other Defects   163
Tenderization of Meat   164
   Proteolytic Enzyme Preparations   164
Microbiology of Poultry   165
Selected References   167

In general, the microflora of beef and pork mimics that of the barn-yard or feedlot. Microorganisms from soil, water, air, and manure predominate. During slaughter, organisms found on the outside of the animal contaminate the meat by direct contact, through air and water, and from the hands and tools of workers. In addition, the microflora of the slaughtering area, which include contaminants from previously slaughtered animals, contribute to the overall microbial load of red meats.

Unhealthy animals are generally removed at antemortem inspection, but animals with no visible signs of disease are slaughtered and processed in the plant. A few of these animals are likely to be diseased and thus contribute to the general microflora of the plant. Healthy animals can also carry pathogens without showing signs of the disease itself. For example, contamination of meat with salmonella organisms is a problem. Fortunately, these organisms are killed by thorough cooking before consumption; unfortunately, some of these organisms may contaminate areas in the kitchen and ultimately end up in foods that are not cooked as thoroughly as meat.

Several factors influence the microbial load of meats. The bacterial load in the gut of animals before slaughter and certain physiological changes have been shown to have an influence. Withholding feed for 24 hr before slaughter is recommended to empty the digestive tract. Some evidence suggests that when animals are excited or fatigued, bacteria can enter their tissues more easily. One reason for this may be that such animals use up muscle glycogen forming lactic acid, which changes the pH of the tissues; this in turn may encourage the invasion of microorganisms. Besides these antemortem conditions, the method used for killing and bleeding and the rate at which the carcass is cooled influence the microbial load of meat. Other factors that influence the numbers and types of microorganisms present in meat are age of animal at slaughter, type of rations, conditions of storage, and whether the meat is ground or not. Sometimes the bung and esophagus are tied off during processing to avoid contamination.

MICROORGANISMS COMMONLY PRESENT IN MEAT

Among the pathogenic bacteria found in meat as a result of infection of the animal prior to slaughter or as a result of cross contamination are those from the genera *Brucella, Salmonella, Streptococcus,* and *Mycobacterium,* as well as some anaerobic bacteria such as the clostridia. Two spoilage-producing psychrophilic genera, *Achromobacter* and *Pseudomonas,* often make up the predominant microflora, especially after storage. Bacilli and staphylococci are also present frequently and may contribute to the slime that forms on the surface of spoiled meats.

Achromobacter and some yeasts can produce lipases, which cause rancidity.

Molds cause trouble in the meat-packing industry, especially with preserved meats. For example, although the growth of molds on hams may not cause any decomposition, molds still represent an economic loss because some of the meat must be trimmed. Lack of proper handling and sanitary practices and excessive humidity in storage rooms all contribute to the growth of molds. Pickling vats also provide an especially favorable environment for the growth of yeasts and molds, especially when good sanitation practices are not used.

Aspergillus, Mucor, Penicillium, Alternaria, Monilia, and *Rhizopus* are the principal mold genera associated with meats. These molds are most often encountered in fresh beef, pork sausage, wieners, and dried beef when the temperature and humidity in storage are favorable for growth.

MICROBIAL CONTENT OF MEAT PRODUCTS

The numbers and types of organisms present in meats depend on the microbial load of the carcass and the environmental conditions in the meat. Typical microbial counts on the surface of animals range from 100 to 100,000/g in beef and from 5,000 to 1,000,000/g in swine.

Moisture, nutrients, pH, time and temperature, oxidation–reduction potential, level and types of contamination, surface area and inhibitors influence the growth of microorganisms in meats, as they do in other foods. Since meat is an excellent source of nutrients, microorganisms—given the other conditions required for growth—multiply rapidly and cause spoilage in a short time. In fact, meat is often included as a component of artificial culture media used to grow microorganisms. The effect of temperature on meat stored for 24–96 hr is illustrated by the data in Table 11.1.

Chopped meat (hamburger) often has quite high microbial counts ranging from 5,000,000 to 10,000,000/g of sample. Minced or spiced meats show a relatively low bacterial count because of the preserving action of heating during processing . The microflora in smoked or dried meats is usually very low primarily because the chemical and physical environmental conditions are detrimental to the growth of organisms. Salting or brining of meats has the same effect as smoking on the number of microorganisms present in the meat; however, halophilic bacteria may be able to thrive in a salt environment. Canned meats are as a rule sterile.

Frozen Meats

A very large amount of meat is frozen. The so-called quick-freezing method tends to minimize "drip" after the meat is thawed. During slow freezing, the free water forms large ice crystals throughout the meat; in this case, many of the soluble nutrients are lost during thawing, which degrades the meat in taste and in physical appearance.

Frozen meats stored at appropriate temperatures may be of good quality after a long period, provided the product was properly chilled

Table 11.1. Effect of Temperature on Growth of Microorganisms in Meat

Temperature		Percent increase of microflora over the initial count at the end of each period (Hours)			
(°F)	(°C)	24	48	72	96
35	1.7	2	4	8	9
40	4.4	3	7	10	20
45	7.2	8	20	25	30
50	10.0	20	30	40	45
55	12.8	25	40	45	60

and handled before freezing. As discussed in Chapter 7, cold temperatures are not an effective means for destroying bacteria; they merely slow the metabolism and reproduction of organisms. Thus, bacteria tend to die off slowly during cold storage. Psychrophilic bacteria, however, are able to survive in cold-storage temperatures for relatively long periods.

Cold-storage meats often are covered with molds, which are not destroyed any more readily than bacteria by cold temperatures. If mold or bacterial spores are present on the surface of the meat prior to cold storage, they may germinate, although the rate of multiplication would be very slow. Even if spores do not germinate during cold storage, they are likely to remain viable. For example, mold spores have been kept frozen for 3 years and retained their viability.

Trichinella spiralis, a worm that lives endoparasitically in the muscle tissue of swine, can be a health hazard when raw pork is eaten (Fig. 11.1). Quick freezing at 30°F (−1°C) followed by 24-hr storage at

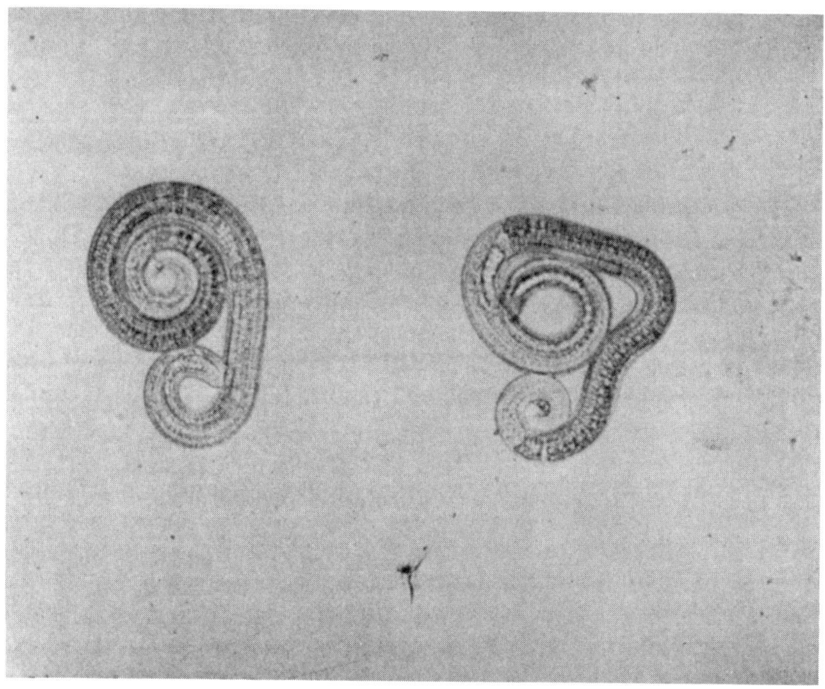

Fig. 11.1. *Trichinella spiralis,* a parasitic worm that infects the muscle tissue of swine and may cause trichinosis in humans.

0°F (−18°C) destroys the larvae. Federal regulations in general require refrigeration below 5°F (−15°C) for 20 days to destroy the parasite. However, these requirements vary with the size of the cut. This parasite is also destroyed by cooking at 137°F (58°C).

SPOILAGE OF MEATS

A number of different types of meat spoilage are caused by microorganisms. Surface slime is characterized by a shiny, viscous, moist covering on the surface of the meat. It is generally caused by *Pseudomonas, Achromobacter, Streptococcus, Leuconostoc, Bacillus,* and *Micrococcus* species. Some lactobacilli can also produce slime.

Some species of the *Lactobacillus* and *Leuconostoc* genera cause a green discoloration of meat when nitrite is one of the curing ingredients. Greening in sausage is an example of this. Some lipolytic genera such as *Pseudomonas* and *Achromobacter* as well as yeasts cause rancidity with resultant off-flavors in fats.

Photobacterium and some *Pseudomonas* species cause phosphorescence on meat. Certain microorganisms under some conditions will produce pigments, which are visible on the surface of the meat. Examples of this are the red spot condition caused by *Serratia marcescens* and the blue color caused by *Pseudomonas syncynea.* Other pigmenters are *Micrococcus* and *Chromobacterium.*

Molds cause stickiness, a white fuzzy growth called whiskers, black spot, white spot, green patches, and off-odors and off-flavors depending upon the types of molds present. Anaerobic organisms cause souring, putrefaction, and "bone taint." All microorganisms, if present in high enough numbers, can cause souring, rancidity, and foul odors.

Ham Sours

The scientific literature offers ample support for the view that tissues of normal healthy animals are sterile (i.e., free of microorganisms) and for the opposite view that they are not. In practical terms, it may make little difference because contamination of the carcass is quite common during the slaughtering and dressing process.

The term *sour* refers to putrefaction of the muscle and not the formation of acid. Ham sours are caused by the growth of bacteria, particularly of those that grow at 38°F (3°C), and are classed into six types depending on the location of the spoilage. Ham sours are relatively rare.

The following practices or conditions have been suggested as contributing to the development of ham sours:

- Faulty exsanguination of the animal
- Improper chilling
- Use of contaminated salt in pickling
- Use of previously contaminated brines
- Excessive mechanical injury by dehairing machine
- Injury or bruises during processing of hams
- Holding carcass too long at high temperatures
- Failure of salt to penetrate into marrows
- Use of insufficient salt to inhibit bacterial growth
- Poor quality hogs

Spoilage of Sausage

Although microorganisms grow on the surface of sausage, most spoilage is caused by growth of organisms inside the casing (Table 11.2). For example, some lactic acid-producing bacteria can multiply to the point where lactic acid is produced. This is considered a form of spoilage except when fermented sausages are desired. In such cases, the sausage is deliberately inoculated with a culture of lactic acid-producing organisms.

Some of these same organisms can cause a green discoloration on the surface or in the core of sausages when nitrites are present. This condition is caused by growth of these microorganisms and production of a heat-stable peroxide before smoking and cooking. Apparently,

Table 11.2. Some Sources of Bacterial Contamination in Sausage Manufacture

Ingredient	% of raw emulsion	Count per gram	Approx. load contributed
Bull beef	41.20	1,590,000	656,000
Ground pork	13.75	88,500,000	12,170,000
Pork jowls	13.75	193,000	26,560
Ice	21.30	28	6
Dried milk	3.30	88,000	2,900
Corn sugar	1.65	595	10
Seasoning mix	.34	260	0
Rework (cooked)	1.37	190,000	2,600
Rework (raw)	1.37	8,150,000	111,700
Total	98.03		12,969,776

Source: Price (1965).

heating and smoking complete the process and the green discoloration appears. *Lactobacillus viridescens* and certain *Leuconostoc* spp. seem to be the principal offenders.

Spoilage of Cured Meats

The common curing agents—salt, sodium nitrate and nitrite, sugars, spices, and smoke—all influence the growth of microorganisms. Consequently, the microflora of cured meats and uncured meats are different.

Smoke contains many antibacterial substances such as creosols, aldehydes, ketones, and acids. Nitrates and nitrites can actually be used by some microorganisms as a source of nitrogen. Salt inhibits the growth of many microorganisms, but in inhibiting bacteria it also removes competition for salt-tolerant organisms.

Bacon is susceptible to microbial spoilage. Lipolytic bacteria may cause rancidity when bacon is held in storage. Also sulfide-forming bacteria may alter the color and produce an oxidized flavor. The presence of high concentrations of nitrite, although it inhibits growth of *Clostridia*, may cause undesirable color changes.

Other Defects

In addition to microbial spoilage, several quality defects occur in meat. One problem of particular concern is *soft-watery pork*. It consists of an extremely pale watery condition of the muscles. Functional properties such as water-holding capacity are adversely affected by this defect. *Boar odor* is the perspirative or urine-like odor from the meat of uncastrated, sexually mature, male hogs. Nitrate or nitrite discoloration, which results in pink meat after cooking, although not harmful is objectionable to consumers except in bacon and hams.

Various kinds of meat tend to dry out during cold storage. The thin moisture film on the surface of the meat is removed by natural desiccation; the fat is exposed to oxidation and ultimately rancidity will develop. Desiccation is speeded up if large amounts of salt are present in the product. It is known that seasoned sausage develops rancid odors and flavors and even loses color when stored at $0°F$ ($-18°C$), whereas ground pork can be stored at a comparable temperature for several months with no noticeable change in the product. Cooking the sausage will destroy the endogenous enzymes that cause deterioration.

TENDERIZATION OF MEAT

The tenderness of meats is influenced by the age and sex of animals, by various production practices such as the feeding of hormones and castration of males, and by postmortem tenderization treatments. After slaughter, tenderization occurs as a result of (1) autolysis by cellular enzymes that digest the cells, (2) treatment with proteolytic enzymes, which break down the tissues, and (3) cooking, which coagulates protein, melts fat, and hydrolyzes collagen. Although postmortem tenderization is primarily an enzymic, not a microbial, process, microbiological considerations can be significant.

Generally beef is held for about 10 days at 36°F (2°–3°C) after slaughter. This treatment permits autolysis to proceed at a slow rate, while the low temperature retards the growth of microorganisms. During this holding or aging period, mold growth may appear on the cut surfaces of the meat unless some precautions are taken to minimize it. Improved sanitation in processing plants has reduced the problem of mold and bacterial growth on carcasses.

In addition to molds, several genera of psychrophilic bacteria are capable of growing very slowly on meat at refrigerator temperatures. *Achromobacter, Pseudomonas,* and *Proteus* are the principle genera present. There is some evidence to suggest that these bacteria may play an important role in the flavor development of beef.

In the Tenderay process for aging beef carcasses, UV lights are used in coolers to reduce the rate of microbial growth. As a result, the storage temperature can be increased; this accelerates enzyme action and thus reduces the time required for tenderization.

Two kinds of enzyme may occur in meats: endogenous enzymes and those produced by microorganisms associated with meats. Undesirable flavors may be developed by microbial enzymes that act on amino acids and their derivatives, converting them into highly odorous substances such as amines, mercaptans, and skatole.

It is interesting to note that enzyme action comes to a halt when the water content of meat is dropped to 2–3%. This is the principle involved in the preservation of dried meats of various kinds.

Proteolytic Enzyme Preparations

The tenderization process often can be accelerated or enhanced by treating meat with proteolytic enzyme preparations, which supplement the endogenous autolytic enzymes present in meat.

The following proteolytic enzymes have been isolated and partially purified for use as meat tenderizers:

- *Macin* from osage orange
- *Asclipian* from milkweed
- *Ficin* from figs
- *Papain* from papaya fruit
- *Bromelin* from the juice of fresh pineapple

In addition to these, an unknown enzyme similar to macin and asclipian has been obtained from edible mushrooms, and a protease-like substance has been obtained from *Aspergillus* species. Papain and bromelin have shown the greatest potential as commercial meat tenderizers.

Proteolytic enzyme preparations may simply be sprinkled over the surface of meats. This is the common practice with preparations that are intended for home use. In commercial operations, proteolytic enzymes may be injected into an artery of the animal at the time of slaughter. This procedure tends to distribute the enzyme throughout the meat and results in better and more uniform tenderization.

The control of enzyme action is very important. An ideal enzymic tenderizer is one that can be inactivated during cooking. If the enzyme is not inactivated, the meat tends to become mushy during heating or cooking.

Enzymes are not generally used in tenderizing hams or in the preparation of ready-to-eat hams. These hams are cooked and tenderized in the smokehouse to an inside temperature of 137°F (58°C) or higher.

MICROBIOLOGY OF POULTRY

The sources of contamination and typical microflora of poultry and poultry meat are similar to red meats. In the past, poultry and eggs have served as a reservoir for *Salmonella,* and some workers believe that poultry have a species affinity for *Salmonella.* Some studies, however, suggest that the reason for the high incidence of salmonella infection in poultry is contaminated feed ingredients, especially meat scrap and fish meal.

Walker and Ayres (1956) in a study of microorganisms found on poultry in six Iowa processing plants reported that live poultry generally had 600–8100 organisms per square centimeter of skin. After processing and eviscerating, the numbers increased to between 11,000

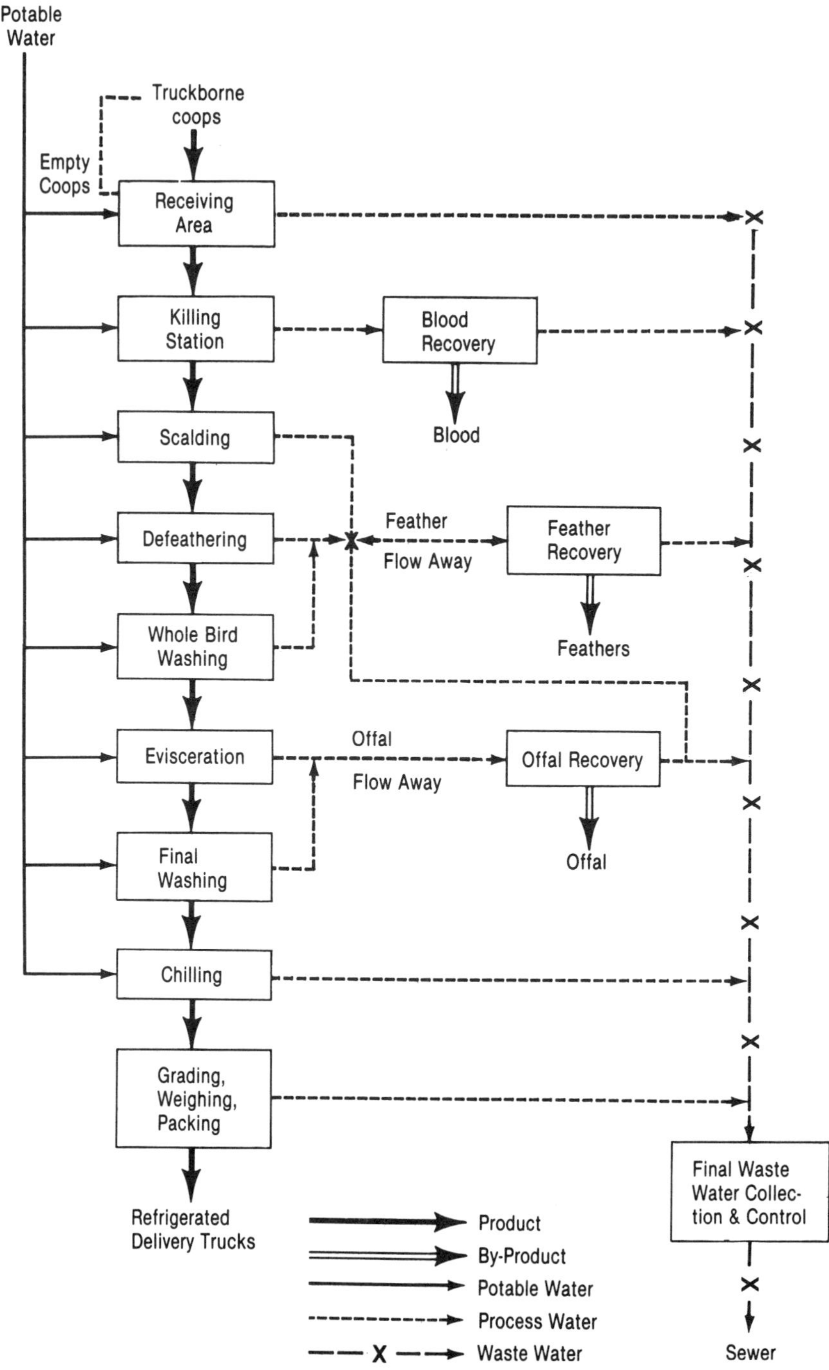

Fig. 11.2. Flow chart of poultry processing plant. High standards of sanitation are required to control microbial contamination.

and 93,000 per square centimeter. The flow chart of a typical poultry processing plant in Fig. 11.2 indicates the many points at which microbial contamination may occur.

The principal genera found on cut-up poultry are *Pseudomonas, Micrococcus, Achromobacter, Flavobacterium, Alcaligenes, Proteus, Bacillus, Sarcina, Streptococcus, Eberthella, Salmonella, Escherichia, Aerobacter, Streptomyces, Penicillin, Oospora, Cryptococcus,* and *Rhodotorula.*

SELECTED REFERENCES

Augustine, D. L. 1933. Effects of low temperature upon encysted *Trichinella spiralis. Am. J. Hyg.* **17**, 696–710.

Ayres, J. S. 1955. Microbiological implications in the handling, slaughtering and dressing of meat animals. *Adv. Food Res.* **6**, 109–161.

Dawson, L. E., and Stadelman, W. J. 1960. Microorganisms and Their Control on Fresh Poultry Meat. NCM-7 Tech. Bull. 278. Michigan State University, East Lansing, Michigan.

Dawson, L. E., Chipley, J. R., Cunningham, F. E., and Kraft, A. A. 1979. Incidence and Control of Microorganisms on Poultry Products. North Central Regional Publ. 260. Michigan State University, East Lansing, Michigan.

Jenson, L. B. 1954. "Microbiology of Meats." 3rd ed. Garrard Press, Champaign, Illinois.

Kotula, A. W., Murrell, K. D., Acosta-Stein, L., and Tennant, I. 1982. Influence of rapid cooking methods on the survival of *Trichinella spiralis* in pork chops from experimentally infected pigs. *J. Food Sci.* **47**, 1006–1007.

Mountney, G. J. 1976. "Poultry Products Technology." AVI Publishing Co., Westport, Connecticut.

Parkhurst, C. R., and Mountney, G. J. 1987. Poultry, Egg and Meat Production. Van Nostrand Reinhold Company, New York (in press).

Price, J. F. 1965. Materials and Methods for Extension Programs in Meat and Poultry Processing. Mich. State University, East Lansing, Michigan [Mimeo].

Walker, H. W., and Ayres, J. C. 1956. Incidence and kinds of organisms associated with commercially dressed poultry. *Appl. Microbiol.* **4**, 345–349.

Winegarden, M. W., Lowe, B., Kastelic, J., Kline, E. A., Plagge, A. R., and Shearer, P. S. 1952. Physical changes of connective tissues of beef during heating. *Food Res.* **17**, 172, 184.

12
Fish and Shellfish

Common Fish Preservation Methods 169
Microbiology of Fish 170
Microbiology of Lobsters 173
Microbiology of Oysters 174
  Oyster Production 174
  Bacterial Flora of Oysters 176
Microbiology of Other Mollusks 177
Paralytic Shellfish Poisoning 178
  Sources of Shellfish Poison 178
  Symptoms and Treatment 179
  Characteristics of the Poison 180
  Coastal Quarantines 180
Selected References 180

COMMON FISH PRESERVATION METHODS

Although numerous methods can be used to preserve fish, some of them are no longer permitted. The basic principle in the preservation of fish is to prolong the quality of the product so that there is little change in its appearance and flavor.

The icing of fish is a well-established method of preservation. However, ice may carry many kinds of microorganisms; even ice produced artificially is not sterile. Storing fish in several times the normal concentration of carbon dioxide also is an effective preservation method. A 0.02% concentration of refined sodium nitrite has proved very effective as a curing agent for fish. However, when hydrochloric acid and sulfur dioxide are used, the product develops an unpleasant odor, flavor, and appearance, probably caused by a chemical reaction between the acid and the oil of the fish.

Salt is an old standby in preserving fish. Usually, the immersion of fresh gutted fish in a 20% salt brine for 30 min prolongs its keeping quality. Fortunately, most pathogenic bacteria are not salt tolerant, and many organisms are destroyed by this treatment, although there are several exceptions.

The use of smoke as a means of curing and preserving fish is a well-established practice. Dehydration of the fish during the smoking process is one factor in controlling microbial growth. The formation of certain chemical compounds, such as formaldehyde, induces a chemical reaction between the constituents in the smoke and the food material, producing a chemical inhibiting agent for most organisms. Moreover, the formation of certain flavor characteristics in the product is usually desirable from a consumer acceptance standpoint. Smoking of fish is limited in use and is suitable only for certain varieties of fish.

Today, freezing is one of the most common ways of preserving fresh fish during transport to distant points for consumption. However, freezing involves considerable loss of water, so that when the fish are thawed, much of the flavor is lost. Ice glazing of fish, (i.e., the formation of a thin coating of ice around the product) minimizes dehydration and is very effective in controlling microbial spoilage. The method has not been used very extensively because it is fairly expensive.

Fish preserved by canning are heat-treated prior to canning or are sterilized in the container and sealed. The problems involved in canning fish are about the same as those encountered in canning other foods. The storage life of the product is entirely dependent upon the efficiency of the heat treatment.

MICROBIOLOGY OF FISH

Fish tissue, like other animal tissue, is not free of microorganisms. Fish muscle is readily susceptible to enzyme action and decomposition by bacteria. It is a common practice to eviscerate fish after catching and then ice the edible portion (Fig. 12.1). Surface contamination and sealing depends upon the degree to which sanitary practices are employed. The addition of chlorine and calcium hypochlorite has been recommended to keep the wash water sterile. The use of ice containing calcium or sodium hypochlorite has given satisfactory results and prolongs the keeping quality of fish packed in ice.

All kinds of edible fish carry microflora in the intestinal tract. The microorganisms present vary according to the food and the water to which the fish have been exposed. Both aerobic and anaerobic organisms have been isolated from the intestinal contents of fish. Obviously, many of these microorganisms can be associated with spoilage of the fish. Good sanitary practices in handling, processing, and storage of fish are major factors in preserving the quality of fish products.

Sewage-polluted waters from which fish may be taken are often a

Fig. 12.1. Unloading halibut from the cold hold of a fishing vessel.

significant public health problem. Studies have shown that many enteric bacteria associated with polluted waters can survive in the gut of fish. Unless steps are taken to destroy these organisms, fish can be a menace to public health.

Fish that are preserved by canning may present problems of spoilage. However, like all food products that are preserved in closed containers, canned fish should remain sterile and keep for long periods. The sanitary practices employed during processing, the efficiency of the heat tratment, and storage conditions will largely determine the shelf life of these products.

Clostridium botulinum was reported in the slime of tuna by Lang (1935). He found type A toxin in cultures from fish slime and cleaned fish just prior to canning. *Clostridium botulinum* was also found in the washings from empty sardine cans. The heat treatment required to destroy this organism in various fish products is reported in Table 12.1. Apparently, the rate of heat penetration into oil-packed products is lower than that into water-packed products. Thus, sauces and oil tend to protect *C. botulinum.*

An attempt has been made to preserve fish in vinegar sweetened with sugar. The fish are prepared by removing all inedible portions including the scales. They are thoroughly washed in water, placed in 6% vinegar pickle with 10% salt, and left in the vinegar brine for 4 days. The fish are then packed in sterilized cartons with the addition of mustard seeds and peppercorns between the layers. Lastly, about 2–3% vinegar containing sugar is added and the cans immediately sealed. Although canned products processed in this way occasionally swell, examination of the contents has identified only acetic acid bacteria with no evidence of sporeformers or putrefactive bacteria. Usually, the swells result from the action of acid on the metal of the container.

**Table 12.1. Time Required to Destroy C. *botulinum*
in Various Fish Products at 250°C (121°F)**

Product	Lethal treatment time (min)
Abalone	1.57
Kamaboko	8.97
Mackerel in brine	2.44
Sardines in brine	2.34
Sardines in mustard sauce	6.55
Sardines in oil	8.00
Sardines in tomato sauce	3.66
Shad	1.58
Shad roe	5.61
Squid in ink	0.69
Tempera	6.17
Tuna in oil	6.95

Source: Lang (1935).

MICROBIOLOGY OF LOBSTERS

Lobsters constitute an important delicacy in the American diet. Several investigators claim that freshly caught lobsters are not free of bacteria and that these organisms are always present in the digestive glands and intestinal tract. Just what role these organisms may play in spoilage is not known.

Lobster meat contains a very active autolytic enzyme system. The enzymes are not active as long as the lobster remains in its natural environment and alive. When lobsters are caught, killed, and packed in ice, these enzymes seem to be activated and the meat undergoes marked changes. This is one reason why lobsters are kept alive until they are cooked.

Lobster is commonly canned, and the keeping life of such products is generally good. During processing, the autolytic enzymes are inactivated and thus do not cause spoilage in the canned products. However, hydrogen sulfide may be liberated from lobster meat and react with iron in the can to cause a black discoloration caused by formation of iron sulfide. Bacteriological studies indicate that several species of bacteria are capable of producing hydrogen sulfide in live lobster.

Lobsters harvested at different times during the year exhibit marked differences in the chemical composition of the meat. For example, the pH of lobster meat is correlated with the season of the year. It is also known that a low pH seems to increase the black discoloration of canned lobster meat. It may be possible that an acidic pH favors the growth of H_2S-forming bacteria, which cause the discoloration. This would explain why discoloration is more of a problem at certain times of year than at others.

As with fish, the microflora of lobsters, crabs, and other shellfish depend largely on the water in which they grow. Contamination of shellfish from polluted waters can pose significant public health dangers. The primary way to guarantee that shellfish are free from enteric organisms is to hold them in clean unpolluted water for a period of time, during which they self-purify themselves, and then to cook them thoroughly before eating or processing them. Shellfish handled in this fashion generally are safe to eat. An important exception is shellfish that are contaminated with the toxin that causes paralytic shellfish poisoning. Because this to toxin is not removed by washing or completely inactivated by cooking, shellfish suspected of being contaminated must not be consumed. Paralytic shellfish poisoning is discussed in more detail later in this chapter.

MICROBIOLOGY OF OYSTERS

Oysters can present dangerous public health problems because they are eaten raw or nearly raw most of the time. Some outbreaks of hepatitis can be traced directly to oysters grown in contaminated waters.

Oyster Production

To understand properly the bacteriological problems of the oyster industry, it is necessary to have a clear understanding of all the processes and conditions involved, from the very beginning of the seed oyster through its growth, harvesting, and final preparation for sale. In the United States, oyster production varies slightly with the different regions and with different kinds of oyster.

Natural vs. Artificial Propagation

In areas where healthy natural oyster beds are plentiful, seed oysters can be gathered from natural beds. This involves bringing large numbers of oysters to the surface, keeping the seed, and returning the large oysters and shells to the water. The gathered seed oysters are then taken to pools used as oyster feeding beds.

Several factors determine the usefulness of these pools for oyster beds: (1) the presence and prevalence of pathogenic microorganisms, (2) the population of surrounding cities, and (3) the amount of industrial wastes draining into the area, as well as the size of the drainage area. Bacteriological tests for the presence of *E. coli* in the water and oysters, as well as chemical tests for oxygen availability, should be made. Such tests can identify waters that should not be used as oyster beds because of excessive factory or other wastes.

The mortality rate of embryonic oysters, called *spats*, in natural breeding beds is very high. Furthermore, in some parts of the country few natural beds still exist because of overharvest in the past, degradation of the environment, and other reasons. In some places, oysters are propagated artificially on concrete floors constructed at a point between high and low tide. Spats or breeding stock obtained from feeding beds are seeded into these artificial beds. The spats have a natural tendency to attach themselves to rocks, shells, or other objects during their growing period. It is common to dump the shells from mature oysters back into the water, a practice that increases the yield of oysters from a bed. The shell or rock to which the spat attaches itself is called the *cultch*. When oysters reach maturity, they free themselves from the cultch, which may then be used by other embryonic oysters.

Growth Requirements

In many northern oyster beds, unfavorable weather conditions contribute to the failure of oyster culture. Southern beds rarely experience such failures. Oysters grow much better in an environment that is less saline than seawater. A salinity one-half of normal seawater approximates ideal growing conditions for oysters. A mixture of fresh water and seawater seems to offer an ideal environment, provided a minimum of sewage pollution in the fresh water can be maintained to reduce the public health hazard.

Purification of Contaminated Oysters

There are at least two ways oysters may be handled to eliminate organisms harmful to man. Both depend on the process of self-purification by which oysters tend to eliminate microorganisms and waste products.

One good way to purify oysters is to transplant them to a sewage-free bed and in 1 or 2 days they will be relatively free of coliform organisms. The rapidity of this cleansing, of course, depends largely on the conditions under which the oysters are placed. The best results are attained when feeding conditions are optimal because then oysters pass water through their bodies most rapidly. The rate of water movement determines the necessary time of cleansing. Temperature, salinity, and season of the year also play an important role. It has been estimated that an oyster can imbibe and expel water weighing two times its own weight every minute.

Because of the lack of clear, natural water in which to purify oysters by natural self-purification, chlorination has been tried. When 6 ppm of chlorine is added to water, the surface or shell of the oyster is partially sterilized. However, oysters exposed to this concentration of chlorine do not imbibe or expel water because of the irritation caused by the chlorine. The use of calcium hypochlorite rather than free chlorine in the water greatly reduces the irritation to oysters, so that they continue to imbibe and expel water. Since the hypochlorite destroys many of the harmful organisms in the water, oysters allowed to remain in a hypochlorite bath for approximately 24 hr practically wash and sterilize themselves.

Although purification of oysters is quite effective, this practice is not used extensively because present state and federal laws do not permit the removal of oysters from polluted beds, except for seed oysters. The tendency at the present time is to prohibit the dumping of raw sewage into any surface water supply.

Shucking and Cleaning

Oysters are commonly shucked soon after harvest and then shipped in buckets or containers. Cleaned oysters, handled improperly, may become so contaminated as to be unfit for eating.

If harvested by dredging, oysters should be cleaned before bringing them into the shucking house. After oysters are shucked, they are always washed. This process, depending on the care that is taken, may reduce or increase the bacterial content. Rather than hosing or just immersing them in tubs of water, oysters are often cleaned by blowing air through them and the water. The oysters are then placed in tins and iced. If they are kept properly iced, there is only a minimum danger of spoilage.

At shucking houses, there are inspectors both for the state and federal governments whose duty it is to enforce sanitary regulations. The state inspector has the most authority, while the federal inspector merely supervises the work.

Bacterial Flora of Oysters

As mentioned already, oysters have an efficient self-purification mechanism. A viable oyster rapidly imbibes and expels water in order to obtain food. If the oyster is located in unpolluted water, this process also insures that it will be bacteriologically clean. The problem, of course, is that the waters oysters are grown in are likely to contain some bacteria; furthermore, the oysters may become contaminated during processing. It is no surprise, then, that various types of bacteria have been found in oysters despite their natural self-purification mechanism.

Among the bacteria that have been reported most commonly found in oysters are micrococci, gram-negative bacilli, and gram-positive aerobes and anaerobes.

Since *Escherichia coli* is a common inhabitant of the intestinal tract of man and animals, its presence in large numbers in water indicates fecal pollution. Therefore, *E. coli* has been used as an indicator for detecting contaminated water supplies. Likewise, public health recommendations emphasize the importance of coliform organisms as indicators of possible pollution in the waters of oyster beds.

The value of any indicator organism depends on its survival compared with that of the pathogens of interest. Preliminary studies have shown that *E. coli* survive in raw river water for 100–180 days and that *Salmonella typhi* survive approximately the same length of time. There is sufficient experimental evidence to demonstrate the transmission of enteric diseases by shellfish, and it has been shown that *S. typhi* will survive for several months in both shell and shucked oysters. Workers at the Public Health Service Shellfish Santitation Laboratory at Woods Hole, Mass., found that *S. schottmuelleri* remains viable as long as *E. coli.* Both of these organisms survived in shell oysters and soft clams when stored for periods comparable to those that might be used in transit from the point of harvesting to consumption.

The danger of polluted oysters is particularly acute because a large percentage of oysters are eaten raw or in oyster soup. In studies made on the effect of cooking on the microbial content of artificially inoculated oysters, it was found that the usual methods of cooking oysters do not necessarily sterilize them. Probably the reason for this is that to retain their fine taste and texture, they must not be cooked too long nor at too high a temperature.

MICROBIOLOGY OF OTHER MOLLUSKS

Other mollusks such as mussels, clams, and scallops also obtain their food by rapidly imbibing and expelling water. When they are grown in polluted waters, the risk of contamination is also very high. Clams, like oysters, are sometimes eaten raw, and generally none of the mollusks are cooked very much because extensive cooking makes them tough and alters their delicate flavor. Thus, the microbiology of oysters is similar to that of the other mollusks, and the same public health concerns apply to all of them.

In recent years, mussels have become quite popular and their annual consumption in the United States has steadily increased. However, acute toxemia has been attributed to mussels harvested from polluted water. The California State Department of Health has repeatedly warned the public of the danger from eating mussels obtained along the California coast. Sanitarians have reported many cases of "mussel poisoning" caused primarily by *Salmonella typhi.* This pathogenic organism has been known to survive for long periods in mussel flesh.

The microbiological quality of all shellfish depends largely on the quality of the environment in which they grow. The most sanitary processing, handling, and distribution cannot assure that marine foods are safe to eat if they have been harvested from polluted waters.

PARALYTIC SHELLFISH POISONING

Shellfish may become contaminated with an extremely potent toxin for which there is no known antidote. Ingestion of toxic shellfish by humans produces a variety of symptoms and sometimes leads to death. The lethal properties of shellfish toxin are comparable to those of the toxin causing botulism.

There are two other types of poisons that resemble shellfish poison but are generally not as toxic to humans. One is *waterbloom poison,* which may be elaborated by the profuse growth of freshwater algae or by their subsequent decay. Farm animals, deer, duck, and other wild birds have been killed by this poison. There are no reported cases of humans being poisoned through drinking water containing large quantities of algae.

The other related group of poisons is found in the flesh and organs of fish. Three different types of fish poison have been investigated, but little is known concerning the origin or nature of these poisons. One of the types, ciguatear, is common in the Carribean and in several instances has been reported as causing intoxication in Florida. The great barracuda is one of several fish species involved and is the most frequently associated with fish poisoning in Florida. A second type of fish poison is found throughout the South Pacific areas and is generally associated with the coral belt. The third type of poison, tetrodon, is found primarily in the Japanese areas. Of the three types of fish poison, tetrodon is the most deadly and causes symptoms most similar to those of shellfish poisoning in man and animals.

Sources of Shellfish Poison

The original source of the poison in shellfish is certain species of unicellular microscopic marine organisms of which the dinoflagellate *Gonyaulax catenella* is perhaps the best known. It is a free-swimming organism, which multiplies by formation of chains of two, four, or even eight individuals. These are dark orange or greenish brown and live like true plant cells by photosynthesis. Like all plankton organisms, this species is most abundant in the summer. At times it may multiply to 40 million per liter. At such times the water is a deep rust color—the so-called *red water* or red tide—in the daytime and a beautiful luminescent spectacle at night.

Scallops may become highly toxic, but they are of limited public health significance because ordinarily only the nontoxic adductor muscle is eaten. The principal types of shellfish that may reach dangerous levels of toxicity and be consumed by man are mussels and clams. The poison evidently does not harm the shellfish. If a large number of *G. catenella* are present in the water, the toxicity of mussels and clams may rise to dangerous levels within a few days. In the absence of the organisms, the stored poison is slowly eliminated.

Symptoms and Treatment

The primary symptom of toxic shellfish poisoning in human beings is peripheral paralysis, which may vary from a slight tingling and numbness about the lips to a complete loss of strength in the muscles of the extremities and neck, to ultimate death by respiratory failure. In a severe case of poisoning, the tingling, stinging sensation around the lips, gums, and tongue develops about 5–30 min after the consumption of the toxic shellfish. This is regularly followed by a numbness or a prickly feeling in the fingertips and toes; and within 4–6 hr, the same sensation may progress to the arms, legs, and neck, so that voluntary movements, as for example raising of the head, can be made only with great difficulty.

In all cases of moderate severity, this weakness and stiffness of locomotion is accompanied by a peculiar feeling of lightness. Vomiting is an inconstant symptom, and diarrhea and abdominal pain have not been recorded in untreated cases. In fact, there is a tendency to constipation, which persists for several days. Records of carefully observed cases show the average temperature to be slightly subnormal (mean 98°F). The pulse is firm and slightly accelerated (80–100/min). During recovery some patients have chilly sensations in their limbs and may feel stupefied or become easily fatigued for a number of days. The longest recorded interval between eating poisonous shellfish and death is 10 hr and the shortest is 3 hr.

Symptoms of poisoning may be mild or fail to develop in those who have consumed toxic shellfish in conjunction with a heavy meal or who have boiled them with rice and garlic or fried them in oil. When toxic shellfish are taken into an empty stomach, the intoxication rate seems to be higher. As soon as intoxication is recognized, emptying of the stomach by an emetic and purging by brisk laxatives has been the usual practice, but in severe cases, these measures cannot be relied upon to prevent the absorption of a fatal dose of poison. If difficult breathing develops, artificial respiration should be applied. Clinical observations indicate that, even in severe cases, this procedure extended over several hours may prevent death.

A number of physiological and pharmacological studies have been made on shellfish poison to investigate its mode of action. In autopsies on humans no significant findings were recorded in the abdominal or chest cavities except a slight pulmonary congestion. In every case, the folds of the stomach were studded with small hemorrhages.

Characteristics of the Poison

McFarren *et al.* (1956) concluded that shellfish poison is a neurotoxin with a central effect upon the cardiovascular and respiration centers and a peripheral effect upon nerve endings, both motor and sensory.

Ordinary cooking of shellfish in a little water, even for brief periods, provides the consumer with some degree of protection. In one study, steaming, boiling, or pan-frying toxin-containing meat reduced its toxicity by 70%. Cooking may actually reduce the toxicity of contaminated meat even more because a part of the poison is expressed in the juices. However, when clams are steamed or boiled, varying amounts of the juices may be ingested with the meats. The temperatures involved in pan frying are considerably higher than in steaming or boiling and seem to be somewhat more effective in destroying the poison, even though none of the extractives are discarded.

Coastal Quarantines

In some coastal areas, stretches of the beach are periodically marked off and quarantined to prevent commercial harvesting of mussels or clams and to discourage local residents and tourists from taking them. Such quarantines are usually imposed when the toxicity of the shellfish in an area is shown by laboratory tests to exceed 400 mouse units/g of meat.

SELECTED REFERENCES

Kelley, F., and Arcizz, C. S. 1954. Survival of enteric organisms in shellfish. Public Health Rep. 69 (12).

Lang, O. W. 1935. Studies on the presence of *Clostridium botulinum* in canned fish. Univ. of Cal. Public Health Rep. 2, 26.

McFarren, E. F., Schaffer, M. L., Campbell, J. E., Lewis, K. H., Taft, R. E., Jensen, E. T., and Schantz, E. J. 1956. Public Health Significance of Paralytic Shellfish Poison. A Review of the Literature. Robert A. Taft Sanitary Engineering Center, Cincinnati, Ohio.

Novak, A. F. 1973. Microbiological considerations in the handling and processing of crustacean shellfish. *In* "Microbial Safety of Fishery Products," C. O. Chichester and H. D. Graham (Editors). Academic Press, New York.

Shewan, J. M. 1971. The microbiology of fish and fish products—a progress report. *J. Appl. Bacteriol.* 34, 299–315.

Speck, M. L. (Editor). 1984. Compendium of Methods for Microbiological Examination of Foods. Am. Public Health Assoc., Washington, DC.

13
Milk and Milk Products

Spoilage of Milk and Milk Products   185
Pathogens and Quality Control   188
Dairy Fermentations   190
   Starters   191
   Butter Cultures   192
   Sequence of Microbial Growth   196
   Effects of Bacteriophages and Bacterial Inhibitors   197
Survival of Organisms in Fermented Dairy Products   197
Microbial Inhibitors in Milk   198
Fermented Milks   200
   Ripened Skim Milk   200
   Bulgarian Milk   200
   Acidopholus Milk   201
   Yogurt Milk   202
   Taette   203
   Kumiss   203
   Kefir Milk   203
Butter   204
Cheese   205
Selected References   210

Milk is the white or yellowish fluid produced by mammals to feed their young. It contains small fat globules suspended in a solution of casein, other proteins, lactose, and inorganic salts. The formal definition according to U.S. standards is as follows: "Milk is the whole, fresh, clean lacteal secretion obtained by the complete milking of one or more healthy cows, excluding that obtained within 15 days before and five days after calving, or such longer period as may be necessary to render the milk practically colostrum-free. The name 'milk' unqualified means cow's milk." In other countries, milk is obtained from goats, sheep, horses, water buffalo, camels, and other species of mammals. Cream is the sweet, fatty, viscous liquid separated from milk. To be considered cream, it must contain 18% butterfat.

Milk may be processed in various ways to change its composition so that it has different flavor, odor, viscosity, texture, or other functional properties suitable for specific uses. Table 13.1 lists many common

Table 13.1. Common Dairy Products

Product	Characteristics	Source/production method
Acidophilous milk	Somewhat acrid but pleasant-tasting milk	*Lactobacillus acidolphilus* fermentation
Bulgarian buttermilk	Curdled milk with a distinctive cheeselike taint	*Lactobacillus bulgaricus* fermentation
Buttermilk	Milk poor in content with sour taste	Supernatant in butter manufacture
Colostrum	Bitter early milk	Cow shortly after calving; this milk is nutritive for the calf but unacceptable for human use
Condensed milk	Thick, sweet canned milk	Concentrated to 25% moisture and sucrose added to give 40% sugar
Dried milk	Powdered whole milk that can be reconstituted with water	Spray drying of milk
Evaporated milk	Thick, creamy milk with cooked taste	Same as condensed but without sugar addition and usually fortified with vitamin D through irradiation
Fresh milk	Milk taken from cow under normal circumstances, cooled, and supplied within a day or two	Normal milking
Homogenized milk	Milk of consistent content (no cream layer)	Filtering and treating of milk to homogenize structure
Instant milk	Fat-free powdered milk that reconstitutes to make a good cooking milk	Freeze drying of milk
Irradiated milk	Milk sterilized by radiation or milk rich in vitamin D	Passage of milk through a radiation sterilizer; irradiation with ultraviolet to increase vitamin D content
Junket	Characteristic dessert with flavoring	Addition of impure rennet enzymes from animal origins
Kefir	Fermented milk with traces of alcohol and carbon dioxide	Mixed fermentation involving *S. lactis*, *L. bulgaricus*, and lactose-fermenting yeasts
Kumiss	Fermented alcoholic mare's milk	Mixed fermentation similar to kefir
Leben	An alcoholic fermented milk	Cows, goats, European bison, etc.
Pasteurized milk	Low in fermentation organisms and free of heat-sensitive pathogens	Heating milk to 145°–150°F (63°–66°C) or heating circulating milk to over 161°F (72°C) for at least 15 sec. (high temperature–short time method)
Peptinized milk	Curdled milk	Addition of pepsin enzyme

Table 13.1. Continued

Product	Characteristics	Source/production method
Skir or Skyr	Semisolid milk	*S. lactis* and *L. bulgaricus* fermentation
Sterilized milk	Milk in sealed containers free of most microorganisms	Heating to 212°F (100°C) for prescribed time
TT milk	Milk with reduced tubercule risk	From cows tested for presence of TT antibodies
Taette	Soured and often ropy milks with characteristic flavors	*S. lactis* and other microorganisms in a mixed growth
Yogurt	Distinctive buttermilks, which may be flavored with fruit	*L. bulgaricus* and *Streptococcus thermophilus* fermentation

types of milk available throughout the world. In addition to these liquid and semisolid milk products, butter and several hundred different varieties of cheese are made from milk. Cheese can be classified into about twenty major types including the following: Parmesan, provolone, Cheddar, pasteurized process American, domestic Swiss, pasteurized process Swiss, Roquefort, blue, brick, Limburger, Camembert, Neufchatel, cream, and cottage creamed.

The microbiology of milk and milk products has been studied more extensively than that of any other foodstuff for several reasons. Milk and milk products are probably consumed by more of the human population throughout the lifespan than any other food. Such a pattern of consumption includes high-risk infant, aged, and infirm populations. Individuals with lactose intolerance must consume milk in the form of fermented milk or cheese if dairy products are used. Milk is an animal product produced and collected in a barnyard or pastoral environment and frequently is mishandled. A number of diseases common to milk-producing animals and humans can easily be transmitted through milk to humans. Because milk is rich in nutrients and is liquid, it provides an excellent medium for the growth and dissemination of bacteria. Pathogenic spoilage and fermentative microorganisms are all capable of rapid growth in milk under the right conditions.

SPOILAGE OF MILK AND DAIRY PRODUCTS

The principal forms of milk degeneration and contamination, along with their causes, are listed in Table 13.2. The most common bacterial genera found in spoiled milk and milk products are *Streptococcus, Lac-*

Table 13.2. Etiology of Milk Spoilage and Contamination

Common name	Description	Cause	Comments
Abortus milk	Normal appearance	*Brucella abortus* and *B. melitensis*	Causes undulant fever (brucellosis) in man
Blue milk	Milk develops blue tint	*Pseudomonas* sp.	Clean and sterilize utensils; discard contaminated milk
Curdled milk	Milk develops semisolid consistency	Fermentative bacteria	Poor storage conditions (e.g., too warm)
Gassy milk	Bubbles form	Yeast or *E. coli*	Reject and check for contamination
Mastitis milk	Normal appearance	Mastitis-causing bacteria	May produce sore throat if consumed
Red milk	Red color	*Serratia marcescens*	Discard; check for source of contamination
Ropy milk	Ropy, stringy texture	*Alcaligenes viscolactis; Bacillus lactis* var. *viscosus*	Reject milk
Slimy milk	Unpleasant mucoid texture	Encapsulated bacteria or heavy white cell concentration	Reject milk
Soapy milk	Foams when shaken	*Bacillus saponicus*	Reject milk
Stringy milk	Similar to ropy milk but ropiness less noticeable	Early growth of *Alcaligenes* or fibrous protein in cows udder	Check before use
Tainted milk	Peculiar taste	Contamination with odors or chemicals; heat treatment	Usually discard
Tubercular milk	Normal appearance	*Mycobacterium tuberculosis*	Discard
Other pathogens	Normal appearance	*Salmonella,* streptococci, and staphylococci	Discard; check source and pasteurization procedure

tobacillus, Microbacterium, Achromobacter, Pseudomonas, Flavobacterium, and *Bacillus.*

The natural souring of milk caused by *Streptococcus lactis* and *Lactobacillus* sp. is undesirable in market milk. However, souring is desirable in fermented milk and cheese production. Gas production by coliform organisms and members of the *Clostridium* genus is undesirable, causing a "stormy fermentation" of milk. Lactose-fermenting yeasts may play a role in gas production especially in gassiness in cream. The souring of cream greatly favors these yeasts.

The proteins in milk can undergo both coagulation and proteolysis. Coagulation is caused by acid and the action of the enzyme rennin. Enzymic proteolysis is usually accompanied by acid formation. Micrococcus in the udder of a cow, *Streptococcus faecalis*, and *Bacillus cereus*, a sporeformer that survives pasteurization, all cause acid proteolysis of milk proteins. *Streptococcus liquefaciens* coagulates milk by producing a rennet-like[1] enzyme followed by acid production. However, proteolysis is much in evidence. *Alcaligenes, Pseudomonas, Achromobacter, Micrococcus,* and *Proteus* are genera capable of growing at low temperatures and causing proteolysis. A bitter flavor in milk is usually associated with these organisms.

Alcaligenes viscosus is responsible for ropiness in milk held at $10°C$ $(50°F)$. The ropiness is caused by the formation of gums and mucins. Other organisms may cause a fake type of ropiness. Coliforms and *Streptococcus lactis* var. *hollandicus,* for example, cause ropiness in butter cultures, thus reducing the acid production of the starter culture.

Certain organisms in raw milk produce the fat-splitting enzyme lipase, which, regardless of its source, hydrolyzes butter fat to fatty acids and glycerin. *Achromobacter, Alcaligenes, Proteus,* and *Pseudomonas* are genera of bacteria capable of producing lipase. Some varieties of molds also are lipolytic. Their presence may be desirable in mold-ripened cheeses such as blue cheese, which is ripened by *Penicillium roqueforti.*

Milk is very susceptible to flavor changes. Some off-flavors are related to certain production practices (e.g., type of feed) and condition of the cow (e.g., presence of mastitis; stage of lactation). Acid flavors are caused by lactic acid-producing organisms. Volatile fatty acids such as acetic or butyric are produced by coliform bacteria. Bitterness may result from lipolysis and also is influenced by the stage of lactation of the cow. *Actinomycetes,* if present in large numbers, produce a musty-flat flavor. *Streptococcus lactis* var. *maltigenes* is responsible for a cooked or caramel flavor in milk and cream; pasteurization will not remove this maltlike flavor.

[1]Rennet is a preparation from the stomach lining of calves that contains impure rennin.

Red milk may be caused by *Serratia marcescens* pigment, while *Torula glutinis* may cause red or pink colonies on the surface of sour cream or milk. Bovine mastitis results in a red coloration of the milk.

Clostridium sporogenes may cause a swelling of evaporated milk cans because of gas formation. The spores survive the heat process and reproduce in the can under anaerobic conditions producing gas, coagulation of proteins, and a bitter flavor. *Bacillus calidolactis*, a typical thermophile, may produce similar conditions to those produced by *C. sporogenes*, providing a small amount of air is available in the can. An unsealed can may swell, and the contents become bitter; coliforms and certain yeasts may be largely involved.

Undesirable flavors in butter may come from absorption of volatile compounds in the air, but microorganisms are largely responsible for off-flavors. Surface taint is caused by *Pseudomonas putrefaciens* present in wash water and on equipment. Fishiness may be due to *P. ichthyosmia*. Rancid, esterlike flavors come from *P. fragi*.

Chemical reactions independent of microbial activity may be associated with tallowiness caused by oxidation of unsaturated fats or exposure to copper in the equipment. Lecithin may form trimethyl amine, giving a characteristic fishy odor.

Anaerobes such as *C. sporogenes* and *C. tyrobutyricum* produce foul-smelling compounds in Swiss cheese. Coliforms, micrococci, and yeasts may contribute their share of proteolysis causing off-flavors.

Oospora lactis, a typical dairy mold, may be present in soft cheeses; the mold may cause some proteolysis, thus making a soft mushy cheese. Species of *Pencillium* may produce green spots in cracks or trier holes of cheddar and other cheeses. *Monilia nigra* produces black spots on the surface of hard cheeses.

PATHOGENS AND QUALITY CONTROL

The common pathogens in milk are generally considered as a group, called milkborne infections. The organisms pass through the mammary gland and are secreted in the milk, which if not pasteurized can transmit the disease to humans. Organisms in this group are *Brucella* spp., *Myobacterium tuberculosis*, beta-hemolytic *Streptococcus* spp., *Streptococcus agalactiae*, *Staphylococcus aureus*, coxiella, tickborne encephalitis virus, foot and mouth disease virus, and vaccina virus. Other common organisms that can be transmitted by drinking contaminated milk are *Salmonella* spp., enteropathogenic *E. coli*, and *Vibrio parahaemolyticus*.

Fortunately, the use of suitable quality control procedures makes milk a safe, healthful food, which can be consumed by high-risk populations including infants, the elderly, and sick people.

Table 13.3 lists the basic steps involved in processing fluid milk. Quality control begins with healthy animals tested and vaccinated for diseases transmitted through the host. Good management and clean, sanitary environments are also important. Rigorous sanitation in milking parlors (Fig. 13.1), especially of the udder and milking equipment before milking, is vital. The collected milk is strained, rapidly cooled in refrigerated stainless steel storage tanks, and refrigerated during transport to a processing plant. At the plant, sediment and fat are adjusted by centrifugation, and the milk is pasteurized at 161°F (72°C) for 15 sec. Generally, milk is also homogenized to prevent creaming.

Fermented milk products have the additional safeguard of a high

Table 13.3. Steps in Processing of Fluid Milk

Operation	Purpose	Remarks
Milking	To obtain raw product	Healthy cows, good sanitation, good ventilation are required
Mechanical cooling to 40°F	Minimize microbial growth	Refrigerated bulk milk tanks used to cool to 50°F in 1 hr, and 40°F or below in 2 hr
Storage at 40°F for up to 48 hr	Economics	Care must be taken to maintain temperature and minimize agitation to avoid off-flavors
Tank truck transport	Economics	Trucks insulated to keep temperature below 50°F
Transfer to plant storage—40°F	Accumulate for processing	Milk may be stored up to 72 hr in raw state
Centrifugal clarification; standardizing	Removal of sediment; fat control	Sediment and some bacteria removed and some cream removed to meet 3.5% fat standard
Pasteurization 161°F for 15 sec	Protection of health and improved shelf life	Generally by means of plate heat exchangers, using regeneration
Homogenization 1500–2500 psi	Prevent creaming	Changes the physical-chemical nature of the product
Packaging; storage at < 45°F; distribution	To meet consumer requirements	Diverse packages and distribution systems in use

Source: Harper and Hall (1976).

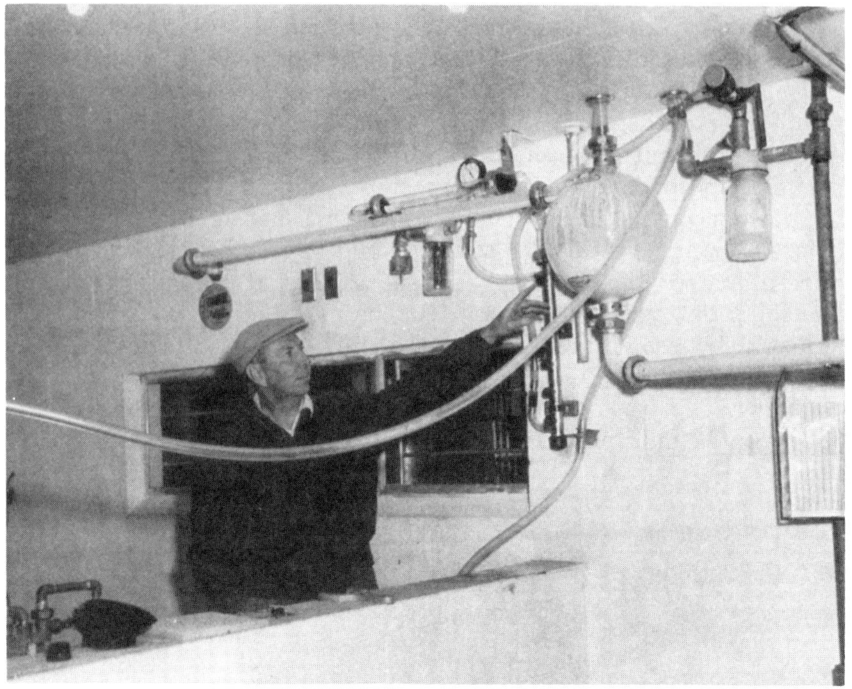

Fig. 13.1. In a properly designed milking parlor, the dairyman can milk a cow in a minute. As the milk flows, it passes through sanitary pipes to a bulk milk tank for colling. Courtesy U.S. Dept. of Agriculture.

lactic acid content, which effectively inhibits the growth of most bacteria.

DAIRY FERMENTATIONS

The characteristic flavor, odor, and/or texture of numerous dairy products depend on specific fermentations. Common fermented dairy products include various types of fermented milk, ripened cheeses (e.g., Swiss, Limburger, Cheddar, blue), butter, sour cream, yogurt, and junket.

Various factors that influence dairy fermentations and the general sequence of events during normal fermentations are discussed in this section. The characteristics of various fermented products are covered in later sections.

Starters

Cultures of lactic acid-producing and citric acid-fermenting organisms used to produce acid in milk are referred to as *starters*. They fall into three general categories. Lactic acid-producing species are generally *Streptococcus cremoris* or *Streptococcus lactis* (Figs. 13.2 and 13.3). Citric acid-fermenting species are *Leuconostoc cremoris* and *S. paracitrovorus (L. dextranicum)*. The third group includes microorganisms that produce both lactic acid and diacetyl. The most common of these is *Streptococcus lactis* var. *diacetylactis*.

Common starter species can be purchased as lyophilized, liquid, or frozen cultures. Frozen concentrated cultures are the most widely used. The starter culture is added to heat-treated milk that has been cooled to room temperature. The inoculated milk is then incubated at 70°–72°F (21°–22°C) for 14–16 hr.

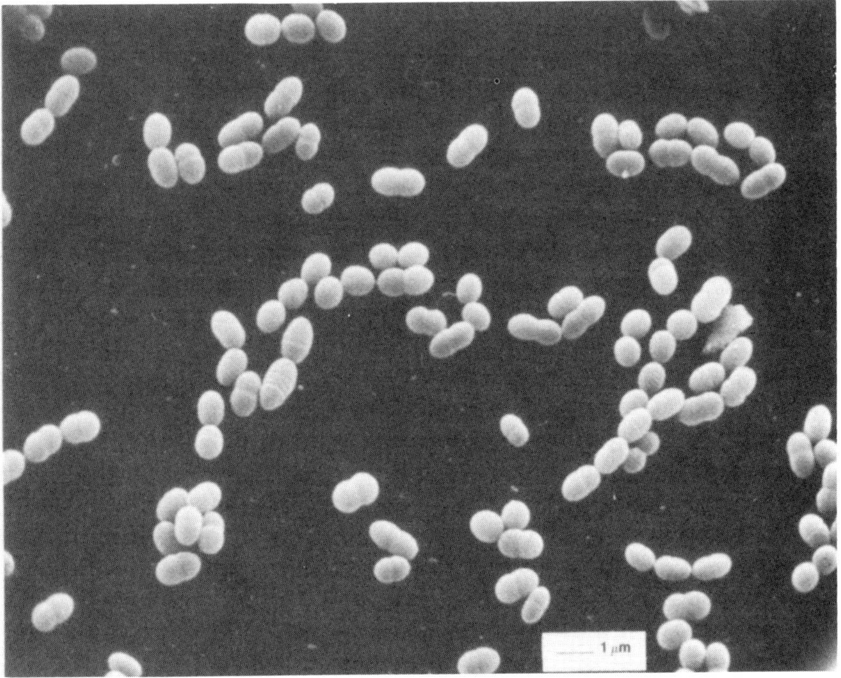

Fig. 13.2. *Streptococcus cremoris*, one of the species in butter cultures. Courtesy E. A. Zottola, University of Minnesota.

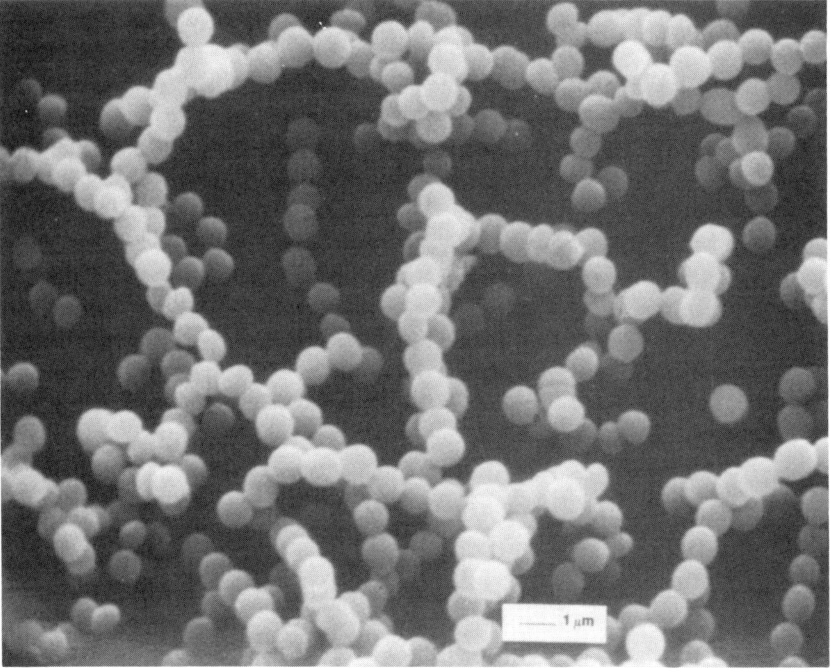

Fig. 13.3.   *Streptococcus lactis,* another lactic acid-producing species common in many dairy fermentations. Courtesy E. A. Zottola, University of Minnesota.

Butter Cultures

In 1780, Scheele isolated and identified lactic acid (hydroxypropionic acid) in sour milk. Storch in Denmark and Conn in the United States in 1890 were among the first scientists to use lactic acid cultures, obtained from ripened or sour cream or buttermilk, in butter production. Pasteurization of cream destroyed many organisms that caused defects in the butter and inhibited the bacteria that caused natural souring, which imparts much of the flavor and aroma to butter. To take the place of natural souring, lactic acid cultures were added to pasteurized cream and allowed to develop prior to the churning process.

As the acidity of milk increases during fermentation, the casein precipitates. Precipitation usually occurs when the pH reaches 4.6–4.8. The acidity will continue to increase, even after coagulation of casein, until the acid prevents further growth.

Although milk contains about 5.1% lactose, it is the limiting factor in acid production. When fresh milk is inoculated with *S. lactis* and *A. aerogenes* and incubated at room temperature for 60 hr, 22% of the lactose disappears and 88% is converted to lactic acid. As acid is produced, it combines with free casein, forming a casein salt, which is called the *curd*.

The lactic acid formed by microorganisms during milk fermentations usually prevents the growth of acid-sensitive putrefactive bacteria. However, molds can use the acid as a source of energy, thus decreasing the acid concentration and creating a more favorable environment for acid-sensitive bacteria to grow and break down the casein.

Although there are many species of lactic acid bacteria, the discussion in this section is limited to butter cultures, i.e., starter cultures used in the manufacture of butter, cultured buttermilk, oleomargarine, and many kinds of cheese. Other species are discussed in a later section dealing with fermented milks.

Three types of streptococci are associated in a good butter culture: *Streptococcus lactis*, *S. citrovorus* (*Leuconostoc citrovorum*), and *S. paracitrovorus* (*L. dextranicum*). *Streptococcus lactis* is often called the sour milk organism because it is normally associated with raw milk and grows very rapidly in milk when conditions are favorable for the formation of lactic acid. *Streptococcus citrovorus* and *S. paracitrovorus*, sometimes called citric acid fermenters, are largely responsible for the characteristic flavor and aroma of the product. For instance, the flavor of butter is caused by diacetyl produced by microorganisms.

Streptococcus lactis produces D-lactic acid in amounts ranging from 0.65 to 1.00% calculated as titratable lactic acid (pH 4.3–4.7). Although lactic acid is odorless, it definitely contributes to flavor because of its acid taste. *Streptococcus citrovorus* produces so little acid that it cannot be detected when the organism is growing in a tube of litmus milk. However, *S. paracitrovorus* produces enough acid in litmus milk to change the litmus to red but usually not enough to coagulate the milk.

An active butter culture produces appreciable amounts of volatile acids. Acetic and propionic acids are produced during an active fermentation. These volatile acids give a distinct flavor and aroma to the product. They can be increased by adding small amounts of citric acid to the milk, which is fermented by citric acid-fermenting bacteria. The addition of citric acid also increases the production of acetylmethylcarbinol and diacetyl; these compounds greatly intensify the aroma and flavor in butter.

When *S. lactis*, *S. citrovorus*, and *S. paracitrovorus* are grown in pure culture, each culture produces very small amounts of flavor compounds. However, when they are grown in a mixture as they normally occur in a good active starter, flavor compounds appear in appreciable quantities. A symbiotic relationship seems to exist in which these organisms are mutually beneficial, and it is only under these conditions that the flavor compounds are formed.

The mother substrate(s) and the mechanism by which acetyl-methylcarbinol and diacetyl are formed by the butter culture organisms are not very well understood. It has been suggested that the compounds may be formed from the citric acid in the milk by a series of oxidation and reduction reactions or they may be formed from the lactose.

One scheme that has been suggested for the breakdown of citric acid by butter culture organisms is as follows:

$$
\begin{array}{c}
\text{OH} \\
| \\
\text{COOH.CH}_2\text{.C.CH}_2\text{.COOH} \; + \; 2\text{H} \\
| \\
\text{COOH} \\
\text{Citric acid}
\end{array}
\quad \xrightarrow{\text{reduction}} \quad
\begin{array}{l}
\text{CH}_3\text{.CHOH.CO.CH}_3 \; + \\
\qquad\qquad 2\text{CO}_2 \; + \; \text{HOH} \\
\text{Acetyl methyl carbinol}
\end{array}
$$

$$
\begin{array}{l}
\text{CH}_3\text{.CHOH.CO.CH}_3 \; + \; 2\text{O} \\
\text{Acetyl methyl carbinol}
\end{array}
\quad \xrightarrow{\text{oxidation}} \quad
\begin{array}{l}
2\text{CH}_3\text{.COOH} \\
\text{Acetic acid}
\end{array}
$$

$$
\begin{array}{l}
\text{CH}_3\text{.CHOH.CO.CH}_2 \; + \; 2\text{H} \\
\text{Acetyl methyl carbinol}
\end{array}
\quad \xrightarrow{\text{reduction}} \quad
\begin{array}{l}
\text{CH}_3\text{.CHOH.CHOH.CH}_3 \\
\text{2,3-Butylene glycol}
\end{array}
$$

Another suggested mechanism of fermentation and formation of aroma and flavor compounds by butter cultures involves the following reactions:

$$
\begin{array}{l}
\text{C}_{12}\text{H}_{22}\text{O}_{11} \; + \; \text{HOH} \\
\text{lactose}
\end{array}
\quad \xrightarrow{\text{hydration}} \quad
\begin{array}{l}
\text{C}_6\text{H}_{12}\text{O}_6 \; + \; \text{C}_6\text{H}_{12}\text{O}_6 \\
\text{glucose} \qquad \text{galactose}
\end{array}
$$

$$
\begin{array}{l}
\text{C}_6\text{H}_{12}\text{O}_6 \\
\text{galactose}
\end{array}
\quad \xrightarrow{\text{dehydration}} \quad
\begin{array}{l}
2\text{CH}_3\text{.CO.CHO} \; + \; 2\text{HOH} \\
\text{methyl glyoxal}
\end{array}
$$

$$
\begin{array}{l}
\text{CH}_3\text{.CO.CHO} \; + \; \text{HOH} \\
\text{methyl glyoxal}
\end{array}
\quad \xrightarrow{\text{hydration}} \quad
\begin{array}{l}
\text{CH}_3\text{.CO.CH(OH)}_2 \\
\text{methyl glyoxal hydrate}
\end{array}
$$

$$
\begin{array}{l}
\text{CH}_3\text{.CO.CH(OH)}_2 \\
\text{methyl glyoxal hydrate}
\end{array}
\quad \xrightarrow{\text{glyoxalase}} \quad
\begin{array}{l}
\text{CH}_3\text{.CHOH.COOH} \\
\text{lactic acid}
\end{array}
$$

$$
\begin{array}{l}
\text{CH}_3\text{.CHOH.COOH} + 2\text{H} \\
\text{lactic acid}
\end{array}
\quad \xrightarrow{\text{reduction}} \quad
\begin{array}{l}
\text{CH}_3\text{.CH}_2\text{.COOH} \; + \; \text{HOH} \\
\text{propionic acid}
\end{array}
$$

Or, with H acceptor, methyl glyoxal yields

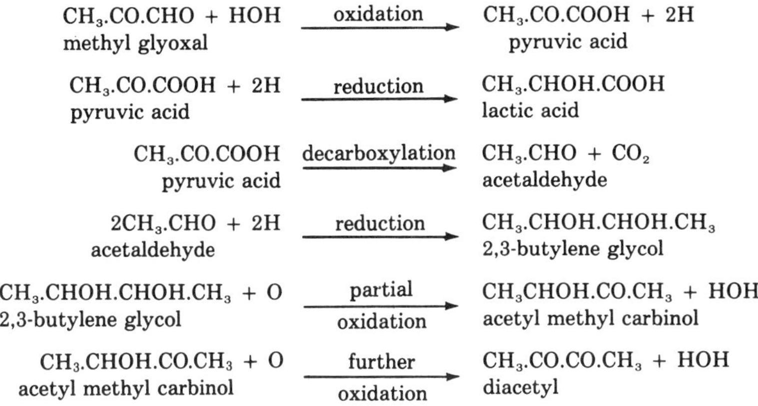

Butter cultures are no longer used to any great extent in buttermaking because good results are not always obtained. If butter cultures are not properly handled and transferred regularly, poor results can be expected. Starter distillate has come into general use in buttermaking. The distillate is prepared by steam-distilling a good butter culture so that the volatile compounds are not lost. It is prepared commercially, standardized, and sold to the dairy industry as *starter distillate*. It can be added to butter directly just before it is removed from the churn.

Cultured buttermilk is prepared by inoculating pasteurized skim milk with a good active butter cutlure. The milk should be heated at 180°–190°F (82°–88°C) for at least 30 min. Sufficient starter should be added to produce an acid concentration in the milk of around 0.75–0.90% in 12–16 hr. Usually an active starter will produce this level in 12 hr using 0.5–1% inoculum. Butter cultures are sensitive to any marked changes in temperature. Therefore, the optimum temperature of 68°–72°F (20°–22°C) should be maintained throughout the fermentation.

The oleomargarine industry still uses butter culture to ripen the milk in which the vegetable fats are churned. Obviously, this product acquires the delicate flavor of butter from the fermented milk. In fact, oleomargine manufacturers are more concerned about maintaining good butter cultures than the butter industry because of the satisfactory use of starter distillate in buttermaking. Starter distillate cannot be used satisfactorily to obtain flavor compounds in the manufacture

of oleomargine because some unexplained chemical reaction between
the vegetable fats and the distillate produce a marked off-flavor. However, some of the recently developed flavor preparations are giving
very good results.

Although starter organisms play an important role in cheese ripening, the characteristics of many types of cheese depend largely on the
processing techniques employed, the temperature of storage, and the
kinds of organisms added to the cheese during its manufacture.

Sequence of Microbial Growth

Raw milk is not free of microorganisms. The animal may contribute
organisms during the formation of milk in the mammary gland, but the
greatest source of microorganisms in milk is external contamination.
Obviously a heterogeneous microflora exists in raw milk. If the milk
is allowed to undergo a normal fermentation, the microbial sequence
illustrated in Fig. 13.4 occurs.

The phase represented from A to B is the lag phase in which microorganisms are destroyed or unable to multiply. *Streptococcus lactis,* usually present in all raw milk, grows first and produces acid up to approximately 0.9% (phase B to C). Eventually the acid restricts the
growth of these organisms and then the lactobacillus group, also normally present in milk, becomes predominant (phase C to D). Being
more acid-tolerant than *Streptococcus,* they are able to survive the
acidity already produced and in turn produce more acid, up to 3.5–

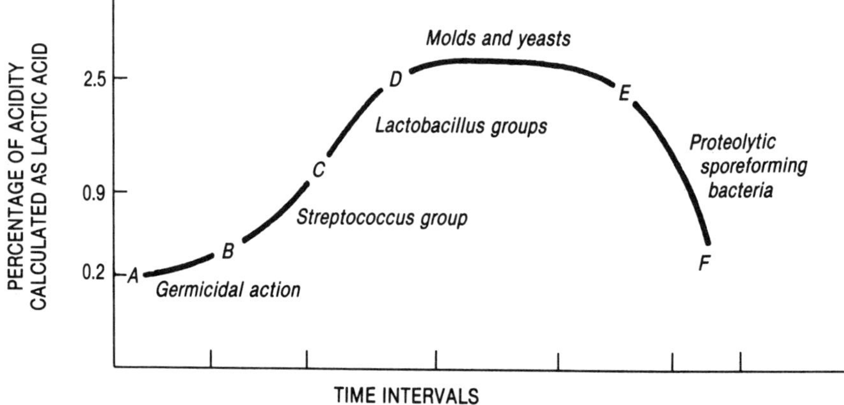

Fig. 13.4.   Sequence of microbial growth in dairy fermentation and relationship to acid concentration.

4.0%. The *Lactobacillus*, like the *Streptococcus*, cannot survive in an acid environment for a long time. Gradually they are eliminated in this high-acid milk, and typical acid-tolerant molds and yeasts take over (phase D to E). The molds play an important role in using the acid and produce several kinds of basic compounds, which tend to neutralize the remaining acidity. A change from a high-acid product to one that is basic then takes place. Proteolytic bacteria (phase E to F) then become the predominant organisms. Because they cannot survive in an acid medium, these organisms do not grow until conditions in the milk are favorable for them. Proteolytic bacteria, like the molds, produce basic products; as a result, the milk becomes extremely alkaline, with a pH from 8 to 10.

Effects of Bacteriophages and Bacterial Inhibitors

The presence of bacteriophages in dairy fermentations will prevent or slow down the production of lactic acid by lactic acid-producing organisms in butter and cheese. The problem can be reduced by selecting starter cultures that are resistant to bacteriophages.

Some microorganisms also produce substances that can inhibit the growth of other microorganisms used as starter cultures in the manufacture of fermented dairy products.

SURVIVAL OF ORGANISMS IN FERMENTED DAIRY PRODUCTS

The microorganisms responsible for acid formation in fermented dairy products and other fermented foods may survive for long periods in the acid environment they create. The acid tolerance of a species as well as the amount of acid formed determines the extent and length of survival. In certain fermented milks produced by various *Lactobacillus* species, a pH of 3.0–3.5 may be reached during fermentation; this is comparable to a titrable acidity of around 3.5–4% as lactic acid. Although these organisms exhibit considerable acid tolerance, they cannot survive indefinitely at such pH levels; eventually the milk becomes sterile because of its high acid content. The survival of organisms normally associated with various food fermentations is also influenced by the buffering system, type of processing, temperature, and other environmental factors. In general, the conditions present during fermentations favor the survival of desirable acid-tolerant organisms and discourage growth of undesirable ones.

Public Health groups have proposed enactment of laws requiring that all dairy products be made from pasteurized milk. This proposal would affect the cheese industry, since many types of cheeses are made from raw milk. Although milk is heated during cheesemaking, the temperature is not sufficient for pasteurization. Moreover, there is the question of how long pathogens survive in cheese, because of the acidity formed by desirable organisms.

For many years food sanitarians have considered fermented milks safe for human consumption. However, the U.S. Public Health Service has assembled sufficient evidence during the past few years to show that some disease outbreaks can be traced to the consumption of fermented milks. Unfortunately many contradictory reports can be found on this subject.

These investigators also studied the survival of *Salmonella typhi, S. paratyphi, S. schottmulleri, Shigella dysenteriae,* and *Shigella paradysenteriae* (Flexner), inoculated into sterile skim milk acidified with lactic acid, into naturally fermented milks, and into milks inoculated with lactic acid-producing bacteria. The results of this study are summarized as follows:

1. The amount of acid produced by lactic acid bacteria in sterile skim milk varies widely. This variation in acidity may contribute to the survival of many pathogenic organisms.
2. The pathogenic test bacteria were destroyed more easily in acidified milk than in milk allowed to undergo natural fermentation.
3. Larger numbers of pathogenic organisms survived in milk when the biological acidity increased gradually than when it increased rapidly. The presence of acid-tolerant strains of the pathogens may account for their resistance to the biological acidity produced during natural fermentation.
4. When the acidity of fermented milks was 0.95–1.15% (pH 3.9–4.2), many pathogens survived during an incubation period of 7 days. Likewise, when the acidity was 0.50–0.86 (pH 4.3–5.0), there were many viable pathogens after 9 weeks' incubation.

The authors concluded from this study that the acidity of naturally fermented milks is not sufficient to destroy all pathogenic bacteria.

MICROBIAL INHIBITORS IN MILK

Antibiotics and sulfa drugs commonly used as effective therapeutic agents against bovine mastitis have been detected in the milk from treated cows. These drugs impede the lactic acid fermentation necessary in the manufacture of cheese and cultured dairy products. Although these antibiotics are known to occur in milk, their effects on the naturally occurring microflora and enzymes in milk, and the effects of processing and storage of milk on their stability, have not been adequately studied. Drug residues are now prohibited in milk above certain levels.

When milk containing 0.13–0.53 units of penicillin per milliliter was pasteurized for 30 min at 143°F (62°C), about 8.2% of the penicillin was inactivated; at 160°F (71°C) about 10.1% was inactivated. After storage for 7 days at 36°–41°F (2.2–5°C) and subsequent pasteurization at 143° and 160°F, 34.8 and 20.6%, respectively, of the penicillin was inactivated. When penicillin was added to raw and pasteurized milks and both samples stored in a refrigerator, the quality of the raw milk was not acceptable within 10 days, whereas the pasteurized milk did not develop a rancid flavor until after 17–30 days. Obviously, penicillin in combination with regular pasteurization can appreciably increase the keeping quality of the milk.

Results from studies on streptomycin were comparable to the results obtained with penicillin under similar conditions. Approximately 14% of the streptomycin was lost from milk pasteurized at 143°F (62°C) for 30 min; at 160°F (71°C), 22.8% was inactivated. Pasteurization at 167°F (75°C) for 15 min resulted in a 20.6% loss of potency. However, when the milk was autoclaved at 15 lb for 15 min, no trace of streptomycin could be detected. When raw milk containing streptomycin was stored, there was an average 18.0, 27.4, and 35.1% inactivation during the first, second, and third week, respectively.

The results of these studies indicate that in milk, streptomycin is comparatively more heat-labile than penicillin. However, during the same period of storage of raw and processed samples, streptomycin is inactivated to an appreciably lesser extent than penicillin.

Several other inhibitory agents may be found in milk. These include sanitizers and bacteriophages all of which in certain concentrations tend to cause lowered plate counts and retard lactic acid development. However, none of these agents is satisfactory as a milk preservative and none is effective against all types of organisms. To avoid starter difficulties caused by bacteriophage action, one should follow recommended practices for the handling of starters.

Probably the best way to avoid objectionable residues of antibiotics in milk is to insist that dairy producers comply with the recommendations of the Committee on Antibiotics of the Federal Food and Drug Administration. There is no doubt about the success of antibiotics in treatment of mastitis, but such agents should not be used unless they are needed.

The harmful effects of antibiotics and other inhibitory substances on human beings has not been determined, although it is known that some people show violent and dangerous allergic reactions to certain of them. Until such information is available, milk used for human consumption should be kept as free as is practical of any microbial growth inhibitors.

FERMENTED MILKS

The souring of milk is well known, and some have ascribed therapeutic value to fermented milks, especially acidophilus milk. All of these milks are prepared by growing selected lactic acid organisms in the milk either as one pure culture or as a mixture of these organisms. Fermented milks are important products of the dairy industry and are particularly popular in warm climates.

The therapeutic value of fermented milks has been greatly exaggerated. The underlying principle is based upon the ability of the organisms to perpetuate themselves in the intestinal tract, thereby creating an unfavorable environment for many kinds of undesirable bacteria that may be present. The diet eaten tends to influence the type of microflora that will predominate in the intestine. A high protein intake favors proteolytic organisms.

Ripened Skim Milk

In the United States, the most common type of fermented milk is ripened skim milk. Its composition varies with the type of culture used and modification of the milk. The average composition of skim milk is as follows: water, 90.35%; protein, 3.72%; fat, 0.15%; lactose 4.98%; and ash 0.80%. During the fermentation process, a small portion of the lactose in the milk is converted to lactic acid. In some instances, 1–2% butterfat is added to improve the body and flavor of the product. It is a common belief that the nutritional value of fermented milk is comparable to that of whole milk in many respects, since 40% of the proteins, approximately 50% of the lactose, and practically all the minerals remain.

Bulgarian Milk

Bulgarian milk resembles cultured buttermilk in most respects, but it has a more acid flavor caused by the high acidity produced by the culture. *Lactobacillus bulgaricus* is the starter organism used in the preparation of this product (Fig. 13.5). Sterile skim milk is used because the temperature of incubation, 90°–100°F (32°–38°C), is also favorable for the growth of undesirable organisms that resist pasteurization. Fermentation by *L. bulgaricus* often produces milk with a 2–3% acid content. This high acidity is not acceptable to many consumers. The addition of whole sweet milk aids in masking the acid flavor results in a good refreshing milk beverage.

One part of cultured buttermilk to one part of Bulgarian milk produces a fine-flavored product that has the desirable characteristics of each. Because the growth features of each fermented milk are different, they must be prepared separately and then mixed at the time of consumption for the best results.

Acidophilus Milk

Acidophilus milk is comparable to Bulgarian milk in the acid produced and general growth requirements for the culture. *Lactobacillus*

Fig. 13.5. *Lactobacillus bulgaricus.* Debris in center is coagulated milk protein. Courtesy E. A. Zottola, University of Minnesota.

acidophilus is the organism used in the preparation of acidophilus milk.

The preparation of acidophilus milk is very difficult, as is that of Bulgarian milk, because the temperatures used during fermentation also favor many other bacterial species that are undesirable. Usually sterile skim milk is used. About 1–2% inoculum is recommended. The culture should be freshly isolated from the intestinal tract of man or other animals for the best results. Cultures carried in skim milk for a short time tend to lose their ability to establish themselves in the intestinal tract.

Of all the fermented milks, acidophilus milk has the best claim to real therapeutic value. *Lactobacillus acidophilus* can grow and survive in the intestinal tract for prolonged periods provided the proper carbohydrates are present. The organism produces sufficient acid in the intestinal tract to suppress many types of undesirable organisms and thus is helpful in treating certain intestinal disorders. Since none of the bacterial species associated with other types of fermented milks can survive in the intestinal tract for very long, they have little therapeutic value.

The relatively high acidity produced by *L. acidophilus* limits the popularity of this fermented milk. However, there are many ways to modify the product and still receive the beneficial results. Usually, the addition of small amounts of whole milk to acidophilus milk at the time of consumption makes the product more palatable. The viability of the organisms tends to decrease rapidly as the milk is stored in the refrigerator. One week is a safe period for storage if one wishes to receive the most beneficial results.

Yogurt Milk

Yogurt milk is another popular fermented milk especially in the Balkan countries and more recently in the United States. The product can be made from cow, goat, sheep, or buffalo milk.

To prepare yogurt itself, fresh raw milk is inoculated with an active fermenting batch of yogurt or with a dried starter. The milk culture is incubated at 100°–115°F (38°–46°C) until a thick curd develops, or until a sharp acid flavor and a distinctive aroma is achieved. The principal organisms involved in yogurt fermentations are *Streptococcus thermophilus* and *Lactobacillus bulgaricus*. The high acidity during the fermentation yields an acid curd.

To prepare yogurt milk, the milk is sterilized before inoculation with starter. Usually 1% of the starter culture is added. The milk is then incubated at 98.6°–100°F (37°C) for 10–12 hr. The finished product can be placed in suitable retail containers and held at 50°F (10°C) until it is consumed.

Taette

The principal organism involved in the fermentation of taette is *Streptococcus lactis* var. *hollandicus*. Usually cow's milk is inoculated with a small amount from a previous batch. Pasteurization of the milk is highly desirable to prevent the growth of undesirable organisms. The culture mixture should be incubated around 70°F (21°C) for 24–36 hr. The fermented milk becomes viscous or stringy, with a mild acid flavor. The acidity may reach 1% lactic acid. Taette is a delicacy in the Scandinavian countries, where it is used as a "spread" for bread and other pastry products.

Kumiss

Kumiss, which is usually made from mare's milk, is native to Russia. The product was originally made by early wandering tribes of eastern Russia, who were great horsemen and by selection gradually developed a hardy breed of horse that gave large amounts of milk.

Streptococcus lactis and *L. bulgaricus* are important in the formation of kumiss. A lactose-fermenting yeast, which produces variable amounts of alcohol, also is necessary in the fermentation. The native method of inoculating milk is to add some fermenting or decayed matter, such as a piece of flesh, tendon, or vegetable matter. Usually the lactic acid bacteria and yeast are able to compete with other organisms and thus establish themselves as the predominant microflora in the milk.

The temperature of incubation may vary from 75° to 86°F (24° to 30°C) for 12–72 hr. About 1% acid and 2% alcohol are formed during this period. In some cases, the product is distilled to increase its alcohol content.

Kefir Milk

Kefir milk is perhaps one of the oldest and most interesting of the fermented milks. It was prepared originally by the people of southern Russia, Turkey, and the Balkan countries. The product can be prepared from milk from cows, goats, and sheep. The product goes under varying names including *kefir, hippe, kepi,* and *kaphir,* all of which come from a common root meaning a pleasant or agreeable taste.

A preliminary treatment of the kefir culture is necessary in order to get an active fermentation. Dry kefir "grains," which resemble popcorn and consist of irregular lumps of mixed fermentative organisms, should be soaked in lukewarm water for 3 hr and then placed in a 1% sodium bicarbonate solution for 3 hr. After this treatment, the grains should be swollen; they are then rinsed in cold water. The culture is added to fresh pasteurized skim milk and incubated at 70°–80°F (21°–27°C) for 24 hr until the milk curdles. Daily transfers to fresh pasteurized skim milk are absolutely essential so the grains will increase in size and rise to the surface during the fermentation.

The final product is prepared by adding 1 part of kefir grains to 3 parts of pasteurized whole or skim milk. The milk culture should be incubated at 57°–61°F (14°–16°C) for 8–10 hr with frequent stirring. The container should be loosely stoppered. The kefir grains are removed by straining the fermented milk through coarse cheesecloth. The grains are washed with lukewarm water and preserved for subsequent batches of kefir milk. The fermented milk is usually transferred to clean milk bottles, stoppered securely, incubated at 57°–60°F for 24 hr, and then placed in the refrigerator until used. The fermented milk contains 0.5–1.0% lactic acid, carbon dioxide, and 0.3–1.0% alcohol, which are formed after 3 to 5 days' incubation.

The kefir fermenting organisms are *S. lactis, L. bulgaricus,* and lactose-fermenting yeasts. These organisms exhibit a symbiotic relationship. No doubt during the active fermentation of milk, a synergistic action takes place that may account for the characteristic flavor and aroma associated with kefir milk.

One suggested procedure for preparing a substitute for kefir involves adding from 2–4 teaspoons of sucrose and a small amount of prepared yeast to a pint of cultured buttermilk. The yeast can be obtained at most food markets. The milk mixture is incubated for 3–4 days at 65°–70° (18°–21°C). The fermentation should be carried out in heavy bottles with securely fastened rubber stoppers because a strong gas pressure is built up. The kefir substitute can be kept in a household refrigerator for several days providing the containers are tightly stoppered to prevent a loss of carbon dioxide.

BUTTER

Butter is made from milk, cream, or both and contains at least 80% butterfat. It is made by churning the milk or cream until the butterfat globules in the milk adhere. Frequently, a special prepared culture of *S. lactis* and associated species are added to develop desirable acidity,

Table 13.4. Steps in Traditional Churning Process for Butter

Operation	Purpose	Remarks
Separation	To concentrate the fat to 25–40% fat to facilitate the churning operation	Milk is warmed to 90°–110°F to improve separation efficiency
Neutralization	To facilitate churning	Used only for sour cream, not employed for sweet cream; acidity of cream should not exceed 0.15% calculated as lactic acid
Pasteurization	To eliminate pathogens and enzymes that cause off-flavors	Cream heated to 160°–170°F for 30 min or equivalent
Cooling	To solidify part of milk fat to permit churning	Cream is cooled to 40°–50°F and held cold for several hours before churning
Churning	To invert the emulsion from oil in water to water in oil; to form granules	Cream is tempered to 50°–55°F; rotated in a cylindrical drum, cube, or cone
Washing	To firm up butter granules	
Salting	For flavor and to reduce microbial activity	Salt added to give 1–1.5% salt
Working	To distribute moisture and salt; composition control	Controls final body characteristics of the butter

Source: Harper and Hall (1976).

flavor, and aroma. The characteristics and uses of butter cultures were discussed in an earlier section.

The various steps involved in the churning of butter are outlined in Table 13.4.

CHEESE

Cheese, which is made from the coagulated portion (curd) of milk, cream, skim milk, whey, or buttermilk, consists mainly of casein, fat, and water. Cheese has been manufactured for human food for at least 11,000 years. The early stimulus for the production of cheese was to change highly perishable milk into a solid, nutritious, flavorful, and relatively stable foodstuff.

In the United States, about 65% of all the milk used in cheese making is made into Cheddar. The various steps in Cheddar cheese manufacture are listed in Table 13.5 and illustrated in Figs. 13.6–13.11.

Because the ripening of cheese is a complex microbiological process, it is very difficult to evaluate the exact role of the different kinds of

Table 13.5. Steps in Traditional Process of Cheddar Cheese Manufacture

Step	Purpose	Remarks
Temperature adjustment	To bring temperature to optimum for starter organisms	Milk added to vat and temperature adjusted to 86°–88°F
Starter addition	To develop about 0.01–0.02% acidity to assist rennet coagulation	Lactic starter used at about 1% for 1/2–1 hr
Settings	To coagulate the milk	Rennet, or suitable substitute, added at a rate of about 3 oz/1000 lb of milk to coagulate in 30 min
Cutting	To promote whey removal	Curd is cut into 1/4- to 3/8-in. cubes
Cooking	To remove whey and firm curd	Temperature raised to 102°–104°F in about 45 min and held about 45 min until firm
Draining	To remove whey	Whey may be pumped through a separator to recover fat and stored for further processing
Cheddaring	To mat the curd and to develop the typical body of cheddar curd	This is the characterizing step in Cheddar cheese manufacture; curd is cut into slabs, turned every 15 min and piled every 30 min to give 3 to 4 high piles of matter curd. The curd is matted until the acidity reaches 0.5% (pH about 5.2)
Milling	To prepare curd for pressing	The curd is cut into pieces about 1 × 2 in.
Salting	To stop further acid production	Salt added to give 1.5–1.8% in finished curd. Salt is mixed and allowed to become absorbed into curd for 20–30 min
Pressing	To remove additional whey and arrive at final desired moisture content	Cheese is hooped into forms or round or square hoops; usually 20 to 40 lb of finished cheese (21½ to 43 lb of curd) Cheese is pressed at about 20 psi for 1 hr; then the cheese is removed from the press and dressed by placing a cloth around the curd to insure a smooth surface. The curd is then pressed for an additional 14–16 hr
Curing	To ripen the curd to give the characteristic body and flavor of cheddar cheese	Cheese is placed in a 45° to 60° curing room at 80% humidity for ripening. For a fully cured cheese, curing requires about 6 months

Source: Harper and Hall (1976).

Fig. 13.6. Rennet being added to milk to make cheese. Courtesy Canadian Dept. of Agriculture.

Fig. 13.7. Cutting the curd. Courtesy Canadian Dept. of Agriculture.

Fig. 13.8. Cheddaring. Courtesy Canadian Dept. of Agriculture.

Fig. 13.9. Hooping the cheese. Courtesy Canadian Dept. of Agriculture.

Fig. 13.10. Pressing the cheese. Courtesy Canadian Dept. of Agriculture.

Fig. 13.11. Packaging Cheddar cheese in wax. Courtesy Canadian Dept. of Agriculture.

organisms present in the milk and in the cheese during the processing. No one knows the role of the initial microflora and enzymes in the raw milk, but these endogenous biological agents are important in the ripening of cheese, as are the starters that are added to the milk. When milk is thoroughly cooked, many of these biological agents are destroyed. Under these conditions it is very difficult to manufacture certain varieties of cheese regardless of the activity of desirable starters. The major commercial varieties of cheese and the ripening organisms involved in their production are listed in Table 13.6.

The formation of curd is a principal step in cheesemaking regardless of the variety. Curd may be formed in two ways: (1) by the action of rennet and (2) by the action of lactic acid produced by fermentative organisms. Differences among varieties of cheese are related to many factors such as kind of milk used, method of curd formation, amount of moisture retained in the curd, amount of salt added, size of finished cheese, temperature and conditions of ripening, length of ripening period, kinds of microorganisms used as starters, and the general microbial content of the raw milk.

The production of lactic acid from lactose in the milk begins early in the cheesemaking process and continues as long as lactose is available. Cheese has a high buffering capacity in relation to the sugar content; therefore, the high acid content does not seriously inhibit the growth of lactic acid-producing bacteria. Lactic acid favors curdling of the milk with rennet. Apparently this enzyme is more active in the presence of some acid than at neutral pH. The presence of acid also helps to control many putrefactive bacteria, which may alter the quality of cheese.

The production of acid during cheese fermentations also affects the

Table 13.6. Classification of Commercial Varieties of Cheese

Class	Variety	Ripening organisms
Hard	Cheddar	Bacteria with or without eye-forming bacteria
	Swiss	
Semi-hard	Brick	Bacteria
	Blue	Mold
	Roquefort	Mold
	Gorgonzola	Mold
Soft	Limberger	Bacteria
	Camembert	Mold
	Cottage	Unripened by bacteria
	Cream	Unripened by bacteria
	Neufchatel	Unripened by bacteria

texture of cheese. Acid production favors expulsion of whey from the curd. Retention of too much moisture in the curd interferes with normal biochemical changes. The fusion of curd particles is very necessary if the cheese is to have the proper body. Insufficient acid formation will cause the cheese to crumble instead of having a smooth firm mass.

The general use of antibiotics for the treatment of mastitis and other diseases in dairy cows has presented a serious problem in the manufacture of certain varieties of cheese. Many of the well-known antibiotics such as penicillin and terramycin have a marked inhibitory effect on the organisms involved in a desirable fermentation, particularly the starter cultures that are used to ripen milk for the various varieties of cheeses. A cow treated with penicillin by injection into the teat canal will eventually build up a penicillin residue in her milk. While the concentration may be very small, it is high enough to alter the metabolic activities of many desirable bacteria.

Many cheesemakers object to the use of antibiotics in treating dairy cows unless the treatment is conducted by a veterinarian. It has been recommended that milk from cows treated with antibiotics be retained on the farm from 3 to 10 days before being delivered to the cheese factory. This procedure seems to be satisfactory for insuring a milk supply that is free of antibiotics, but it is very difficult to enforce.

SELECTED REFERENCES

Frandsen, J. H. 1958. "Dairy Handbook and Dictionary." Nittany Printing and Publishing Co., State College, Pennsylvania.

Harper, J. W., and Hall, C. W. 1976. "Dairy Technology and Engineering." AVI Publishing Co., Westport, Connecticut.

Hobbs, B. C., and Christian, J. H. B. 1973. "The Microbiological Safety of Food." Academic Press, New York.

ICMSF. 1980. Milk and milk products. *In* "Microbial Ecology of Foods," Vol. II: Food Commodities. Academic Press, New York.

Kosikowski, F. 1966. "Cheese and Fermented Milk Foods." Edwards Brothers, Inc., Ann Arbor, Michigan.

Shahani, K. M., Gould, I. A., Weiser, H. H., and Slatter, W. L. 1956. Observations on antibiotics in a market milk supply and the effect of certain antibiotics on the keeping quality of milk. *Antibio. Chemother.* **6**, 544–549.

Webb, B. H. and Johnson, A. H. 1974. "Fundamentals of Dairy Chemistry," 2nd ed. AVI Publishing Co., Westport, Connecticut.

14
Eggs and Egg Products

Formation and Structure of Eggs   212
Microbial Content of Shell Eggs   213
  Factors Influencing Microflora   215
  Contamination during Handling   216
Antibacterial Substances in Eggs   217
Enzymes in Fresh Eggs   218
  Lysozyme Activity   219
Storage and Treatment of Shell Eggs   220
  Shell Treatment   220
  Heat Treatment   220
Contamination with *Salmonella*   221
Dried and Frozen Egg Products   222
  Pasteurization of Liquid Eggs   222
  Dried Eggs   223
Selected References   226

A number of egg characteristics influence penetration, growth, and spoilage by microorganisms, as well as the transmission of human pathogens. These characteristics include the following:

- Eggs have a shell, which offers limited protection from bacteria.
- Just before expulsion, the egg and shell pass through the end of the intestinal tract where they are exposed to intestinal microflora.
- Improper handling of eggs can result in penetration of microorganisms through the shell and into the egg where they can cause spoilage or present health hazards to humans.
- Eggs are excellent food for microorganisms and are sometimes used in culture media.
- Egg white contains several bacteriostatic substances.
- Salmonella organisms are often found in or on chickens.

Because eggs are consumed uncooked in meringues or egg nog or partially cooked in some types of scrambled eggs or omelets, any microorganisms present are not destroyed. In addition, eggs are used in diets for high-risk populations. In spite of these potential hazards, high-quality, normal, sound eggs are considered a safe, nutritious, healthful food.

FORMATION AND STRUCTURE OF EGGS

The yolk is the first part of an egg to develop. This takes place in the ovary, where there are numerous small yolks (ova), each contained in a separate sac. As the yolk develops in the vitelline membrane, it passes through the oviduct where the egg may be fertilized. As the yolk passes along the oviduct, albumen is added along with the shell membrane. The egg finally reaches the uterus where the final shell membrane is added. Then, before the egg is expelled from the cloaca, the shell receives a waxlike coating, which aids in protecting the egg from external contamination.

It is possible for hens harboring certain bacterial infections to contaminate the egg in the process of formation. It is also possible for the egg, during the process of formation, to pick up organisms such as *Salmonella pullorum,* which causes pullorum in chicks.

The structure of a hen egg is shown in Fig. 14.1. In a normal, newly laid egg, the white, or *albumen,* appears to be composed of two distinct parts: a thick viscous portion surrounding the yolk and a thinner portion. Actually there are four parts, or layers, of white. The innermost (chalaziferous) layer immediately surrounds the yolk and ends in twisted strands of thick white resembling cords; these are called *chalazae.* There are usually two of them, one on each side of the yolk, provid-

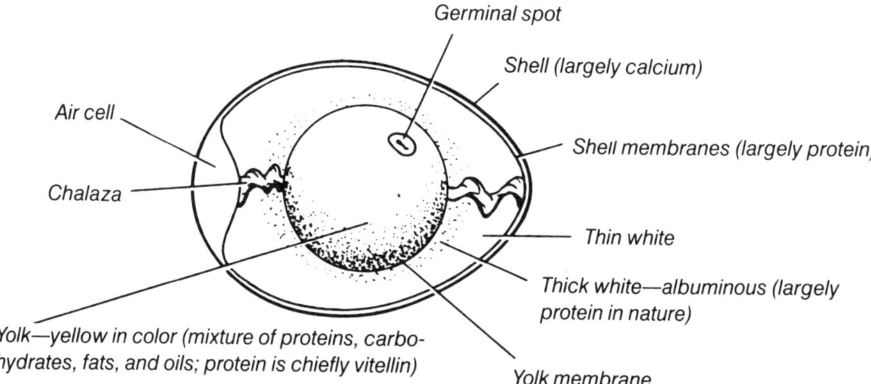

Fig. 14.1. Structure of a hen egg.

ing an axis upon which the yolk may rotate but which restrict the movement of the yolk away from the center of the egg.

Next to this layer is one of thin white contained in an envelope of thick white. The fourth layer is thin white and fills the space between the thick white and the shell membrane. Except for the chalazae, all layers of white are generally clear or transparent. In newly laid eggs and in some older eggs, the white may be slightly cloudy. Also, certain types of deterioration or certain feeds may cause the egg white to be discolored.

The yolk is suspended in the white near the center of the egg and is contained by the yolk (*vitelline*) membrane. On the surface of the yolk, and usually on its upper side when an egg is broken, is the *germinal spot*, or disk. The germinal spot in a fertilized egg, under proper temperature conditions, develops into the embryo. The size of the germinal spot in a fertilized egg varies depending on the development that has taken place. As an egg ages, the yolk tends to take up water and becomes larger, the yolk membrane weakens, and the yolk assumes a flattened or spread-out shape.

When an egg is first laid, the contents entirely fill the shell. As the egg cools, the contents shrink and air is drawn through the pores of the shell. A small air cell forms between the two shell membranes, usually at the large end of the egg where these membranes separate most easily. Some loss of water takes place with the aging of the egg, and the air cell gradually grows larger.

Blood spots on the surface of the yolk or suspended in the white lower the grade of an egg. The spots are caused by rupture of blood vessels in the yolk follicle (saclike membrane) during ovulation. Some spots, called meat spots, degenerate and change in color to reddish brown, tan, or white. Heredity and breed seem to be factors associated with blood spots.

Eggs in which the blood has diffused throughout the albumen are known as bloody whites. This may be caused by a blood clot breaking and spreading through the white as the egg passes down the oviduct. Eggs with bloody whites are classed as inedible.

MICROBIAL CONTENT OF SHELL EGGS

Several genera of molds have been isolated from shell eggs. They are *Alternaria, Cladosporium, Mucor, Penicillium,* and *Thamnidium.* Spoilage is caused by the mycelium of the mold entering through cracks or pores in the egg.

Whiskers is a form of fungal spoilage that covers the surface of the egg shell, especially when eggs are stored in high humidity. Different molds cause spots of different colors. *Penicillium* sp. cause blue to green spots in the egg white. *Cladosporium herbarum* is one of the common molds causing spoilage of eggs. It produces black spots on eggs and may be present in other foods. The yolk is ruptured by the hyphae, releasing an abundance of food for more mold growth.

Age, moisture, temperature, general sanitation under which the hens are kept, and the condition of the storage facilities are important factors that determine the general microflora of shell eggs. Sometimes mold infection may originate from handling and packing after the eggs have reached the storage plant. Several genera have been isolated from packing material, although not all the molds studied could penetrate eggs during cold storage.

In commercial practice, spoiled eggs are classified according to the color of the rots. Winter (1942) made an exhaustive study of the microflora in black rot eggs. *Alcaligenes, Escherichia,* and *Proteus* appeared to be the predominant genera in black rot eggs, and *Alcaligenes* and *Proteus* produced this defect very quickly when inoculated into fresh eggs (Table 14.1). These genera are widely distributed in feces, soil, and water and are common on dirty eggs. The best way to control black rot in eggs is to produce clean eggs and handle them in accordance with accepted sanitary practices.

Green rot is caused by some *Psuedomonas* species, red rots by the

Table 14.1. Bacteria Isolated from Black Rot Eggs and Their Ability to Cause the Defect When Inoculated into Fresh Eggs

Kind of bacteria	No. lots found in	No. eggs found in	No. eggs inoculated	No. eggs developing rot
Aerobacter aerogenes	5	10	39	2
Alcaligenes bookeri	3	20	24	13
Alcaligenes fecalis	2	8	8	5
Alcaligenes recti	3	12	22	17
Eberthella oedematiens	2	5	4	4
Eberthella pyogenes	1	1	6	0
Escherichia coli	5	16	88	21
Escherichia freundii	2	3	27	0
Micrococcus epidermidis	1	3	4	0
Micrococcus varians	1	1	4	4
Proteus ichthyosmus	4	20	39	28
Proteus vulgaris	1	1	4	4
Serratia marcescens	2	2	18	18
Salmonella pullorum	1	1	4	0

genus *Serratia,* pink rots under some conditions by *Pseudomonas,* and white rots with threadlike white areas by *Pseudomonas.* With the exception of green rots, spoilage is rare in normal trade channels because shell eggs are not held long enough for organisms to multiply sufficiently to cause spoilage even though such organisms are on the surface of the shell.

Yeasts are rarely found in shell eggs probably because their size makes penetration through the shell difficult. At least one virus, which causes a form of avian leukosis, can be transmitted through eggs, but does not infect humans.

In general, eggs pose few public health problems. In a few instances, eggs have been thought to be the source of organisms causing intestinal disturbances. It has been found that soft boiling, coddling, or frying on one side does not always render an egg free from *Salmonella.* Eggs artificially infected with *Salmonella pullorum* required 5 min in boiling water to kill all the test organisms.

Van Oijen and Wedeman (1940) recommended that duck eggs be cooked in boiling water for 10 min before use and that broken-out liquid be heated at 149°F (65°C) for 20 min to destroy *Salmonella enteritidis* and *Salmonella paratyphi,* which are sometimes present. Bringing custard to a second boil, after adding a little thickening, was found to render it free from *S. enteritidis.* In one study, reconstituted egg powders heavily contaminated with several *Salmonella* species were used in the preparation of scrambled eggs, omelettes, sponge cakes, custards, and muffins. The scientists were unable to detect any viable *Salmonella* in these cooked foods.

Factors Influencing Microflora

Since the composition of eggs offers ample opportunity for microorganisms to grow, the shell plays an important role in restricting, but not entirely stopping, their entry. The membrane lining the inside of the shell is also semipermeable and permits the entrance of microorganisms from the exterior. Some workers maintain that the shell membrane offers very little resistance to organisms, while a few investigators believe that bacteria do not enter an egg until its natural resistance is broken down with age.

Moisture apparently plays an important part in the transmission of microorganisms through the shell. Excessive moisture facilitates the passage of microorganisms, while lack of moisture retards their passage. Clean, dry eggs are essential to the production and maintenance of good quality.

A deficiency of shell-forming food in the diet of the hen results in thin, defective shells, which are more permeable to microorganisms. Eggs laid under unsanitary conditions or improperly handled are quite certain to show greater contamination. The temperature at which eggs are stored is very important. Since the egg is a living cell with a complex enzyme system all its own, a favorable temperature favors the growth of organisms already present in the egg and also activates endogenous enzymes, all of which materially influence the quality of eggs. Cracked and dirty eggs are invariably high in microorganisms. The season may also influence the microbial content. For example, in one study eggs from hens housed on winter range under average conditions produced eggs with microbial counts of 500–12,000/ml of mixed egg content; eggs from hens on summer range had 1500–350,000 organisms/ml.

Contamination during Handling

In general, over 95% of chicken eggs are sterile inside when produced. Salmonella organisms are sometimes found inside eggs from hens infected with this organism.

Egg shells contain 7,000–17,000 pores, many of which do not go completely through (Fig. 14.2). At the time of laying, an egg is coated with a mucous-like material, which dries immediately. It is called the *cuticle*, or *bloom*. In addition to acting as a coating, it plugs the pores partially and acts as a mechanical barrier to microbial invaders.

The cuticle degenerates during storage, making it easier for microorganisms to penetrate the shell. Microorganisms do penetrate egg shells even though the eggs receive good treatment. Proper handling, however, can reduce spoilage. Generally, all commercial eggs are washed, and improper washing is the most common source of contamination.

The most common means of contamination is by immersing warm eggs in cold, contaminated water. When an egg is immersed in cold water, the contents contract and contaminated water is aspirated through shell pores into the white. Haines and Moran (1940) demonstrated this phenomena by placing freshly laid eggs in a cold suspension of bacteria; as the eggs cooled, the egg meats contracted and the bacteria were drawn through the shell.

The following factors influence the extent of contamination during washing: (1) porosity of egg shell; (2) temperature differential between water and egg (the greater differential, the more contamination); (3) length of time of immersion; (4) concentration of microorganisms in water; and (5) age and condition of the interior of the egg.

Fig. 14.2. Photomicrograph of an egg shell. Courtesy Hy-Line Poultry Farms.

ANTIBACTERIAL SUBSTANCES IN EGGS

Freshly laid eggs are normally low in bacterial content. A number of investigators report evidence of the presence of antibacterial substances in fresh eggs (Table 14.2). Although there is conflicting evidence relative to this point, it has long been established that healthy tissues in any animal exert a germicidal effect. Analogously, it would appear reasonable to expect that freshly laid eggs, like freshly drawn milk, possess at least a transitory germicidal action. The hydroxyl ion concentration within the egg tends to increase this germicidal property. Hydroxyl ions increase during the first few days after an egg is

Table 14.2. Antibacterial Substances in Eggs

Substance	Action
Lysozyme	Lysis of cell walls of gram + bacteria; flocculation of bacterial cells
Conalbumin	Chelation of iron
Ovomucoid	Inhibits trypsin
Avidin	Inhibits biotin
Riboflavin	Chelation of cations
Other uncharacterized proteins	Inhibit trypsin, chymotrypsin, and fungal protease; combine with riboflavin or B_6

laid when stored in a ventilated room. This increase in hydroxyl ions means greater alkalinity, which in turn means a greater inhibitory effect upon microorganisms.

Since many workers have found very few, if any, bacteria in egg white, they have concluded that this portion of the egg has some bactericidal properties. Haines and Moran (1940) claimed that 98% of the egg whites examined were sterile. Other workers showed that egg white contains bacteriolytic, bactericidal, and bacterior-inhibiting substances, which they called *lysozyme*. When this substance was diluted 50,000,000 times, it showed marked lytic power on a susceptible test organism. The egg yolk has much weaker bactericidal power.

Egg white has a definite lethal or lytic effect on *Staphylococcus, Streptococcus, Meningococcus, Salmonella typhi,* and *Bacillus anthracis.* Egg white added to blood *in vitro* maintains its antibacterial activity. For example, when it is injected into the blood stream of small animals, the blood shows marked antibacterial powers, which is evident for several hours.

ENZYMES IN FRESH EGGS

Lineweaver *et al.* (1948) found only small amounts of a few enzymes in hen eggs. The low enzyme content of hen eggs is comparable to that of dormant seeds. It is reasonable to assume that developing embryos control the enzyme activity of the egg or seed and that the fresh egg or dormant seed would, therefore, possess very little enzyme activity. These authors concluded that the deterioration of processed eggs, frozen or dried, is not caused by endogenous enzymes. Moreover, autolipolysis of egg yolk has been reported to occur only in the presence of contaminating microorganisms.

Although tributyrinase occurs in both yolk and white, the yolk contains much larger quantities. Esterases are present in the livetin fraction of the yolk and are capable of hydrolyzing methylbutyrate, benzylbutyrate, acetylcholine, benzylcholine, acetyl-β-methylcholine, triacetin, tripropionin, and tributyrin. Amylase is usually associated with the livetin fraction. Peptidase is present in small amounts in both white and yolk. Likewise, phosphatase is present in the yolk, while catalase more often occurs in the white. Lipase capable of attacking fatty acids with six or more carbons, phenol oxidase, cytochrome oxidase, and peroxidase could not be detected. It is extremely doubtful that endogenous egg enzymes play an important role in egg deterioration.

However, some investigators believe that several endogenous enzymes are associated with normal eggs and that their activity varies with the state of maturity and general conditions under which eggs are stored. Such enzymes as pepsin, trypsin, lipase, catalase, and reductase have been demonstrated in eggs. Lipase activity is low in fresh eggs but considerable in stale eggs. The catalase content varies widely in fresh eggs and apparently occurs in equal amounts in the white and yolk. In a putrid egg, this enzyme is present in large amounts. No doubt the enzyme systems differ in eggs from different species. Geese and duck eggs have a pronounced flavor and odor, and some workers claim this is caused by particular enzymes present in them but absent in hen eggs. It appears that very little work has been done with respect to enzymes in eggs other than hen eggs.

Lysozyme Activity

As mentioned already, lysozyme is one of several bacteriostatic substances in eggs. Many investigators believe that lysozyme plays an important role in deterioration of egg quality during storage.

Kraft and Bryant (1956) showed that lysozyme activity in shell eggs increased during storage at 78°F (25°C) and then gradually declined. Lysozyme was not related to the incidence of bacterial infection in shell eggs stored at specified temperatures, or to the pH of the albumen. These authors observed that pseudomonas bacteria responsible for green fluorescence in shell eggs were not lysed or inhibited in growth by either lysozyme or egg albumen. Apparently the pH of the albumen did not affect the percentage of shell eggs infected during storage regardless of the temperatures used.

STORAGE AND TREATMENT OF SHELL EGGS

The quality of stored eggs gradually decreases because of a decrease in moisture; absorption of off-odors; continuous loss of carbon dioxide, which increases the alkalinity of the egg white; and finally a decrease in hydrogen ion concentration caused by the loss of carbon dioxide. The latter factor greatly favors the activity of proteolytic enzymes present in the thick white. Moreover, there is a diffusion of water from the white through the yolk membrane to the yolk, thus raising the pH of the yolk. Egg white usually has a pH of 7.4–7.6 when it is laid, but at room temperature, the pH may reach 9.4–9.5 within a week. At cold-storage temperatures with little or no ventilation, the pH may reach 8.8–9 in 3 months. Any procedure that stops or slows down chemical and physical changes in the egg will help to maintain its quality during storage.

Shell Treatment

Treating egg shells with oil to control microbial invasion and to decrease the loss of moisture and carbon dioxide is not new. Dutch farmers early in the 19th century preserved eggs for the winter season by dipping them in linseed oil.

Oiling eggs as soon as they are gathered improves their quality. Eggs must be oil-treated before too much carbon dioxide escapes from the egg. A light paraffin oil (viscosity 50–60) is sprayed on the eggs after they have been collected or packed in filler flats.

Heat Treatment

Pasteurization has been applied to shell eggs to inactivate endogenous enzymes and to destroy the bacteria present on the shell, in the shell and shell membrane, and in the albumen and yolk.

There is a critical temperature range at which coagulation of the albumen occurs if either the time and/or temperature are increased (Fig. 14.3). The point at which coagulation begins is considered the upper limit for satisfactory results. Obviously several factors must be considered such as temperature of eggs before pasteurization, size, age, grade, and type of heating medium. The best results have been obtained by adjusting the eggs to room temperature around 75°F (24°C) before immersing and rotating them in oil. The temperature of the oil is held at 140°F (60°C) and the eggs rotated for 10 min.

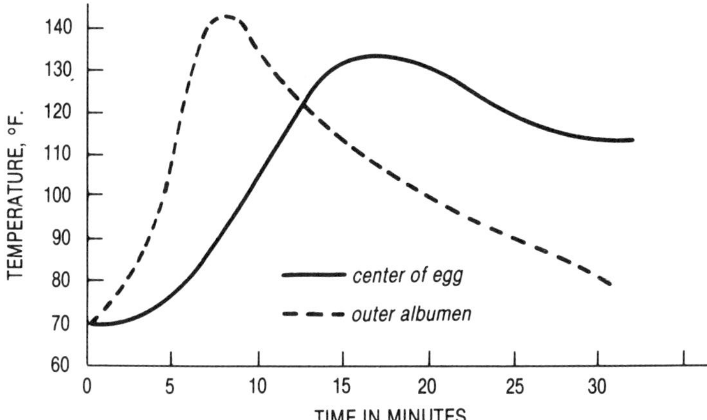

Fig. 14.3.   Temperatures of a shell egg during pasteurization.

This process is primarily of academic interest since it is no longer practiced in the trade.

CONTAMINATION WITH *SALMONELLA*

In the past, organisms of the genus *Salmonella* have been a major problem in eggs. The situation has been brought under control by an intensive campaign by the industry itself, more stringent federal and local regulations, pasteurization, and a better knowledge of the sources of contamination.

Surveys have implicated feed, particularly meat scrap and fish meal, as one of the sources of *Salmonella* in eggs. A typical case history involving menhaden fish meal illustrates the chain of contamination. The fish caught in the ocean are unloaded at the dock. After unloading, the hold and decks of the ship where the fish are held are washed down with water pumped from the harbor. The water is contaminated with *Salmonella,* which contaminates the hold, which contaminates the next load of fish.

After unloading, the fish are processed by cooking at times and temperatures that destroy all *Salmonella,* but the cooked meal is dumped out on the same area where the uncooked fish were held prior to cooking, thus recontaminating the cooked meal. This cooked meal, which contains *Salmonella,* eventually is mixed with other feedstuffs and fed to poultry. A few birds become infected with *Salmonella* and spread the disease throughout the flock. Finally the birds begin to lay occasional eggs contaminated with *Salmonella.* These eggs are generally of poor quality because the infection also affects the hens' reproductive organs. Eventually such birds stop laying altogether.

The cycle can be stopped at this point if the eggs go to a breaking plant and are pasteurized. If they are marketed as shell eggs, only clean, normal, sound shell eggs should be sold, since it has been established that high-quality eggs almost never harbor *Salmonella.*

Other sources of *Salmonella* are from contaminated poultry flocks, wild birds, insects, soil, and water.

DRIED AND FROZEN EGG PRODUCTS

The usual sequence of operations in the manufacture of frozen and dried egg products is as follows: (1) candling eggs to remove cracked and inedible ones; (2) washing eggs (no quarternary sanitizers are permitted); (3) breaking eggs and removing egg meats; (4) churning and/or milling, pressure filtration; (5) clarifying or filtering; (6) cooling to 40°F (4°C); (7) fermentation or enzyme treatment to remove sugars and proteins (whites for dehydration only); (8) pasteurization; (9) processing (adjust pH, additives, etc.); (10) packaging for freezing *or* dehydrating and packaging of egg solids.

Pasteurization of Liquid Eggs

Eggs are pasteurized primarily to eliminate *Salmonella,* but the reduction in other microflora is also of considerable value. The temperature at which egg white proteins are denatured is very close to the temperature required to kill *Salmonella.* For that reason, accurate temperature regulation is essential to prevent the loss of functional properties in the white. To help solve this problem several different methods of pasteurization have been developed. Table 14.3 shows the typical time–temperature schedules for ordinary heat treatment of various types of liquid eggs. Other methods are referred to as the Western Regional Research Laboratory method, the Armour method, and the Ballas method.

To pasteurize eggs by the Western Regional Research Laboratory method, egg whites are acidified with lactic acid to pH 6.8–7.3, and then aluminum sulfate is added to prevent damage by heat to the egg

Table 14.3   Suggested Temperatures for Pasteurization of
Liquid Egg Meats in 3.5 Minutes

	Temperature	
Product	(°F)	(°C)
Whole egg	140.0	60.0
Yolk plain	142.0	61.1
+ 10% sugar	146.0	63.3
+ 10% salt	146.0	63.3
Whites		
pH 7	140.0	60.0
pH 9	134.0	56.7

Source: Anon. (1967).

whites. After this treatment, the eggs are pasteurized at 140°F (60°C) for the usual 3.5 min.

In a process patented by Armour and Co., egg whites are heated to 125°F (52°C) for 1.5 min to inactivate the enzyme catalase. Then hydrogen peroxide is added, and the pasteurization process is continued at the same temperature for an additional 2 min. After cooling, catalase is added to remove the hydrogen peroxide.

A process in which the white is heated to 134°F (57°C) for 3.5 min under vacuum has been developed by the Ballas Egg Products Co.

Dried Eggs

As Fig. 14.4 indicates, three types of dried egg products are made: dried whole egg meats, dried egg yolks, and dried egg whites. The following steps are involved in the production of dried whole egg meats:

1. Shell eggs are stored at 40°F (4°C) for at least 72 hr prior to breaking. Note: Federal requirements specify that the broken egg meats shall not be over 60°F (16°C).
2. Egg meats are thoroughly mixed in a paddle churn.
3. Extraneous matter and particles of egg shell are removed by clarification.
4. Eggs are held in a holding vat, comparable to a milk pasteurizer, at 40°F (4°C) for not longer than 48 hr. If they are held beyond 48 hr they must be frozen.
5. Eggs are pumped to a spray drier by a high-pressure pump at a pressure of around 3000 lb. Egg from the pump is broken up into a fine spray as a result of passing through special spray nozzles. The drying temperature is 340°F (171°C).

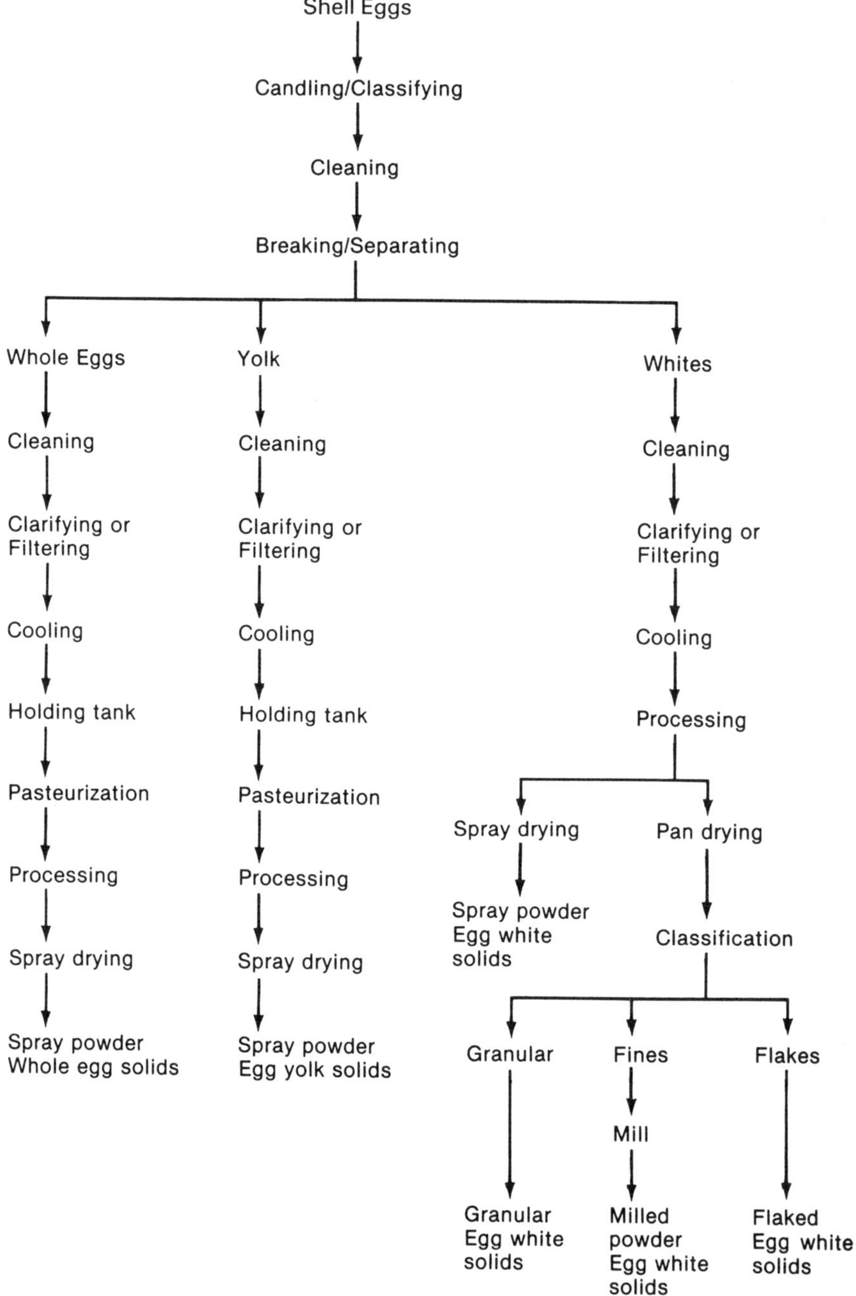

Fig. 14.4. Schematic outline showing the various processing steps involved in the production of dried whole egg, egg yolk, and egg white solids.

6. Powdered eggs are collected at the bottom of the sprayer and passed through a system of cooling coils, so that the temperature of the powder is around 90°F (32°C) or less.
7. Whole egg powder is packed in barrels or 14-lb cartons and stored at 50°F (10°C).

Microorganisms in Whole Egg Powder

The heat treatment involved in the production of dried whole egg powder does not produce a sterile product. Many enteric organisms have been isolated from the finished product. Obviously many heat-resistant and typical thermophilic organisms are present, usually as the result of unsanitary practices in the egg drying plant. The U.S. government specifies that the bacterial content of dried whole egg meats shall not be over 100,000/g and the moisture content must be 5% or lower.

The microbial content of spray-dried whole egg powder is raised markedly at first when the product is stored at high temperatures. However, as the storage time is increased, the total count is decreased. Conversely, when the temperature is lowered, higher counts are obtained and a more diversified bacterial flora can be detected. The genus *Bacillus* seems to be the predominant microflora in spray-dried egg powder held for 3 months regardless of the temperatre of storage. Likewise the storage of spray-dried whole egg powder in tin cans or grease-proof cartons does not affect the total bacterial count or the type of bacterial flora.

Fermentation and Drying of Egg Whites

As noted earlier, in the production of dried egg whites, the whites are fermented, which liquefies the thick white portion. During the fermentation, carbon dioxide is formed and ammonia is produced by de-aminating enzymes, which makes the whites become alkaline. Bacteria then grow and break down the proteins and carbohydrates. This fermentation is allowed to continue for 30–60 hr at 70°F (21°C). A scum rises to the top and a sludge settles to the bottom of the vat. Both are discarded as unfit for drying and make up about 8% of the total. At the present time, commercial enzyme preparations (e.g., trypsin) are used to accelerate this fermentation, then at the proper stage the fermentation is stopped by addition of ammonium hydroxide and ethyl alcohol.

The time required for thinning egg whites is determined by the amount of trypsin added and by the temperature at which the enzyme is allowed to act. A temperature of 40°–43°F (4°–6°C) retards bacterial development.

Egg whites hydrolyzed by trypsin resemble thin white more closely than does "fermented" white. Bacterial growth during fermentation treatment produces changes in the white besides breaking down the water-insoluble proteins. Furthermore, a uniform standard product can be produced over a long period of time. Trypsin-hydrolyzed white tends to darken at room temperature and must be stored at low temperatures because it is difficult to control the activity of the enzyme. The enzyme glucose oxidase is also used.

The fermented liquor is drawn off into pans about 12 in. in diameter and 1.5 in. deep. Pans are placed on racks in drying room at a temperature of 120°F (49°C) for 20 hr then raised to 140°F (60°C) for 48 hr. The albumen is removed from the pan and allowed to cure for 24 hr, then it is packed. Some plants are using the spray-dry method. The liquid egg white is forced through nozzles under 1500–3000 lb of pressure. The yield is not good with the spray method.

Salmonella in Egg Whites

Ayres and Slosberg (1949) studied three methods for destroying *Salmonella* in egg whites and found all three processes satisfactory. Two methods were concerned with the destruction of the pathogens in the fermented liquor albumen prior to drying. The other method involved the storage of the dried albumen for controlled periods of time.

Vat pasteurization at 139°F (59°C) on a commercial scale destroyed *Salmonella*. In the second method, hydrogen peroxide at levels of 0.1 and 0.2% followed by the removal of the residual H_2O_2 with excess catalase destroyed the test organisms, but produced an excess of foam. The storage of dried fermented albumen at room temperature was the least successful in destroying the organisms. However, a storage sequence of 120°, 130°, and 135°F for 20, 8, and 4 days, respectively, destroyed potential enteric pathogens. Increasing the storage temperature of dried fermented albumen eliminates the danger of recontamination during the drying process encountered when liquid albumen is sterilized.

SELECTED REFERENCES

Anon. 1967. Egg Pasteurization Manual. U.S. Dept. Agriculture, Western Utilization Res. Dev. Div., Albany, California.

Ayres, J. C., and Slosberg, H. M. 1949. Destruction of *Salmonella* in egg albumen. *Food Technol.* **3**, 180–183.

Board, R. G. 1966. Microbiology of the hen's egg. *J. Appl. Bacteriol.* **29**, 319–341.

Gibbons, N. E., and Moore, R. L. 1944. Dried whole egg powder. *Can. J. Res.* **22F**, 48.

Haines, R. B., and Moran, T. 1940. Porosity of and bacterial invasion through the shell of the hen's egg. *Hygeia* **40**, 453–461.

Kraft, A. A., and Bryant, A. W. 1956. The influence of environment on the lysozyme activity in shell eggs. *Food Technol.* **10**, 45–47.

Lineweaver, H., Morris, H. J., Kline, L., and Bean, R. S. 1948. Enzymes of fresh hen's eggs. *Arch. Biochem.* **16**, 443–472.

Mountney, G. J. 1976. "Poultry Products Technology," 2nd ed. AVI Publishing Co., Westport, Connecticut.

Stadelman, W. J., and Cotterill, O. J. (Editors). 1986. "Egg Science and Technology," 3rd ed. AVI Publishing Co., Westport, Connecticut.

Winter, A. R. 1942. Black rot in fresh shell eggs. *U.S. Egg and Poultry Mag.* **48**, 506–509, 520.

Wrinkle, C., Weiser, H. H., and Winter, A. R. 1950. Bacterial flora of frozen egg products. *Food Res.* **15**, 91–98.

15

Fermented Foods

Sauerkraut 230
  Sequence of Microbial Growth and End Products 230
  Effect of Temperature 232
  Salting 232
  Use of Starters 233
  Spoilage and Defects in Sauerkraut 234
Pickles 236
  Sequence of Microbial Growth 237
  Defects in Pickles 237
  Processing of Sweet and Sour Pickles 237
  Processing of Dill Pickles 238
Vinegar 238
  Raw Materials and Preparation of Fruit 238
  Reactions in Vinegar Fermentation 239
  Vinegar Fermentation Methods 240
  Starter Organisms and Fermentation Conditions 243
  Processing of High-Quality Vinegar 243
  Spoilage and Defects in Vinegar 244
Olives 245
  Green Olives 245
  Ripe Olives 246
  Spoilage of Olives 246
Citric Acid 247
Selected References 247

Controlled fermentations are used to produce sauerkraut, pickles, vinegar, olives, various dairy products (discussed in Chapter 13), and other products. In general, fermented foods are acidic due to the formation of lactic and other acids by the fermenting microorganisms. Since acid in foods destroys or retards the growth of microorganisms, fermentation can be viewed as a preservation method. Indeed, the original impetus for the production and use of fermented foods may have been their relative stability. The effect of acidity on the growth and survival of microorganisms was discussed in Chapter 5.

SAUERKRAUT

A series of desirable microorganisms plays a major role in transforming fresh cabbage tissues into the fermented product called sauerkraut, which means "acid cabbage." Under suitable conditions, normal cabbage fermentation is a spontaneous process in which bacteria attack sugars present in cabbage and convert them into acids, alcohol, and carbon dioxide. At the same time, the cabbage proteins undergo biochemical changes. The primary end products of this fermentation are lactic acid and smaller amounts of other acids. Lactic acid bacteria are primarily responsible for the normal sauerkraut fermentation. The species involved can be classified into three groups:

* *Leuconostoc mesenteroides,* an acid- and gas-producing coccus
* *Lactobacillus plantarum* and *L. cucumeris,* bacilli that produce acid and a very slight amount of gas
* *Lactobacillus pentoaceticus (L. brevis),* acid- and gas-producing bacilli

In addition to the desirable bacteria in kraut fermentation, there are many kinds of microorganisms present on cabbage that may interfere with or alter the course of fermentation. The quality of kraut depends very largely on how well these bacteria are controlled during the normal fermentation. Spoilage bacteria are typical soil organisms; they attack the protein and to a lesser extent the sugars in cabbage producing substances that reduce the quality of sauerkraut. Fortunately, the growth of these organisms is inhibited in normal kraut fermentation. Yeasts are also present on cabbage, but they seldom develop early enough in the fermentation to do any particular damage. However, these organisms are not affected by the acids formed and may even use them as a source of energy, reducing the acidity or neutralizing it by the end products resulting from their metabolism. Molds and transitional fungi are troublesome organisms. They, like yeasts, can use the acids formed during fermentation and may seriously reduce the quality of kraut.

Sequence of Microbial Growth and End Products

After shredded cabbage is placed in a suitable jar and salted, mechanical pressure is applied to expel the juice from the cabbage. The cabbage juice contains fermentable sugars and other nutrients suitable for microbial activity.

Within a period of hours after the cabbage is placed in the jar, gas-producing cocci (*L. mesenteroides*) begin to form acids. When the acidity reaches 0.25–0.3% (calculated as lactic acid), the cocci begin to slow down and gradually die, though their enzymes continue to function. When the acidity reaches 0.7–1%, the majority of the cocci have disappeared. However, two types of lactobaccilli (*L. plantarum* and *L. cucumeris*) are active in increasing the acidity up to 1.5–2% in spite of the inhibitory effect of salt and low temperature. Finally, *L. pentoaceticus* completes the fermentation, thus bringing the acidity up to 2 and sometimes 2.5% acid. Figure 15.1 shows the acid formation during normal fermentation.

The end products formed during normal kraut fermentation include appreciable amounts of lactic acid (nonvolatile) and smaller amounts of acetic and propionic acids (volatile); a mixture of gases, of which CO_2 is the principal one; small amounts of alcohol; and a mixture of aromatic esters. The acids in combination with the alcohol form esters, which contribute to the characteristic flavor of good kraut. The acidity formed by the desirable bacteria helps to govern the course of the fermentation by stopping putrefactive changes in the cabbage. Fermented kraut owes its keeping qualities to the acids formed.

The final or completed kraut contains only traces of sugar; the lactic acid content is high, but the amounts of acetic acid and alcohol are about the same as those produced by *Leuconostoc mesenteroides*, the first species in the series. Hence, the bulk of the lactic acid fermentation is carried on by the non-gas-producing bacilli. Any disturbance in

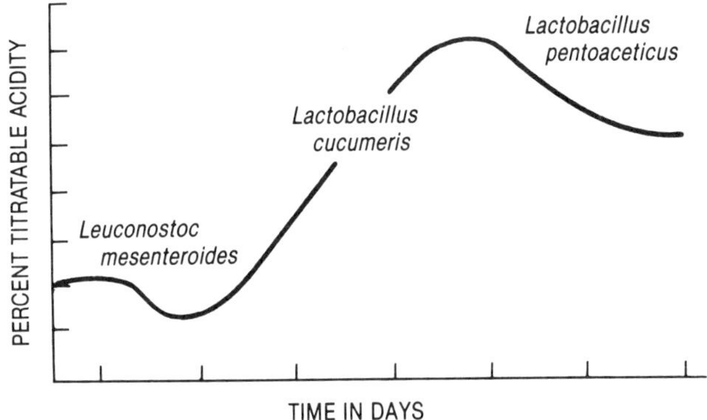

Fig. 15.1. Acid production during sauerkraut fermentation.

the sequence of the order of desirable bacteria will change the flavor and quality of the kraut.

Effect of Temperature

The optimum temperature for kraut fermentation is around 70°F (21°C). A variation of just a few degrees from this temperature alters the activity of the microbial process and affects the quality of the finished product. Therefore, temperature is one of the most important factors in kraut fermentation.

A temperature of 65°–72°F (18°–22°C) is most desirable for initiating fermentation. *Leuconostoc mesenteroides* is the first organism to start the fermentation and is rather sensitive to temperature changes. Temperatures above 72°F (22°C) favor the growth of the *Lactobaccillus* species. Although it is desirable that they complete the fermentation, if these species replace the *Leuconostoc* too early, harmful effects result. If the initial temperature is too low, the *Leuconostoc* will produce about 1% acid, which is too high at this phase of the process. The fermentation virtually ceases and the kraut will lack flavor and will also be of poor quality.

Salting

Salt plays an important role in starting the kraut fermentation and affects product quality. The addition of too much salt tends to inhibit the desirable bacteria, although it may contribute to the firmness of the kraut. The principal function of salt is to withdraw the juice from the cabbage, making a more favorable medium for the proper kind of bacteria to develop.

Generally, salt is added to give a concentration of 2.0–2.5%. This concentration of salt is slightly inhibitory to the lactobacilli but does not affect the cocci. The usual concentration of salt actually has a greater inhibitory effect on the desirable fermenting bacteria than on the organisms responsible for spoilage in kraut. The latter organisms are not inhibited in a salt concentration less than 5–7%. Thus, the acidity produced by the fermentation is chiefly responsible for controlling spoilage organisms and not the added salt.

In the manufacture of sauerkraut, dry salt is usually added at the rate of 2–3 lb/100 lb of cabbage. The use of salt brines is not recommended in sauerkraut making but is common in the fermentation of vegetables with low water content. The effect of salt concentration on the texture of finished kraut is illustrated by the data in Table 15.1.

Only pure salt should be used in kraut making. Special salts containing alkali should never be used because the alkali tends to neutralize the acids formed. The amount of salt to be added should be determined by weight rather than volume because salt from different sources may vary in density.

Use of Starters

Starters similar to those used in the dairy industry have been recommended for making sauerkraut. Starters have been used to some extent in Europe but have not been used on a large scale in the United States.

The one beneficial effect derived from the use of starters is that they are acid and thereby slightly acidify cabbage, causing undesirable soil bacteria to be inhibited. Kraut produced by addition of starters prepared from sour milk, such as *Streptococcus lactis,* is usually of good quality, but its flavor is no better than that of kraut from a normal spontaneous fermentation. Since sour milk organisms grow only for a short time in the kraut, they do not disturb the natural sequence of bacterial growth that takes place in a typical fermentation. Starters prepared from other types of bacteria that do not grow in kraut have the same effect as those prepared from sour milk.

The addition of *Leuconostoc mesenteroides,* which is normally predominant in the early stages of kraut fermentation, gives a good flavor to the final product but apparently alters the sequence of bacterial growth and results in a final product that is not as completely fermented as normal kraut. The addition of gas-producing rod types such as *L. pentoaceticus* results in production of more acetic and less lactic acid than is normal, and the fermentation never reaches completion. When lactic acid, non-gas-producing rods (*L. cucumeris*) are used as

Table 15.1. Effect of Salt Concentration on Texture of Sauerkraut

Salt (%)	Kraut texture
2.8–3.4	Soft to medium
2.2–2.5	Medium to very firm
1.8–2.1	Medium to firm
0.7–1.7	Soft

Source: Pederson (1979).

starters, the resulting kraut is not completely fermented; it is often bitter and apparently more readily attacked by yeasts.

The juice from a normal kraut fermentation may be used as a starter for other fermentations. The effect of this practice depends to a great extent on the types of organisms present in the juice and its acidity. If starter juice has an acidity of 0.3% or more, the kraut produced is of poorer quality than normal kraut. In this case, the cocci that normally initiate the fermentation are suppressed, so that only the bacilli carry on the fermentation. Such starter juice is comparable to a pure culture of *Leuconostoc,* which predominates in kraut juice at this stage. If starter juice has an acidity of 0.25% or less, the kraut produced is normal, but no definitely beneficial effects are shown. Usually the organisms present in old kraut juice have a detrimental effect upon the normal fermentation of cabbage. Use of old juice as starter is likely to produce soft kraut, which is discussed in the next section.

Spoilage and Defects in Sauerkraut

Sauerkraut spoilage is due very largely to aerobic soil microorganisms. As a rule, they do not attack sugars to any extent but act on the protein to produce undesirable changes. Fortunately, their growth and biochemical activity are quickly inhibited by a normal fermentation.

Yeasts, which are not affected to any appreciable extent by acids, seldom develop early enough in the fermentation process to do damage. Molds and transitional fungi require air and cannot grow in covered vats. All of these organisms can metabolize acids. For example, mycoderm, a false yeast, grows very readily on the surface of kraut and can use the acids present, thus altering the normal fermentation process.

Rotted kraut is usually caused by bacteria, molds, and yeasts resulting from unsanitary conditions in harvesting and processing the kraut subsequent to fermentation.

Slimy kraut is not too common. The flavor and odor of the kraut are not seriously impaired. The product is edible, but consumers do not like it. Generally, cooking or canning will cause the sliminess to disappear. This type of spoilage usually is caused by a slimy capsular material formed by certain kinds of bacteria. High temperatures during the fermentation process favor the development of these undesirable bacteria.

Soft kraut may result from several conditions such as large amounts of air, poor salting procedure, and varying temperatures. Indeed, whenever the normal sequence of growth of the desirable bacteria is altered, soft kraut is the inevitable result. The lactobacilli, which seem to have greater ability to break down cabbage tissue than the cocci, contribute to this defect. High temperatures and a lowered salt content favor growth of the lactobacilli, which are sensitive to higher concentrations of salt. The usual concentration of salt used in kraut fermentation slightly inhibits the lactobacilli but does not affect the more salt-tolerant cocci. If the salt content is too low initially, the lactobacilli grow too rapidly at the beginning, thus upsetting the normal sequence of the fermentation.

The continuous use of the same vat without periodic cleaning often causes succeeding batches of kraut to have a soft texture. The reason for this is that lactobacilli in large numbers often adhere to fermentation vats under commercial conditions, especially during the rush season when tanks of kraut are emptied and immediately refilled with very little attempt to clean the vats. Their presence in large numbers may alter the normal fermentation of the succeeding vats of kraut, resulting in soft kraut. This is especially common in warm weather because the growth of lactobacilli is favored by warm temperature.

Dark kraut is one of the more common defects in sauerkraut and is caused by growth of spoilage organisms. Several conditions may favor growth of these organisms. For example, an uneven distribution of the salt tends to inhibit the desirable organisms but allows spoilage microorganisms to grow because many of them are salt tolerant. If insufficient juice is present to cover the kraut, caused primarily by lack of sufficient weights, certain undesirable aerobic bacteria and yeasts will grow on the surface of the kraut and cause darkening as well as off-flavors. Too high a fermentation temperature also encourages undesirable microflora to grow and results in dark kraut.

An even distribution of kraut juice is essential to inhibit the undesirable microorganisms that cause dark kraut. Repacking kraut in containers for shipping may result in unequal distribution of the juice, which may lead to spoilage in some containers.

Pink kraut is caused by the growth of certain types of yeasts. One of these is *Torula glutinis*, which produces an intense red pigment in the kraut juice and on the surface. Usually an uneven distribution of salt or too high a salt content are favorable for the development of these yeasts. A chemical analysis of pink kraut often shows an increased alcoholic content and a slightly lower than normal acid content. Conditions that are optimal for a good normal fermentation will usually suppress the activities of these yeasts.

PICKLES

Cucumber pickles are produced by a fermentation comparable to the one that occurs in making sauerkraut except that brine rather than dry salt is used. Pickle cucumbers should be mature, and approximately 1/8 in of the stem should be retained in harvesting.

Because cucumbers grow on vines in close contact with the soil, they should be thoroughly washed before use. The washed cucumbers are placed in large tanks or barrels, and 15–20% salt brine is added. In this concentration of brine, many of the cucumbers will float; some provision must be made to keep them submerged, otherwise the pickles will spoil. Usually board lids weighed down with stone or brick are used.

The strong brine draws the water and sugar out of the cucumbers. Since cucumbers contain about 90% water, the salt concentration is quickly reduced and more salt must be added as osmosis and fermentation proceed. If the concentration of salt falls below 12%, putrefaction or softening of the pickles will occur. In practice, the salt concentration is gradually raised to 15–18% by adding salt slowly, insuring a brine of uniform strength. Usually, the addition of 1 lb salt/15 gal of brine each week for 5 consecutive weeks is satisfactory.

A spontaneous fermentation usually takes place within a few days after the cucumbers have been placed in the brine. Vigorous boiling indicates fermentation is going on with the formation of acids, which tend to inhibit many undesirable spoilage microorganisms.

During fermentation, certain visible changes take place that are important criteria in judging the progress of the curing process. The change in surface color from bright green to dark olive green is probably caused by the action of the acids formed during fermentation on chlorophyll. The cross-section of the cucumber changes from snow white to a waxy transparent color primarily because of the expulsion of a large amount of air from cucumber tissues. An increase in the specific gravity of the cucumbers also takes place. At first they float on the surface, but the gradual absorption of salt increases their weight until they no longer float.

The salt content in the interior of pickles may be sufficient to inhibit many kinds of undesirable microorganisms. Fortunately, the desirable organisms involved in the fermentation process are not appreciably affected by the salt because they are slightly salt tolerant. Although undesirable proteolytic bacteria initially are more numerous than the desirable acid-forming organisms, they are gradually destroyed by the brine before causing any marked defect in the finished product.

Sequence of Microbial Growth

Leuconostoc mesenteroides, a gram-positive coccus, usually predominates during the first stages of pickle fermentation. This species is more resistant to temperature changes and more salt-tolerant than other species involved in the fermentation. Later, *Lactobacillus cucumeris* or other closely related gram-positive bacilli grow rapidly.

The active stage of fermentation continues for 10–30 days, depending very largely upon the temperature employed. The most favorable temperature range for *L. cucumeris* is 85°–90°F (29°–32°C). During active growth of this organism, the total acidity increases to around 2%. Strong acid-producing types of bacteria reach their maximum growth during this period of the fermentation. The addition of sugar or small amounts of acetic acid to the fermenting mixture increases the production of acid.

Defects in Pickles

If excessive amounts of acid are formed during the fermentation, the acid will tend to shrivel the pickles. Soft or shriveled pickles sometimes are the result of excessive acid formation by *Leuconostoc mesenteroides*. Because of the relatively high salt-tolerance of these organisms, they are able to produce appreciable amounts of acid.

Stirring the brine may introduce air bubbles, making conditions more favorable for the growth of spoilage bacteria. Mycoderm organisms also may be troublesome. They use biological acids as a source of energy and at the same time neutralize the acid by alkaline waste products resulting from their metabolism. This tends to alter the normal fermentation sequence.

In general, though, if the pickles are well covered with brine and the salt concentration and proper temperature are uniformly maintained, a good-quality product can be expected. Pickles usually can be held for a year or longer.

Processing of Sweet and Sour Pickles

In the preparation of sweet or sour pickles, cucumbers undergo essentially a normal lactic acid fermentation. The subsequent pickling operations consist of leaching the excess salt into warm water, firming the flesh or outer coating of the cucumber with a dilute solution of alum, and impregnating the fermented product with plain, spiced, or sweetened vinegar. The final treatment is the addition of a mixture of vinegar and sugar.

Processing of Dill Pickles

In the preparation of dill pickles, cucumbers are fermented in 10–12% salt brine in closed barrels fitted with small holes for the escape of gas during the fermentation. Dill and other herbs and spices are added to the brine with the cucumbers before the fermentation process starts. If conditions are favorable for the fermentation, appreciable amounts of lactic and other acids will be produced. The acids will preserve the pickles for a long time providing air is excluded.

The finished product can be pasteurized very easily. A relatively short heat treatment supplements the natural acidity of the product, resulting in a longer shelf life.

VINEGAR

Vinegar, one of the oldest fermentation products used by man, literally means "sour wine" (from the French *vinaigre* (vin)-wine, *aigre*-sour or sharp). Although any fermentable carbohydrate in foodstuffs may serve as raw material, the fermentation follows a definite sequence. Fruits of various kinds serve as excellent material for spontaneous fermentation. Apples are the most widely used raw material in the commercial production of vinegar.

Raw Materials and Preparation of Fruit

Any raw product that will yield an alcoholic fermentation is acceptable for the production of vinegar. Apples, grapes, pears, peaches, plums, figs, oranges, berries, honey, sugar, syrups, hydrolyzed starchy materials, beer, and wine may be used. Wine and cider are the best raw materials for vinegar production. Malt is widely used in England; wine in France, Italy, and Spain; and cider in the United States. In order to produce a high-grade product, the raw material should be clean, sound, and mature. If wine is used it should be clean and free from preservatives.

It is a common practice to use apples that have been rejected as a No. 1 grade, bruised, small size, and partially spoiled fruits. However, a high-grade vinegar can only be made from good sound fruit. The apples are mechanically crushed, and then pressed. The apple juice is sweet because of endogenous carbohydrates, particularly glucose. The juice then is placed in large containers called fermenters. A spontaneous fermentation takes place in a very short time by microorganisms that are normally present in the raw material.

Reactions in Vinegar Fermentation

The production of vinegar depends on a mixed fermentation involving both yeasts and bacteria. The fermentation is usually initiated by certain species of yeasts, which cause the following overall reactions:

$$C_6H_{12}O_6 \xrightarrow{\text{yeast}} 2C_2H_5OH + 2 CO_2$$

glucose ⟶ ethyl alcohol   carbon dioxide

The alcohol is oxidized by *Acetobacter* to form acetic acid:

$$C_2H_5OH + O_2 \longrightarrow CH_3 COOH + H_2O$$

alcohol   acetic acid   water

The yeasts and bacteria exist in a form known as *commensalism*. The *Acetobacter* are dependent upon the yeasts to produce an easily oxidizable substance. Vinegar could not be produced by the activity of one kind of organism.

A concentration of 10–13% alcohol is necessary for good fermentation. If the alcohol content is much higher, the alcohol is incompletely oxidized to acetic acid. A lower concentration results in the loss of vinegar because of the oxidation of esters and acetic acid. Carbon dioxide and water are formed from acetic acid. In addition to acetic acid, other organic acids are produced in the fermentation. Esters formed from this mixture of acids contribute to the characteristic odor, flavor, and color of vinegar.

Yield of Acetic Acid

Acetaldehyde is an intermediate product in the transformation of the reducing sugar in fruit juice to acetic acid or vinegar. Oxygen acts as the hydrogen acceptor in the conversion of alcohol to acetaldehyde and in the subsequent conversion of acetaldehyde to acetic acid, as indicated in the following reaction scheme:

$$2\ CH_3\!-\!\overset{\displaystyle OH}{\underset{\displaystyle H}{C}}\!-\!H + O_2 \longrightarrow 2\ CH_3\!-\!\overset{\displaystyle O}{C}\!\!=\!\!H + 2H_2O$$

ethyl alcohol   acetaldehyde

$$2\ CH_3\!-\!\overset{\displaystyle O}{C}\!\!=\!\!H + 2H_2O \longrightarrow 2CH_3\!-\!\overset{\displaystyle OH}{\underset{\displaystyle H}{C}}\!-\!OH$$

acetaldehyde   hydrated acetaldehyde

$$2CH_3-\overset{\overset{\displaystyle OH}{|}}{\underset{\underset{\displaystyle H}{|}}{C}}-OH \;+\; O_2 \;\longrightarrow\; 2CH_3-\overset{\overset{\displaystyle OH}{|}}{C}=O \;+\; 2H_2O$$

hydrated acetaldehyde $\qquad\qquad\qquad\qquad$ acetic acid

Generally, 100 parts of sugar will produce 50–55 parts of acetic acid, which is a yield of about 1.2 g acetic acid from 1.0 g ethanol. Since part of the acetic acid and alcohol are lost by evaporation and some of the sugar is used as food for yeasts, the yield is less than 100%.

The overall fermentation reaction can be written as follows:

$$\underset{\underset{\text{glucose}}{C_6H_{12}O_6}}{(180)} \;\longrightarrow\; \underset{\underset{\text{ethanol}}{2C_2H_5OH}}{(2 \times 46)} \;+\; \underset{\underset{\text{carbon dioxide}}{2CO_2}}{(2 \times 44)}$$

$$\underset{\underset{\text{ethanol}}{2C_2H_5OH}}{(2 \times 46)} \;+\; \underset{2O_2}{(2 \times 32)} \;\longrightarrow\; \underset{\underset{\text{acetic acid}}{2CH_3COOH + H_2O}}{(2 \times 60)}$$

Theoretically, 1 g-mol of glucose (180 g) should yield 2 g-mol (120 g) of acetic acid, or 2 parts of acetic acid from 3 parts of glucose. However, if both reactions are only 90% efficient, then the yield of acetic acid from 100 parts of sugar would be 66.7 × 0.9 × 0.9 = 54.0 g, which approximates the yield obtained in fermentations under the usual conditions.

Vinegar Fermentation Methods

Home Process

Vinegar can be made at home by providing an air supply to barrels of cider or wine and allowing it to ferment spontaneously. This does not always produce a high-grade product.

Air is admitted to the barrels through holes (one usually at each end of the barrel placed on its side) above the level of the vinegar medium. These holes should be at least 1 in. in diameter and screened to prevent the entrance of insects. Air may also be admitted through a top bung-hole likewise screened. The acetic acid bacteria form a thin film on the surface of the solution. This film later becomes quite thick and gelatinous. This gelatinous mat, which contains very large numbers of bacteria, is known as the *mother of vinegar*. Eventually, unless supported on a *raft*, or framework, it will sink to the bottom of the barrel and a new film will form.

Although high-grade vinegar can be produced by this method, it is a slow and costly process that involves much attention. The films are easily disturbed by the addition of the alcoholic medium and the withdrawal of vinegar.

Orleans Process

The oldest and also the best of the slow processes for the production of table vinegars is considered to be the Orleans (French) method. Barrels of approximately 200-liters (50-gal) capacity are used for containers. Each barrel is filled with about 65–70 liters of a high-grade vinegar and 10–15 liters of wine. The vinegar serves as the starter culture. At weekly intervals during the next 4 weeks, approximately 10–15 liters of wine are added. Thereafter, about 10–15 liters of vinegar are withdrawn each week, and the same amount of wine is added to replace the vinegar, thus forming a slow but continuous process.

Modified Orleans Process

Although there are other slow methods of making vinegar, most of them are merely modifications of the Orleans process. A problem commonly encountered with these methods is how to add liquids without disturbing the floating bacterial film mat. In one method, wood shavings, which float, are sometimes added to help support the mat. In another, a glass tube is inserted in the top of the barrel so that it reaches almost to the bottom; this tube is then used to add additional solutions, so the mat is not disturbed. A glass tube can also be attached to the bottom of the barrel to serve as a gauge for measuring the level of vinegar in the barrel.

Quick Vinegar Method (German Process)

The objective of the German process is to help the fermenting microorganisms work at a rapid pace. The apparatus used, called a generator, is an upright cylindrical tank filled with beechwood shavings or similar materials. This tank is fitted with devices for allowing the alcoholic solution to trickle slowly down through the shavings on which the acetic acid bacteria are living. The tank is not allowed to fill because that would exclude oxygen, which is necessary. Near the bottom of the generator are holes for allowing air to be drawn in; the air rises through the generator and is used by the acetic acid bacteria for oxidizing the alcohol.

In many respects the generator is like a furnace. Both are contrivances in which active combustion takes place. Coal is added to a furnace to be chemically oxidized; alcohol is added to a vinegar generator to be oxidized by bacterial enzymes. Both yield large amounts of heat. In a furnace the heat is desired; in a vinegar generator too much heat must be avoided and careful attention is necessary to prevent the temperature from rising too high and destroying the bacteria.

Fringe Method

In the Fringe method, vinegar is produced by submerged fermentation. The generator used, called an *acetator,* consists of a tank constructed of stainless steel, wood, and accessory equipment and controls. It is possible to convert 150–200 gal of alcohol per day into 100-grain vinegar by this method. Because of the higher oxidation rate of the submerged process, it is necessary to maintain a residue of alcohol in order to preserve the vitality of acetic acid bacteria. The acid and alcohol mixture can be automatically controlled.

The generator consists of an airtight tank located at the bottom of the apparatus, a collection chamber constructed of stainless steel inside a 1½-in. copper tube water jacket, and above the collection chamber an air intake where oxidation of the vinegar takes place. Above this area is a pile of 10 to 15 feet of wood shavings packed into the generator. On top of the generator is a stainless steel spray in which a rubber tube feed line and automatic controls are located.

Oxygen flow is controlled by a damper located on top of the generator. Air enters the generator through several intakes installed at intervals around the tank near the wooden grate. The incoming air is filtered by special screens.

Vents serve to remove the heat. Otherwise the high temperature would inhibit the growth of acetic acid bacteria. As an acid–alcohol mixture passes through the cooling coils, the temperature is controlled. The cycling of the acetic acid mixture proceeds until the desired acetic acid strength is reached. This may require 8–12 days.

The Fringe method has certain advantages over other methods. It is very economical, the apparatus requires very little space, losses by evaporation are reduced to a minimum, and because it is a continuous process, very little slime forms in the generator.

Synthetic Acetic Acid

Crude acetic acid or pyroligneous acid is the principal source of commercial acetic acid and acetate. It is produced by dry distillation of selected varieties of hardwoods. This is purely a chemical process and no microorganisms are involved.

Starter Organisms and Fermentation Conditions

The organisms involved in the production of vinegar usually grow at the top of the substrate in a thick jelly-like mass, commonly known as mother of vinegar. In many respects, these organisms are comparable to cultures used in the dairy industry. The mother floats on top of the acetifying medium. It is composed of both acetobacter and yeasts which work together harmoniously. *Acetobacter acetic, A. xylinum,* and *A. ascendens* are the principal species of the acetic acid-producing bacteria. *Saccharomyces ellipsoideus* and *S. cerevisiae* are common yeasts involved in producing alcohol from the fermentable substrate.

Maintenance of the initial acidifaction in the production of vinegar is important in order to suppress undesirable organisms and to encourage the presence of desirable acetic acid-producing bacteria. Ordinarily 10–25% by volume of strong vinegar is added to the alcoholic medium to attain a desirable fermentation.

The alcoholic fermentation should be allowed to go to completion before acidification because the sugar in the medium will not be converted to alcohol after the addition of acetic acid. Incompletely fermented juices are usually low in acetic acid. The acetic acid strength of good vinegar should be around 6%.

The optimum temperature for acetic acid fermentation varies with the organism and the process being used. Ordinarily, a temperature of 80°–85°F (27°–29°C) is most favorable. Too low a temperature favors a slow fermentation; a high one accelerates the evaporation of alcohol, acetic acid, and volatile substances that contribute to the characteristic flavor and aroma of vinegar.

Processing of High-Quality Vinegar

The fermentation should be allowed to continue until the acetic acid content is about 6%. After the final stage of fermentation, the acetic acid bacteria will tend to destroy the vinegar by oxidation unless oxygen is excluded. Therefore, during storage the containers should be completely filled and sealed to prevent access of air to the vinegar.

Aging improves the flavor and clarifies the product. During aging many biochemical changes take place in vinegar including the formation of esters, which contribute to the fine flavor and odor of the final product. Aging for 1 year or longer produces a good high-grade vinegar. However, alcoholic vinegars, which are low in acetic acid, are not improved by aging.

Sometimes vinegar is bottled without further treatment, but clarification should be done in most cases. Filtration using filter aids or fining are the two methods used to clarify vinegar. Fining involves the addition of bentonite or Spanish clay, settling, and decantation.

Only clear and aged vinegar should be bottled. The containers should be completely filled and tightly capped. Vinegar may be pasteurized in the bottle at a temperature ranging from 60° to 66°C (140° to 150°F) for 30 min. Sometimes vinegar is pasteurized in bulk, cooled to 21.1°C (70°F), and then bottled.

Spoilage and Defects in Vinegar

Vinegar eels (nematode worms) may cause trouble in the commercial production of vinegar. Unsanitary practices, soil from plants, and insects may be the principal sources of these organisms. Eels have no public health significance, but they reflect the general sanitary conditions in the factory, and their presence in vinegar is objectionable.

Vinegar eels may be controlled by strict sanitary practices in every phase of vinegar production. Sterilization of vinegar barrels destroys these organisms; since they are susceptible to steam or chemical agents, their removal is not difficult. Pasteurizing vinegar to a temperature of 130°F (54°C) or filtration will control the organisms.

Mites entering through the air hole in the barrel are sometimes troublesome. Moisture and warm temperatures favor the growth of mites. Control practices are comparable to those used for vinegar eels.

Decayed fruit, spoiled fruit juices, and vinegar are good breeding places for vinegar flies. Clean sanitary practices along with screens over windows will help to control these flies.

Wine flowers refers to a whitish film on the surface of the liquid. The film is sometimes called *Mycoderma vini* and is composed of yeast-like organisms. These microorganisms grow aerobically and oxidize carbon-containing constituents to carbon dioxide and water, thus altering the flavor and alcohol content of vinegar. This defect may be controlled by adding 1 part vinegar to 3 parts of the alcoholic solution or storing the alcoholic liquid in filled and closed containers.

Darkening of vinegar is caused primarily by iron and tannin or an oxidase. The iron comes from pipe lines and other metal parts, while tannin may come from new barrels. Aeration of the solution helps to remove the darkening caused by iron and tannin. However, if oxidase-producing organisms are present, pasteurization of the vinegar is strongly recommended.

OLIVES

Olive production in the United States began in California in 1869 from seeds brought from Mexico. Comon cultivars for canning and fermentation and their country of origin are 'Mission' (United States), 'Sevellano' and 'Mazanillo' (Spain), and 'Ascolano' and 'Barocini' (Italy). Olives are harvested in California around September 25. They are usually harvested when green.

The production of both green (immature) olives and black (ripe) olives involves treatment in lye solution followed by fermentation in salt brine. The microorganisms involved in the normal olive fermentation are *Leuconostoc mesenteroides* and *Lactobacillus plantarum*. The environmental factors influencing the fermentation are practically the same as those in pickle fermentation.

Green Olives

Green olives are a fermented product preserved by biological acidity formed during the pickling process. In the usual processing procedure, the green olives are placed in a 2% NaOH (lye) solution of 70°–75°F (21°–24°C) until the lye penetrates the flesh or outer tissue of the fruit. The lye is removed by adding cold water, thus diluting the mixture until the fruit is practically free of lye.

The lye treatment of olives is necessary because the outer tissues contain a bitter glucoside called oleuropein, which is highly bactericidal. The compound must be removed because it will retard or alter the course of a normal fermentation of the olives.

The glucoside oleuropein can be removed by (1) dry oxidation; (2) aeration, which involves washing the olives four or five times, then heating the olive mixture in a retort at 242°F (117°C) for 60 min; or (3) enzyme treatment. Certain enzymes are capable of destroying or altering the bactericidal properties of the glucosides. An enzyme obtained from burdock has this property, but its disadvantages outweigh the desirable properties of this enzyme.

After the removal of glucosides, the olives are placed in barrels with 7–10% salt brine and allowed to undergo a spontaneous fermentation. A temperature around 75°F (24°C) is most favorable for maximum activity of the desirable bacteria. Naturally, a heterogenous microflora exists in the raw product, but the salt suppresses the undesirable organisms and favors the lactic acid bacteria. Usually 2–3 months are required for the completion of a normal fermentation of green olives. Since air is practically excluded, the growth of many undesirable acid-destroying organisms is retarded. The exclusion of air also minimizes oxidation reactions, which tend to change green olives to a dirty brown color.

The finished product is packed in jars and sterilized. This procedure insures a good-quality product for a long time.

Ripe Olives

Ripe olives are light brown in contrast to immature green olives. Ripe olives are placed in 5–7% brine to soften the outer tissue before they are exposed to lye solution. This pretreatment in brine allows the lye to penetrate into the fruit more easily. Apparently, the germicidal glucoside is concentrated in the interior of ripe olives. After pretreatment, the olives are placed in a 0.7–2.0% lye solution. The treated olives are washed free of lye, packed in barrels with 2–5% brine, and allowed to undergo normal fermentation, usually for 2–6 weeks depending upon the temperature and other factors.

During fermentation, ripe olives are exposed to air in contrast to green olives, which are kept nearly free of air. Air in the presence of the slight amounts of NaOH left on the fruit oxidizes the polyphenols in the olive tissue to a black color. Hence, the characteristic color of ripe olives. The finished product is usually packed in cans with 3–5% salt brine and then sterilized. This procedure renders a good finished product, which will keep for a long time.

Spoilage of Olives

Aerobacter aerogenes may be involved in gas formation in green olives, and *Clostridium* sp. may cause gas pockets, bad odors, and off-flavors formed by an abnormal fermentation. The production of butyric acid by *C. butyricum* may cause putrefactive odors.

In ripe olives, pectolytic enzymes produced by *Bacillus subtilis* may cause a mushy or soft product. If the wash water used for removing the lye is below 140°F (60°C) many species of bacteria may survive and cause a poor-quality product.

CITRIC ACID

The commercial production of citric acid is accomplished by employing some of the common molds for fermentation. The principal mold used is *Aspergillus niger; Penicillium luterim* and *Mucor pyriformis* also may be used, but they are not as efficient.

Many carbohydrate sources may be used in the production of citric acid. Low-quality molasses is a cheap source of raw material. Cane blackstrap molasses or solutions of glucose or sucrose also may be used provided some sources of nitrogen and mineral salts are available. Conditions most favorable for growth of mycelium tend to increase oxalic acid production, while less mycelium growth increases the citric acid production. Therefore, there is a problem in controlling the growth of the mold in order to harvest the maximum amount of citric acid. The presence of oxalic acid necessitates procedures for separating citric and oxalic acids.

Modern methods of citric acid production use solutions with sugar concentrations between 10 and 20% at a pH around 6.0. Generally, incubation is at 77°–86°F (25°–39°C) for 7–12 days in the surface method.

Citric acid is used in candies, flavoring extracts, and soft drinks. It is sometimes used to preserve fish and other marine foods, and to prevent discoloration of crab meat and browning of sliced peaches.

SELECTED REFERENCES

Beuchat, L. R. 1987. Traditional fermented food products. *In* "Food and Beverage Mycology," 2nd ed. Van Nostrand Reinhold Co., New York.

Fabian, F. W., Bryan, C. S., and Etchalls, J. L. 1932. Experimental Work on Cucumber Fermentation. Tech. Bull. **126**. Mich. Agr. Exp. Stn., East Lansing, Michigan.

Frazier, W. C., and Westhoff, D. C. 1978. Food fermentations. *In* "Food Microbiology." McGraw-Hill Book Co., New York.

Pederson, C. S. 1979. "Microbiology of Food Fermentations." AVI Publishing Co., Westport, Connecticut.

Prescott, S. C., and Dunn, C. G. 1982. "Industrial Microbiology," 4th ed. (Gerald Reed, Editor). AVI Publishing Co., Westport, Connecticut.

Stanton, W. R., and Wallbridge, A. 1967. Fermented food processes. *Process Biochem.* 4 (4), 45–51.

16
Fruits and Vegetables

Microbiology of Fruits 249
  Natural Protection and External Contamination 249
  Reducing the Microflora on Fresh Fruits 250
  Spoilage of Fresh Fruits 251
  Dried Fruits 252
  Fruit Juices 253
Microbiology of Vegetables 253
  Spoilage in Fresh Vegetables 254
  Reducing the Microflora on Fresh Vegetables 255
  Contamination of Vegetables Grown in Polluted Soil 255
  Tomato Products 256
Survival of Microorganisms in Frozen Fruits and Vegetables 258
Selected References 259

MICROBIOLOGY OF FRUITS

The climatic conditions, type of soil, kind of fruit, and stage of maturity are just a few of the factors that may influence the kinds and numbers of microorganisms present on and in fruit.

Natural Protection and External Contamination

Some mature fruits have a hard impervious shell or outer covering, whch protects the fruit or nut meats from microbial invasion, stabilizes enzymatic changes in the product, and minimizes loss of moisture from the fruit. However, bacteria, viruses, and certain pests may invade the fruit while it is in the process of forming on the plant. These may remain inactive for some time, but as the fruit reaches maturity, or shortly thereafter, bring about profound changes in the fruit, usually resulting in spoilage. Spraying with pesticides is recommended during the growth of the fruit because of this.

The skin of a fruit tends to be bactericidal, and the outer covering contains a waxy substance that retards the entry of microorganisms into the interior. When the covering is broken by bruises or other mechanical injury or attacked by insect stings, microorganisms may enter readily. Immature fruits as a rule have a higher germicidal power than mature or ripe fruits. Lemons have very definite bactericidal properties, whereas oranges have very little.

Fresh fruits and vegetables may be sold at fruit stands where sanitary conditions are not satisfactory. The constant handling of products by prospective customers and continued exposure of products greatly increase the possibility of unusual microflora on the fruit. Grading and packaging of fruits in see-through containers is helpful in preserving their crispness, freshness, and water content, and has a tendency to reduce microflora.

The kinds and numbers of microorganisms on fruits depend upon a variety of conditions. Fruit lying on the ground has a different microflora from that on a tree or upright plant. The microbial content and quality of fruits are influenced significantly by storage, packaging, dust in the air, insect contamination, and climatic conditions.

Any mechanical injury to a fruit that ruptures the outer pericarp creates conditions favorable for organisms to enter the product and induce undesirable biochemical changes. Such injury may occur during harvesting and storage or result from an insect sting. Microorganisms have been found in the inner structure of the fruit, in seeds, and in the sap. Thus, once they gain entry, microorganisms may spread throughout the fruit tissue.

Reducing the Microflora on Fresh Fruits

Since fresh fruits generally are not entirely free of pathogenic bacteria and many are eaten raw, they may pose certain public health risks. Several methods have been used to reduce the microflora on fresh fruits.

Mechanical washing, although often done primarily to improve the appearance of fruit, removes large numbers of microorganisms. However, washing is not an effective means of rendering fresh fruit sterile. Rubbing or polishing fruits to bring out their natural color tends to remove the natural wax coating on the skin of a fruit and may increase the risk of internal contamination. The waxy outer coating provides natural protection against the entry of microorganisms and insects and also contains some bactericidal substances. Thus, the wax on the surface of fruits should not be removed.

Another method for reducing the microflora on fresh fruits is treatment with a disinfectant. In one study, for example, treatment with a 0.2% solution of chloride of lime for 30 min destroyed *Escherichia coli* on strawberries, lettuce, and carrots. Chlorinated lime (20% available chlorine) also has been used, followed by washing of the fruit in dilute sodium thiosulfate to remove the chlorine. Immersion in approximately 5% formalin followed by a liberal washing in water also is fairly satisfactory in reducing the microflora on fruits. In recent years, many detergents and detergent sanitizers have been developed. Some of these may prove effective in controlling surface contamination of fresh fruits.

The use of hot water for disinfecting some fruits has also been tried but has not proved particularly effective. Exposure of grapes to water at 176°F (80°C) produced no undesirable effects on the fruit, but many bacteria and viruses were not killed. Likewise, bacilli associated with decaying areas on pears were not destroyed. Presumably, they were sporeforming organisms.

Spoilage of Fresh Fruits

True bacteria do not play an important role in fruit spoilage because the acidity of most fruits prevents their growth. In addition, the presence of bactericidal substances in many fruits has a tendency to destroy many kinds of bacteria. However, handling of fruits by humans may lead to contamination with bacteria.

Bacteria, molds, and yeasts present on strawberries were viable after 3 years' storage at 15°F (−9°C). Moreover, bacteria have been found repeatedly and in appreciable numbers on dates. Grapes sold on sidewalk stands often contain pathogenic organisms and should be considered potentially dangerous. There is no doubt that the human factor plays a major role in the contamination of fruits by bacteria.

Many varieties of fruits contain appreciable amounts of sugars and acids, and their juices are easily attacked by molds and yeasts. Mold spoilage of fruits during storage may be significant. Molds are common on citrus fruits, and spoilage occurs very quickly when a fruit is injured. There are many kinds of spoilage associated with fruits, but the most characteristic one is softening of the flesh followed by rotting, making the product inedible. In certain kinds of fruit rot, the organism invades the fruit while it is attached to the plant. In fact, some species of fungi occur characteristically on each kind of fruit or vegetable. For example, spoilage of cherries, peaches, and plums is caused primarily by *Penicillium expansum*, which causes green to blue mold rot. *Physalospora malorum* causes black rot, although various other molds may be involved.

Plant pathogens may also cause fruit spoilage as well as damaging the tree. The common brown soft spots in stored apples are most frequently caused by *Penicillium expansum.*

As discussed in Chapter 15, yeasts are usually responsible for the characteristic alcoholic fermentation of various fruit juices that is involved in the production of vinegar. Although their presence is desirable in juices that are to be fermented, yeasts may produce an undesirable yeasty flavor in certain fruits, especially dried figs and dates.

Dried Fruits

The moisture content of all fresh fruits is sufficient for microorganisms to grow. However, if the moisture content is lowered by drying, an unfavorable environment for microbial growth is created. Although dried foods are not sterile, drying is one of the oldest methods of preserving fruits and other foods and increases their shelf life significantly.

The microflora present on dried fruits is influenced by the type and age of the fruit, conditions under which it was grown, storage conditions, residual moisture after drying, and the kinds of organisms initially present. The microbial content of dried fruits generally is several hundred per gram. Molds are commonly present, while yeasts and bacteria are usually less numerous.

Sulfuring

During the drying of fruit, discoloration may occur because of the action of endogenous enzymes in the presence of air. Sulfuring prevents this discoloration and helps to preserve vitamin C.

In one method of sulfuring, fruit is exposed to sulfur dioxide (SO_2) fumes, generated by burning sulfur in air. The SO_2 penetrates the moist fruit, forming sulfurous acid on contact with the water in the fruit. This acid inactivates endogenous enzymes, thus preventing discoloration and development of off-flavors and off-odors, and also retards spoilage by microorganisms and insects.

Fruits may also be sulfured by immersion in solutions of potassium metabisulfite or sodium bisulfite. The results with this method, however, are not very satisfactory because penetration into the fruit is very slow and considerable sugar may diffuse out of the fruit. In addition, during the long soaking process, fruits absorb much water, which then must be removed during drying, thus increasing the drying time.

Spoilage of Dried Fruits

If favorable moisture and temperature are available, several types of spoilage may take place in dried fruits. Yeasts can be responsible for souring in prunes and dates. Yeasts also may form a sugar-like substance on some dried fruits, which give the product a yeasty flavor. Sometimes a bitter or acid flavor may be produced.

Fruit Juices

The processing involved in the manufacture of fruit juices can significantly change the composition. The pH, moisture, and solids content influence the type of spoilage organisms present in fruit juices.

A high moisture content favors bacteria and yeasts. The reduction of solids by processing may change the oxidation–reduction potential in favor of yeasts, and in the presence of residual acids and sugar, an alcoholic fermentation may take place, especially at room temperature. However, around 50°F (10°C) *Leuconostoc mesenteroides,* a lactic acid-producing bacterium, may proliferate, creating environmental conditions favorable for molds, which can tolerate acids and may use the acid as a source of energy.

The survival of several bacterial species in frozen orange concentrate has been studied. In one experiment, stock cultures of *Escherichia coli, Salmonella typhi,* and *Shigella paradysenteriae* were inoculated into pasteurized orange concentrate, which was then frozen at 1°F (−17°C). After 48 hr no enteric pathogens could be detected in samples taken from the frozen product. However, in another experiment a freshly isolated culture of *Streptococcus faecalis* remained viable much longer in the frozen concentrate, although it eventually died. The survival of microorganisms in frozen fruit and vegetable products is discussed further in a later section.

MICROBIOLOGY OF VEGETABLES

Since many vegetables are grown near or below the surface of the soil, they are likely to have a heterogeneous microflora, with soil microorganisms predominating. However, there is ample evidence that fresh vegetables grown in infected soil or polluted water can transmit disease-causing bacteria such as *Clostridium tetani* and *Salmonella typhi.* Studies have shown that *S. typhi* can survive in soil for 30–

60 days and in sterile sand for more than 60 days, and so can easily contaminate vegetables. Although several species of intestinal bacteria have been found on the surface of vegetables, the interior of undamaged vegetables usually is free of microorganisms.

The survival of pathogenic bacteria in soil is influenced by several factors. Obviously, the addition of any material (e.g., sewage sludge) containing pathogenic organisms may result in contamination of vegetables grown in the soil. In assessing the potential for such contamination by soil amendments the following should be considered: type(s) of bacteria present, its ability to compete with other organisms in the soil, soil type and pH, soil moisture, and the penetration of air and light into the soil. Especially in the case of vegetables that are eaten raw, the public health risks of using sewage sludge as a fertilizer must be given serious consideration.

Surface contamination by handling in market stands and unfavorable storage conditions also play an important role in the numbers and kinds of organisms present on vegetables. In one study, *E. coli* was isolated from more than 40% of the vegetables purchased at a curbside market. A marked reduction of the microflora is evident when such vegetables are trimmed. The removal of the outer leaves from lettuce and cabbage reduces the changes of enteric organisms, especially *Salmonella typhi*, being present.

Spoilage in Fresh Vegetables

As with fresh fruits, spoilage in fresh vegetables is often associated with rots caused by various molds. Potatoes are particularly susceptible to mold rots. Infected potato tubers may appear shriveled or shrunken and are covered with mold mycelium ranging from dark brown to white. A *Fusarium* species is usually the organism involved.

Bacterial soft rot causes soft mushy condition in potato tubers. *Pythium debaryanum* is largely responsible for this condition.

Erwinia carotovora is a typical bacterium that causes a soft rot, mushy consistency, and bad odors in carrots, radishes, and potatoes. This organism has a unique enzyme system that enables it to attack pectins in these products, thus rendering them inedible.

The presence and distribution of enzymes that may cause deterioration in vegetables have not been studied extensively. Peroxidase activity generally is much greater in some tissues than in others. For example, this enzyme is quite active in the skin of young lima beans but is very weak in the cotyledon. Corn cob tissue contains more peroxidase than the kernel tissue. Likewise, vegetable tissue damage by mechanical injury or improper storage gives a marked peroxidase reaction.

Reducing the Microflora on Fresh Vegetables

In contrast to some fruits, it is generally agreed that green vegetables cannot be effectively disinfected by treatment with chlorine compounds, regardless of the concentration or length of exposure.

The most effective way to reduce the microflora on fresh raw vegetables is blanching. The blanching of vegetables prior to freezing destroys large numbers of surface organisms and at the same time inactivates many of the enzymes in the vegetables. Since freezing *per se* does not destroy microorganisms, blanching is now considered a standard practice in freezing certain kinds of foods. A good blanching procedure should reduce the bacterial count by 90%. After blanching, the product should be quickly cooked in clean water and properly handled before and after freezing to avoid additional contamination and reduce growth of microorganisms.

Contamination of Vegetables Grown in Polluted Soil

As noted earlier, there is some evidence that vegetables grown in polluted soil or irrigated with polluted water may transmit enteric bacterial diseases. For this reason, the use of nightsoil or sewage sludge as fertilizers may pose public health risks, especially in the case of vegetables that are eaten raw.

Considerable research has been done on the bacterial contamination of tomatoes grown in polluted soil; this work is illustrative of the general problem. In these studies, the residual coliform counts on the surface of sound uncracked stems were about the same on tomatoes grown in polluted and unpolluted soil. It made no appreciable difference in the coliform count whether the pollution was added to the soil prior to or concurrent with the growth of the plants.

When the stem ends of the tomatoes were split, the coliform count was greater, regardless of the soil contamination. However, the bacterial microflora was three times higher on tomatoes grown in soil that received sewage irrigation during growth.

Sunlight appeared to reduce coliform contamination when the tomatoes were exposed, but cracks and crevices on the fruit reduced the effect of the sunlight. Dust, wind current, cultivation, and insect movement all influenced the contamination of tomatoes. When salmonella suspensions were sprayed on tomatoes in the field, the organisms could not be recovered after 7 days.

Tomato Products

Raw and canned tomatoes, as well as other tomato products, are widely used in the American diet. Microorganisms play an important role in certain types of spoilage, while others may be a menace to public health.

Canned Tomatoes

Canned tomatoes are sufficiently acid that they are usually sterile when processed and placed in sealed containers. As with any processed food, the quality of the raw product largely determines the quality of canned tomatoes.

Several types of spoilage are associated with canned tomatoes, although the loss through spoilage is not great. *Hydrogen swell* is a physical-chemical reaction between the metal of the can and the acids in the tomatoes, and usually no microorganisms are involved.

Bacterial swell is usually caused by one or more members of the *Lactobacillus* genus of which *Lactobacillus lycopersici* is the principal one. Although this organism is non-sporeforming, it can tolerate the high temperatures used in the processing of the product; it also may gain entrance into the can through a slight defect. Tomatoes and their products are an ideal medium for the growth of these organisms, which produce gas, causing the can to swell.

Flat-sour is characterized by an off-flavor and a cloudy appearance of the product. *Bacillus thermoacidurans* is the principal organism involved. Its growth in canned tomatoes does not cause the cans to swell, thus making it difficult to differentiate between spoiled and unspoiled cans until the can is opened. Catsup and tomato juice are particularly susceptible to this type of spoilage.

Tomato Juice

One of the more popular tomato products is tomato juice. Fresh juice is extracted from the fruit. Temperatures used for extraction vary with the manufacturer. The juice is usually flash-pasteurized at 252°F (122°C) for 42 sec. Salt is added as the product is being filled in the can.

Color and flavor are the principal quality characteristics. If the juice is from green tomatoes, the color and flavor are poor; if the tomatoes are overripe, the flavor may also be inferior. Although a variety of manufacturing procedures are used, the main objective is to obtain a sterile product without affecting the fine delicate flavor of the juice during heating.

Perhaps the most troublesome spoilage in tomato juice is an off-flavor. Overheating can bring about cooked flavors in the juice. *Bacillus thermoacidurans* also may cause a high incidence of spoilage. This organism is very heat resistant and can withstand a processing temperature of 212°F (100°C) for 90 min. This organism not only produces an off-flavor in tomato juice, but may eventually produce a flat-sour type of spoilage.

Catsup

Like tomato juice, catsup is subject to spoilage. Molds and yeasts find a favorable environment in catsup, and their numbers reflect the general sanitary practices used in the processing of this product. In fact, mold counts are a routine laboratory procedure followed during the processing.

Certain fermenting yeasts have caused much trouble in catsup. The yeasts are thermoduric and no doubt survive the heating process used in the preparation of this product. A yeasty flavor and odor along with gas formation are characteristics of this defect. *Lactobacillus lycopersici* may be troublesome in catsup and causes a gaseous spoilage and off-flavor. The organism can be destroyed in 2 min at 167°F (75°C). The presence of these organisms in catsup also indicates that unsanitary methods were employed during processing. Where strict sanitary practices are used, very little trouble is encountered in the finished product.

Salt is usually added to catsup to improve the flavor and not for its germicidal action. However, the final salt content may vary from 0–3.8%; some germicidal action may occur against certain types of microorganisms, especially at the higher concentrations.

Sugar is another flavoring agent like salt. The usual concentration for catsup is not sufficient to inhibit the growth of most microorganisms. However, when 15% sugar and 4% salt were added together to catsup, microbial growth was reduced to a minimum.

Sodium benzoate, a common preservative, is not effective in catsup and other tomato products. Although a concentration between 0.2 and 0.5% stops growth of most varieties of bacteria, yeasts and molds, which are very common in tomato products, are known to be very resistant to sodium benzoate.

SURVIVAL OF MICROORGANISMS IN FROZEN FRUITS AND VEGETABLES

Many variable factors influence the survival of microorganisms during freezing and subsequent cold storage of fruits and vegetables. Of the more than 100 species of bacteria that have been isolated from fruits and vegetables, some are markedly reduced in numbers when stored at $0\,^{\circ}F$ ($-17.8\,^{\circ}C$) for 6 weeks, while others show a high degree of resistance under similar conditions. A few strains of *Escherichia coli* are not able to survive during the first 4 hr of freezing.

Naturally, the substrate in which organisms are suspended for freezing will influence their survival time. For instance, some organisms suspended in vegetable extract and 40% sucrose are not appreciably protected by these substrates, whereas they might be by other substrates. In general, the temperature of freezing may not influence survival of microorganisms significantly, but certain cultures show some differences in the rate of survival at different temperatures. In these instances, $-20\,^{\circ}F$ ($-28.9\,^{\circ}C$) offers the greatest protection. Intermittent freezing is more destructive to microorganisms than continuous freezing. However, a product may deteriorate in quality when subjected to defrosting and refreezing, even though the microbial content may be low. Thus, one should not use microbiological results alone to judge the quality of frozen fruits and vegetables.

As mentioned previously, most vegetables are blanched before freezing, which reduces their microbial content. In contrast, many fruits are not blanched, although they may be disinfected by various treatments discussed earlier. Also, the naturally low pH of most fruits limits the growth of certain types of microorganisms. This offers some assurance that frozen fruit products will maintain their quality and not deteriorate during frozen storage.

The effect of freezing and frozen storage on microorganisms and on various quality attributes of foods are discussed in detail in Chapter 7. It is important to remember that freezing *per se* does not effectively sterilize a product. Although some microorganisms are killed during the freezing process, many simply exhibit a reduction in metabolic activity. In time, some of these will die, but many may survive for quite long periods during cold storage. Thus, the initial microbial count on the raw product should be reduced as much as possible before freezing. In addition, frozen foods, especially those that are eaten raw, should be consumed soon after defrosting to reduce the chances that organisms surviving frozen storage will proliferate.

SELECTED REFERENCES

Goepfert, J. P. 1980. Vegetables, fruits, nuts, and their products. *In* "Microbial Ecology of Foods," Vol. II. Academic Press, New York.

Jones, R., and Fabian, F. W. 1952. The Viability of Microorganisms Isolated from Fruits and Vegetables when Frozen in Different Menstrua. Tech. Bull. 229. Michigan Agric. Exp. Stn., East Lansing, Michigan.

Luh, B. S., and Woodroof, J. G. 1987. "Commercial Vegetable Processing." 2nd ed. Van Nostrand Reinhold Co., New York (in press).

Speck, M. L. 1984. Compendium of Methods for Microbiological Examination of Foods, Chapt. 44. Am. Public Health Assoc., Washington, DC.

Woodroof, J. G., and Luh, B. S. 1986. Commercial Fruit Processing, 2nd ed. AVI Publishing Co., Westport, Connecticut.

17
Flour, Bread, and Cereals

Microorganisms in Grains and Flours 262
Breadmaking 263
  Wrapping of Bread 263
Defects of Bread 264
  Ropy Bread 264
  Sour Bread 264
  Red Bread 264
  Moldy Bread 265
Cereals and Other Grain Products 265
Selected References 266

Some of the microflora associated with grains are referred to as *epiphytic*, that is, they are present on the grain surface. Morphologically, most are rod-shaped bacteria. Some of these bacteria are sporeformers such as *Bacillus mesentericus*, others are chromogenic, and many are gram-negative bacteria of which the genus *Flavobacterium* predominates. In addition, members of the escherichia–aerobacter group may be present. Molds and yeasts also constitute a part of the epiphytic microflora of grains. Mycotoxins (Chap. 22) can also create serious problems in grains.

The assumption that the epiphytic flora of a grain crop are not affected by the microflora of the soil in which the crop grew has been made. However, it is possible for organisms to be transmitted from seed to the new plant. Hence, the practice of treating seed before planting is employed.

The surival of microorganisms on grain depends upon moisture and temperature. The minimum moisture content that will sustain organisms is about 16% in oats, 14% in barley, and 13% in corn. A higher moisture content is favorable for mold growth.

Excessive moisture in grain is an important problem at the present time caused primarily by modern methods of harvesting. The combine has largely replaced the grain binder and threshing machine; the corn picker has taken the place of cutting, shocking, and husking corn.

These developments make it possible to harvest grains with a higher moisture content than was possible with older harvest methods, which required a "drying out" period of the grain. Weather conditions prior to and during the harvesting season play an important part in determining the moisture content of stored grain. Because the moisture content of grain harvested today often is too high for the milling process and favors microbial survival on the grain, artificial drying of grain in storage is necessary.

MICROORGANISMS IN GRAINS AND FLOUR

Studies show that microbial growth occurs more readily in flour than in whole grain providing the moisture content is the same in both products. Moisture content will determine the number and kinds of organisms present. The bacterial count of flour may range from 20,000 to 5,000,000/g. An extremely heterogenous flora is present, which includes many secondary invaders as well as the epiphytic flora of grain.

Serratia marcescens has been shown to cause discoloration and changes in the starch content of damp cereals. *Flavobacterium* spp. are usually very numerous in addition to the *Aerobacter, Staphylococcus,* and *Cellulomonas* genera. If the moisture content is favorable, the cellulose-digesting organisms will break down the outer coating of the grain kernel, and this in turn will accelerate starch hydrolysis by certain organisms.

Flour may contain an appreciable number of mold spores. If the spores are not destroyed in baking, the resulting bread becomes moldy quickly. Moldy bread also can result from external contamination of the wrapping, from use of unsanitary equipment, and from unsanitary conditions in the bakery. The following practices are helpful in controlling molds: (1) strict sanitary practices in the bakery and on the part of workers; (2) use of ultraviolet lights; (3) air conditioning of the building.

Various kinds of yeasts also may occur in flour, but they do not cause any particular harm. Sometimes an epidemic of *ropy bread* is encountered; this is caused largely by *Bacillus mesentericus* in the flour. This organism is very resistant to heat and may survive baking. *Serratia marcescens,* a chromogenic organism, may cause a defect known as *bloody bread.*

Certain anaerobic bacteria such as *Clostridium butyricum* may cause off-flavors in flour.

BREADMAKING

Breads are generally classified into two groups, leavened and un-leavened. Leavened, or raised, bread is the result of gas bubble formation in the dough. Leavened breads can be subdivided according to the mechanism responsible for the formation of gas, as follows:

- Gas from chemical (baking powder): biscuits
- Gas from yeast (regular dough): bread generally used in the United States
- Gas from bacteria (*E. coli* and others): sauerteig

Unleavened breads, in contrast, are not raised and, thus, do not depend on gas formation by chemicals or microorganisms. They include crackers, tortillas, and matzoth.

While some microorganisms are responsible for gas formation, others impart desirable characteristic flavors and odors to particular types of bread. For example, sauerteig dough is inoculated with desirable acid-forming organisms such as *Enterobacter levans* and *Escherichia coli*, which help to give this bread a characteristic flavor.

Black bread is a course wheat or rye bread similar to sauerteig. In the process of making black bread, *Streptococcus mesentericus,* an active peptonizing bacterium, changes the dough from a thick to a more plastic state.

Jamin-bang is a coarse bread prepared by the Indians of Brazil. The leavening action is brought about by a mixture of yeasts and bacteria. The resulting fermentation is an acid-alcoholic type.

Salt rising bread is the result of a fermentation, involving *Escherichia coli, Enterobacter levans,* and *Lactobacillus bulgaricus,* that gives this bread its peculiar flavor. *Clostridium perfringens,* an anaerobic pathogen, has also been used as the starter organism. The source of this culture is its natural presence in the meal used.

Success or failure in breadmaking depends very largely upon the leavening agent used. The yeast *Saccharomyces cerevisiae* is the principal leavening agent used.

Wrapping of Bread

Bread is kept clean and protected from contamination by microorganisms by wrapping it. The National Bakers Association objected to wrapping at first because moisture was held on the surface of wrapped breads, which tended to make them soggy and encouraged mold growth. Now a porous wrap is used, so that moisture accumulation is not a problem. Bread baked and wrapped by machinery is a good, clean sanitary product.

DEFECTS OF BREAD

Just because bread is baked at high temperatures does not guarantee that it will not undergo undesirable changes. The flour and other ingredients used in breadmaking are not free of organisms that can, under certain conditions, survive the baking process. Sanitary practices in the bakery will influence the microbial content of finished bread products.

Ropy Bread

Ropy bread is characterized by discoloration ranging from brown to black and usually has an unpleasant odor. The center of the loaf is soft and discolored, and from this portion long threads can be drawn out when the center of the bread is touched with a glass rod.

Bacillus mesentericus is one of the principal organisms responsible for ropy bread. It is strictly an aerobic, proteolytic, sporeforming organism. Spores present in flour, yeast, malt, and powdered milk are common sources of *B. mesentericus* in bread.

Prevention of ropy bread usually requires a strict sanitary program designed to control the spores that cause this defect. Chlorine, at 1000 ppm in the form of a hypochlorite solution, has been used as an equipment sanitizer.

Sour Bread

A fermentation resulting in the production of acids in the dough is the usual cause of sour bread. Lactic acid bacteria are responsible for this condition and are introduced either in the flour or yeast, or by dirty equipment. Sanitation is the best control method.

Red Bread

The growth of *Serratia marcescens* or *Torula glutinis* on or near the surface of bread results in distinct red patchy areas. Because these organisms are not heat resistant and are not sporeformers, they are destroyed during baking. Hence, their presence in finished bread is due to contamination after baking. Good sanitation throughout the bakery and careful wrapping can help prevent this condition.

Moldy Bread

Bread is particularly susceptible to mold growth during warm humid weather. Wrapping of bread tends to favor mold growth by retaining moisture in the loaf. Slicing bread helps to spread mold spores from the surface of the loaf to the interior. This may be one reason why sliced bread tends to mold more readily than unsliced bread.

Various physical and chemical methods have been used with some success to control mold in bread. Air conditioning the bakery and washing the air along with filtering it certainly have been helpful. Ultraviolet light may not be effective in destroying mold spores and may produce undesirable changes in the bread. Certain fatty acids and organic acids (e.g., citric, lactic, malic, and tartaric) have been used. For example, dilute acetic acid applied to the surface of a loaf of bread is effective in controlling mold.

Mold inhibitors for bread have not been widely accepted. A good inhibitor must be potent in low concentrations, must not be poisonous, and must not cause any undesirable changes in the product. The propionates (sodium and calcium salts of propionic acid) seem to meet these requirements, but they do not completely inhibit mold growth in bread.

Again, a strict sanitary program in the bakery will offer the best assurance against molds in bread.

CEREALS AND OTHER GRAIN PRODUCTS

Most of the cereals are processed and packaged under sanitary conditions, and the products are practically free of organisms. However, cereals containing raisins or figs may contain a few bacteria and yeasts. In most cases the number of organisms is not large.

Flat sour-producing organisms are the predominant type of microflora present in cracked meal, in coarse flour, and in flours from wheat, corn, soybeans, and tapioca. Thermophilic bacteria are more numerous in wheat flour than in any of the other products. The composition of soybean flour seems to favor the development of flat sour-producing organisms, sulfide spoilage organisms, and thermophilic anaerobes.

General sanitary practices in processing the various grains influence their microbial content. The season, stage of maturity, moisture content, and storage conditions are important factors in governing the activity of microorganisms in various grain products.

SELECTED REFERENCES

James, N., and Smith, K. N. 1948. Studies on the microflora of flour. *Can. J. Res.* **26c**, 479–484.

Sultan, W. J. 1986. "Practical Baking," 4th ed. AVI Publishing Co., Westport, Connecticut.

Zottola, E. A. 1973. An introduction to the microbiology of cereal and cereal products. *Bull. Assoc. Oper. Millers* **1977** (July) 3375–3386.

18
Spices

Properties and Classification   268
   Hot (Pungent) Spices   269
   Aromatic Spices   269
   Flavoring Extracts   269
Use of Spices in the Meat Industry   270
Germicidal Action of Spices   270
Storage and Treatment of Spices   271
Selected References   272

Spices, according to the American Spice Trade Association, are tropical plants whose parts are used to season foods. Botanically, spices are the roots, bark, buds, seeds, or fruits of aromatic plants, which usually grow in the tropics. The *true spices,* numbering about 13, include allspice, clove, and black pepper. *Herbs* are leafy parts of temperate-zone plants. *Condiments* are mixtures of spices and other ingredients usually made into a saucelike consistency; catsup and mustard are examples.

Some ingredients, while used in spice blends, are not classed as spices. For example, sugar and salt are used as flavoring or preservatives, while both green and pimento peppers are classed as embellishments.

In the past, spices were thought to have germicidal properties and, thus, were sometimes used in food preservation. In fact, very few, if any, of the spices are germicidal, though some may have a bacteriostatic action if used in large enough quantity. Some commercial spices harbor large numbers of bacteria and molds (Table 18.1).

Spices have been part of history for a very long time. For centuries, the Arabs monopolized the oriental spice trade. The lure of spices and a share of the lucrative spice trade stimulated early European explorations to the Far East.

Spices were also used for purposes other than as seasonings. The Romans would hang an anise plant near their pillow to prevent bad dreams. The natives of the Molucca Islands always planted a clove tree

Table 18.1. Microbial Content of Untreated Spices

| | Number/gram | |
Kind of spice or herb	Bacteria[a]	Molds[b]
Whole allspice	1,000,000	70,000
Ground allspice	64,000	50,000
Sweet basil	525,000	50
Whole cloves	4,400	100
Whole Zanzibar cloves	190	0
Ground China cinnamon	36,000	60,000
Crushed cinnamon	8,000	600
Ground ginger	60,000	2,000
Bay leaves	15,000	350
Ground Bandamace	2,800	400
Ground mustard	1,800	0
Ground East Indian nutmeg	1,200	700
Ground paprika	680,000	5,000
Ground red pepper	2,190,000	1,220,000
Ground white pepper	42,000	9,000
Decorticated pepper	1,780,000	70,000
Ground black pepper	10,400,000	1,300,000
Savory	4,000	450
Ground sage	270,000	20,000
Whole thyme	2,700,000	12,000
Ground thyme	35,000	30,000
Miscellaneous:		
Celery seed	1,150,000	10,000
Onion powder	6,000	0
Garlic cloves	200	20,000
Onion juice	30,000,000	100
Ground garlic powder	90,000	200
Liquid garlic	10,000	10,000
Emulsified spice oil	10	10

[a] Suspensions incubated at 37°C.
[b] Suspensions incubated at room temperature.

whenever a child was born and destruction of the tree was thought to spell doom for the one for whom it was planted.

PROPERTIES AND CLASSIFICATION

Spices themselves have little or no nutritional value, but as seasonings they add zest to foods that would be insipid without them.

Most spices owe much of their flavoring properties to volatile oils, but in some cases the flavor is from a fixed oil. In several instances, no single flavor ingredient can be identified, and the flavor is credited to a natural blending of flavors from many different components. These include alcohols, esters, terpenes, phenols and their derivatives, organic acids, alkaloids, resins, sulfur-containing compounds, and some as yet unidentified compounds.

Spices may be classified into three groups: hot or pungent spices, aromatic spices, and flavoring extracts.

Hot (Pungent) Spices

Pepper is a good example of a spice. Cayenne and red pepper come from *Capsicum frutescens*. Black and white pepper come from *Piper nigrum*. Black pepper is made from immature berries, while white pepper comes from the mature berries, which have the hull removed by fermentation. The condimental properties of pepper are caused chiefly by a volatile oil, which is a hydrocarbon ($C_{10}H_{16}$), and the nitrogenous bases piperidine and piperine.

Aromatic Spices

Allspice or pimento is obtained from an evergreen tree. Allspice, like clove, contains the volatile oil eugenol ($C_{10}H_{12}O_2$).

Cinnamon is obtained from the dried inner bark of the cinnamomum tree.

Cloves are prepared from dried flower buds of the clove plant, an evergreenlike plant, which grows 20–40 ft in height. Their condimental property is caused primarily by the volatile oil eugenol.

Ginger is prepared from root stock of an annual plant growing 3–4 ft high. The root is washed, peeled, and then dried. Sometimes carbonate of lime is added. Preserved ginger is prepared by boiling the root and then adding sugar or honey.

Flavoring Extracts

Vanilla extract is obtained from the vanilla bean, the fruit of a climbing vine. The odor is caused by vanillin ($C_8H_8O_3$). Imitation vanilla is made from coumarin extracted from the tonka bean together with dextrose and glycerin.

Lemon extract is prepared by soaking lemon peel in strong alcohol. Citral ($C_{10}H_{16}O$) is the volatile oil flavoring compound.

USE OF SPICES IN THE MEAT INDUSTRY

In medieval, ancient, and even prehistoric times spices were used because it was believed that they preserved meats and other foods. Actually, the spices served primarily to mask bad flavors and odors, which resulted from varying degrees of decomposition.

Modern meat processors are equally dependent on spices, but for quite different reasons. Refrigeration and modern curing practices have largely displaced the use of spices in meat preservation and removed the need for covering up off-flavors.

Spice blends are now used to make dozens of different products from lower-quality cuts of meat. For example, small pieces of comminuted meat trimmed from primal cuts of beef, pork, lamb, and veal must be made into sausage and canned meats before they can be marketed. Also a large proportion of the meat from older, lean cattle and heavy hogs is more acceptable in the form of sausage than as fresh or cured meat. Many sausages have almost exactly the same meat ingredients but differ markedly in taste because of the kinds and proportions of spices used in their preparation.

The meat processing industry now takes around 20% of the spices consumed annually in the United States. The top five spices used in the meat industry are white pepper, sage, coriander, black pepper, and nutmeg. An estimated one-sixth of the nation's meat is marketed in the form of products whose characteristic taste depends largely on added spices.

GERMICIDAL ACTION OF SPICES

Blum·and Fabian (1943) studied the ability of various spice oils to control microbial growth. They reported that emulsions of spice oils were more effective than the unemulsified oils in retarding the growth of *Saccharomyces ellipsoideus* in all cases except mustard. Emulsified oils from black pepper, angelica, calamus, celery seed, ginger, sweet marjoram, and lime had very little effect upon *S. ellipsoideus.* Free oils from allspice, mace, and sweet orange showed a weak effect; in the emulsified form, they had a strong inhibiting action. The emulsion of an oil was more efficient in germicidal action than its respective components.

The relative penetrating and inhibiting properties of emulsions of various oils were compared. Mustard was concluded to be the most effective, while cassia and cinnamon were almost as good. Black pepper, calamus, ginger, and lime were not effective in their germicidal properties.

Cinnamon and cloves possess greater bactericidal activity than other spices. The active ingredient in cinnamon is cinnamic aldehyde; in cloves it is eugenol.

Extracts of many spices do not effectively inhibit the growth of microorganisms. However, oils obtained from ground mustard, cloves, and cinnamon display a degree of bactericidal activity. In some cases, spices may even accelerate the growth of organisms. In one study, the volatile oil of mustard had a stronger preserving power than cinnamon oil, oil of cloves, thyme, and bay leaves in that order. There is also a great difference in the resistance of different bacteria to the same spice and of the same organism to different spices. Many spices tend to inhibit the growth of *Staphylococcus aureus*. In general, the spice oils appear to be more inhibitory than ground spices.

In another study, the bactericidal properties of horseradish vapors were most effective at 99.5°F (37.5°C), and their potency decreased rapidly at lower temperatures. Horseradish vapors inhibited bacteria better than those of garlic and onion. The active ingredient in crushed horseradish is allyl isothiocyanate, a pungent volatile oil that irritates the eyes and burns the skin. It is very effective against several varieties of microorganisms when tested by different methods.

The bactericidal strength of spice oils can be measured more accurately in liquid form than as vapors.

Generally, molds are most sensitive, yeasts are intermediate, and bacteria, especially spore forms, are least sensitive to the germicidal effects of spices and their oils.

Many spice oils are valued more for their antiseptic properties than for their germicidal action. The germicidal power of any spice or oil is limited because they generally are not present in sufficient quantities to be actively germicidal. However, animal oils, such as seal oil and tuna oil, emit vapors that are germicidal and more active than the oils of spices. If cod-liver oil and sardine oil are exposed to sunlight or ultraviolet light, they become extremely germicidal.

STORAGE AND TREATMENT OF SPICES

Proper storage and processing of spices are important if they are to retain their quality. The highest-quality meats can be spoiled by spices that have been improperly stored. Spices should be stored covered, in a cool, dry place, and away from sunlight. A storage limit of 60 days for natural spices without refrigeration has been recommended. However, at 25°F (-4°C), spices can be kept in excellent condition for an indefinite period without any appreciable volatilization of oil.

As noted already, spices may contain appreciable numbers of organisms. Ethylene oxide treatment of spices has been reported to be effective in reducing their microbial content. Studies on the use of treated and untreated spices in fresh pork sausage showed a marked delay in spoilage when treated spices were used.

SELECTED REFERENCES

Anon. 1985. Spices. *In* "An Evaluation of the Role of Microbiological Criteria for Foods and Food Ingredients," pp. 287–288. National Academy Press, Washington, DC.

ICMSF. 1980. Spices. *In* "Microbial Ecology of Foods," Vol. II: Food Commodities, pp. 731–751. Academic Press, New York.

Blum, H. B., and Fabian, F. W. 1943. Spice oils and their components for controlling microbial surface growth. *Fruit Prod. J.* **22** (11), 326–329, 347.

Cartwright, L. C., and Nanz, R. A. 1948. Comparative evaluation of spices. *Food Technol.* **2**, 330–336.

Chipault, J. R., Mizuno, G. R., Hawkins, J. M., and Lundberg, W. O. 1952. The antioxident properties of natural spices. *Food Res.* **17**, 46–55.

Jones, S. E. 1949. Spices in the essence of geography. *Nat. Geographic Mag.* **96**, 401–420.

Pruthi, J. S. 1980. "Spices and Condiments: Chemistry, Microbiology, Technology." Academic Press, New York.

Yesair, J., and Williams, M. H. 1942. Spice contamination and its control. *Food Res.* **7**, 118–126.

19
Water

Consumption and Uses of Water   274
 Industrial Uses of Water   274
 Water Usage in Agriculture   275
Water Supplies   276
 Movement of Rainfall   276
 Groundwater Supplies   277
 Surface Water Supplies   277
Principles of Water Purification   279
 Sedimentation   280
 Filtration   280
 Storage   281
 Aeration   282
Causes and Treatment of Hard Water   282
 Softening Water   283
 Survival of Bacteria in Lime-Treated Water   285
Waste Treatment   286
 Effects of Industrial Wastes   287
Characteristics of Wastes   288
 Biological Oxygen Demand   289
 Acids and Alkalies   290
 Suspended Solids   290
 Settleable Solids   291
 Oil, Grease, and Immiscible Liquids   291
 Pathogenic Bacteria   292
 Toxic Materials and Heavy Metal Ions   292
 Nitrogen and Phosphorus   294
 Color and Turbidity   294
 Total Dissolved Solids   294
Selected References   295

When water falls upon the land, it either flows into streams, lakes and reservoirs, soaks into the ground, is absorbed by plants, or is evaporated. Water supplies are called *surface supplies* when they are taken from streams, lakes, or reservoirs, and *ground supplies* when they are taken from wells or springs.

Large municipalities make use of surface sources to secure an adequate supply of water, whereas smaller communities, individuals, and industrial plants ordinarily use underground sources. It is almost impossible to find a source of either surface or underground water that meets modern standards for a public water supply without some form of treatment.

A public water should have the following characteristics:

- Contain no organisms that cause disease
- Be sparkling clear and colorless
- Be good tasting, free from odors, and preferably cool
- Be neither scale-forming nor corrosive
- Be reasonably soft
- Be free from objectionable gases, such as hydrogen sulfide, and objectionable minerals, such as iron and manganese
- Be plentiful and low in cost
- Be free of chemical toxicants

CONSUMPTION AND USES OF WATER

Water is one of our most precious natural resources. Sufficient quantities are essential for life. Large amounts of water are used in industry, agriculture, and homes. Many are unappreciative of the small cost of the water we drink and use, considering that it is treated to be bacteriologically and chemically pure.

Industrial Uses of Water

The data in Table 19.1 illustrate the large volumes of water used in the manufacture of many industrial products. Electrical generating plants are also big users of water. The production of electric power by steam requires 80 gal of water per kilowatt-hour of electricity, or nearly

Table 19.1. Gallons of Water Required per Unit of Product in Various Industries

Product	Water required (gal)
Aluminum	960 per lb ingot
Brewing	470 per bbl
By-product coke	3,600 per ton coal
Coal washing	200 per ton
Cotton fiber to fabric	37 per lb of goods
Cotton cloth processing	15 per lb of goods
Distilling	300–600 per bu of washed grain
Oil refining	770 per 42 gal bbl
Papermaking	30 per lb of paper
Steel, finished	65,000 per ton
Tanning	800 per 100 lb hide

1000 times as much water by weight as coal. The water required to generate 15 billion kWh of electricity, a typical annual usage in a medium-size industrial state, is equal to 1200 billion gallons. This would fill a reservoir 1 mile wide and 100 miles long to a depth of 57 ft.

The uses of water in the food industry may be grouped as follows:

- Cleaning, washing, or spraying food products
- Immersion of food in hot water or steam during blanching, scalding, and pasteurizing in a water medium
- As a cooking medium for various products
- As wash water, with or without detergents or other compounds, in the general cleaning of food processing equipment

The amounts of water used in several food processing operations are listed in Table 19.2. Obviously, the water supply for a food processing plant must be free of pathogens and other organisms that might cause spoilage in the products. Water that is used for cooling purposes or for washing food should be chemically treated to render it free of harmful organisms.

Water Usage in Agriculture

High crop yields in modern agriculture depend on adequate water supplies. Large quantities of water are used for crop irrigation in certain regions of the United States. Even where irrigation is not common, adequate rainfall and soil moisture are necessary for good crop yields.

Production of an average U.S. corn crop, for example, requires 860 billion gallons of water. A single corn plant may transpire 1–1½ gal daily when the ear is forming, and as much as 30 gal during the season. High-yielding hybrid corn takes even more water from the soil and transpires it to the atmosphere.

Table 19.2. Water Required in Some Food Processing Industries

Industry	Water required (gal)
Corn canning	1100 per ton of corn in husks
Corn syrup	30–40 per bu corn
Dairying	5 per gal milk
Meat packing	6,000 per ton on the hoof
Tomato canning	60 gal per bushel

WATER SUPPLIES

Although water covers about three-fourths of the earth's surface, much of it is not suitable for human use because of its composition, the presence of pathogenic microorganisms, or its location and availability. The oceans contain a large proportion of the earth's water, but the high salt content of seawater precludes its use for most purposes without desalinization treatment. In the future, as other sources of water become scarcer and the demand for water increases, it is likely that the oceans will be used increasingly as a water supply.

Movement of Rainfall

The basic source of water on earth is, of course, rainfall. Rainwater is approximately as pure as distilled water when it leaves the clouds. By the time it reaches the earth's surface, it has picked up foreign material suspended in the air. The amount and kind of material picked up varies considerably. The most important contaminants are airborne microorganisms. Since the air commonly contains these organisms, their presence in rainfall is usual.

When rainwater reaches the ground, it becomes grossly contaminated with microorganisms, which are always present in large numbers in soil. Runoff water after a rainfall carries large amounts of suspended soil particles and innumerable microorganisms. River water, after a hard rain, becomes muddy because of soil suspensions washed into the river, and its microbial content is extremely high compared with that of the river water prior to the rainfall.

Not all the rainfall runs off as surface water. Large amounts enter the soil to replenish the groundwater supply. As water penetrates through the soil layers, much of the suspended soil matter and organisms are removed. For instance, water leaving an underground tile drain is usually crystal clear, thus demonstrating the filtering effect of soil. The micorbial content of this kind of water is very low. However, as water penetrates through the soil, it picks up soluble mineral salts, such as calcium, magnesium, and iron, which contribute to the hardness of the water. As it moves through soil, water may also come in contact with decayed organic matter, resulting in undesirable flavors and odors.

Even in the absence of human-generated pollutants, surface waters and groundwater may be contaminated with all kinds of microorganisms. Harmful or disease-producing organisms must be removed or destroyed, if a water supply is to be safe for human consumption.

Groundwater Supplies

Groundwater is obtained from wells, springs, and infiltration galleries. Public supplies of groundwater in this country are usually taken from wells. In comparison with water from surface supplies (e.g., lakes and reservoirs), groundwater is usually clearer; contains fewer bacteria; if taken from a single well, usually has a uniform mineral content; and usually has a more nearly constant and lower temperature during the summer.

The temperature of groundwater coming from a depth of 50 ft is the same as the average temperature of the region under which it lies. Water from a depth of less than 50 ft will be a little colder in the winter and a little warmer in summer. Water from a depth greater than 50 ft has a temperature higher than the average temperature of the region under which it lies. The temperature increases an average of 1°F for each 60 ft of depth below 50 ft.

Groundwater supplies, however, generally are insufficient to meet the needs of large water consumers such as municipalities and certain industries. In addition, a good groundwater supply simply may not be available in some locales. Well water has several other disadvantages:

* Calcium and magnesium compounds are present in larger quantities in well water than in surface waters found in the same localities.
* Iron and manganese are present in many well supplies.
* Hydrogen sulfide is often present in well water.
* The cost of pumping well water is usually greater than the cost of pumping surface water.
* The mineral content from two wells may be entirely different even though located in the same plot of ground.

Surface Water Supplies

Surface water, whether from streams, lakes, or reservoirs, usually is contaminated and therefore unsafe and unsatisfactory for human consumption until properly treated. Municipalities sometimes discharge partially treated sewage into a water course that is used as a public water supply. This is perhaps the most dangerous source of surface water contamination.

Soil washings may carry mud, leaves, decayed vegetation, and human and animal refuse into a surface water supply. The turbidity, or muddiness, and mineral content of water in flowing streams vary from day to day. Following heavy rains, it may be extremely muddy and low in mineral content, whereas during dry seasons it may be relatively clear and more highly mineralized. Surface supplies may be muddy or clear, soft or hard, depending on the season. The removal of mud from water taken from surface sources can be an immense task. For example, in many large communities 100–1000 or more tons of mud must be removed from a day's supply of water before pumping it to the consumer.

Surface waters contaminated with human feces and urine may cause typhoid fever, dysentery, and other enteric diseases. Organic wastes furnish food for microorganisms, algae, and lower forms of animal life. Excessive growth of these organisms may impart disagreeable tastes and odors to the water.

The waste liquors from manufacturing plants, mines, and quarries are often discharged into streams. These wastes may be quite acidic, thus rendering the water either unfit for use or too corrosive for distribution through ordinary iron pipe distributing systems. Industrial wastes also may contain excessive quantities of organic material, which after decomposition causes the water to be unpalatable. Substances such as phenols from coke-oven plants are sometimes discharged into streams and lakes. If the water is subsequently treated with chlorine, chlorophenols are produced. These impart to the water a disagreeable medicinal taste, which is extremely difficult to remove.

Farm Ponds

The water requirements in rural areas, especially for large dairy farms, vegetable crop irrigation, and domestic consumption can be substantial even though the population is relatively low. Where underground water supplies such as springs and wells have not been adequate to supply sufficient water, farm ponds often are used to help meet the demand for water. If the topography of the land is suitable, farm ponds can be constructed quite inexpensively. Typical farm ponds range in size from ¼ acre to several acres and can store millions of gallons of water for domestic and livestock use, recreation, and fish culture. The major problem is to remove sediment and purify the water for domestic use. Many publications are available covering the details of the uses of farm pond waters.

The use of contaminated water from farm ponds presents a serious health problem. Waterborne pathogenic organisms may cause disease in man or animals or both. Fortunately, many farm pond operators treat the water adequately to make it safe for human consumption. The contaminated water also carries many food spoilage organisms, which can alter the quality of milk, dairy products, and other foods if the water is used in various ways in food processing.

Farm ponds are the answer in many cases for providing adequate rural water supplies. Practical methods are now available for treating farm pond water both from the chemical and public health standpoints, thus making this source of water adequate for all purposes in rural homes and on the farm.

PRINCIPLES OF WATER PURIFICATION

Although nature purifies surface waters and groundwater by several physical, chemical, and biological processes, such natural purification generally is not sufficient to assure a safe drinking water supply. Nonetheless, many modern water treatment methods are based on natural purification processes.

The example of a stream or river illustrates both the usual sources of contamination and natural purification processes. Bacteria and algae consume and thrive on organic materials in surface waters. Small organisms eat bacteria and algae, larger organisms live on the smaller ones, and fish and higher forms of animal life in turn consume them. Unless the velocity of flow is too great, mud and suspended matter naturally settle out of suspension and organic matter is rendered harmless by oxidation, which occurs much more quickly and effectively if the stream flows over rough beds, riffles, and spillways. In a fast-flowing stream, the dilution of organic matter and microorganisms is a significant factor in purifying the water. Generally, the microbial content of fast-flowing streams is much lower than that of quiet streams. Sunlight has some germicidal effect from ultraviolet rays, but it penetrates only a little below the water surface and is not constant in its action because at night and on some days the sun does not shine at all. As a rule, during the summer months pathogens die off more rapidly than in winter time.

The popular notion that streams purify themselves in 20 miles of flow is untenable. The effectiveness of natural self-purification processes is variable and depends on many factors: flow rate, temperature, time of year, composition of the water, mciroflora present, streambed characteristics, etc. Thus, generalizations regarding the number of miles of flow necessary to accomplish self-purification are not valid. For example, in one study a river that had received sewage for years did not become clear or lose its sewage odor until it had flowed 70–80 miles. It did not test bacterially safe until after many more miles of flow.

Several processes involved in the natural purification of water and their application in modern treatment methods are discussed in the following sections.

Sedimentation

If water in reservoirs or lakes is allowed to stand, much of the suspended solids and many microorganisms attached to them will settle out. In water purification systems this process often is accelerated by addition of certain chemicals to the water that hasten the settling of suspended solids. Such a procedure may remove 80–90% of the bacteria from the water.

Filtration

Percolation of water through the top soil layers tends to remove large amounts of organic matter and many microorganisms. The rate of water filtration depends upon the type of soil. Sand and gravel offer the most efficient type of filtering material, while clay soil offers the least efficient because clay particles are so small that water passes through them very slowly. Natural filtration can remove 95–99% of the microorganisms in polluted water.

Modern treatment plants have modified the filtering process in order to accelerate the treatment of large volumes of water, although the fundamental principle of natural filtration remains the same.

Slow Sand Filtration

Slow sand filters are beds of sand 30–40 inches deep, contained in concrete basins, each about 1 acre in extent. They date from about 1830, are commonly called English-type filters, and are used extensively abroad. Very few slow sand filters have been built in this country since the introduction of American (rapid-type) filters. The filtration rate of English-type filters is about 1–2 million gallons per acre per 24 hours, while that of the American type averages 125 million gallons per acre per 24 hours.

The operation of some slow sand filters in the United States has been supplemented by features characteristic of the rapid sand filter, principally preliminary coagulation and disinfection.

The slow sand filter is adapted to the treatment of water having a turbidity not exceeding about 30 ppm and low color. For high turbidities and high color, coagulation generally is a necessary or desirable adjunct. When such conditions are present, the rapid sand filter is more economical.

The chief disadvantages of slow sand filters are as follows:

- They require a large area.
- Their successful operation is limited to clear water with low color.
- Their initial cost is relatively high.

Rapid Sand Filtration

To obtain a high rate of filtration through sand filters (100–200 million gallons per acre per day), it is usually necessary to treat the water ahead of the filters. This preparatory treatment includes (1) aeration to free the water from dissolved gases and to oxidize iron and organic matter if present, (2) coagulation, and (3) settling.

The addition of the coagulant to the water, followed by a short period of agitation, results in formation of a precipitate that entangles the mud, bacteria, and suspended matter into clumps, which are readily removed by settling. The settled water then passes to the filter for removal of the finely divided suspended matter that remains.

Storage

Storage of water in basins or reservoirs not only helps to provide an adequate water supply during low-rainfall periods but also provides some purification. For such storage, a dam is usually constructed across a valley to form an impounding reservoir. This affords an opportunity for the subsidence of silt and clay, reduction in color, and death of bacteria. Considerable time is required for these results to be accomplished by storage alone, and even then bacteria and finely divided particles of clay are not entirely removed. Even when many days of storage are provided, other methods of treatment usually are required to insure a satisfactory product. The large size of the basins or reservoirs required make this method very expensive. The tendency toward stagnation and the multiplication of lower forms of animal and vegetable growth also make it undesirable.

Water in reservoirs after 30 or more days' storage may become practically free from turbidity and bacteria. However, this condition may be entirely changed in a few hours by the process called *overturning*. This occurs when the surface water becomes chilled or warmed to a temperature of 39.2°F (4°C). At this temperature water has its greatest density. The heavier surface water, of course, does stay at the surface, and convection currents are set up that may become sufficiently active to cause a complete overturn of the reservoir water, resulting in much turbidity being diffused throughout the entire contents of the reservoir.

Aeration

Surface waters sometimes have offensive tastes and odors. Aeration may remove these if they are caused by dissolved gases resulting from the decomposition of organic matter. If, however, they are caused by dissolved organic matter, aeration is not very effective.

Groundwater usually contains carbon dioxide and sometimes hydrogen sulfide, iron, and manganese. Aeration reduces the carbon dioxide and hydrogen sulfide and, in most cases, oxidizes the iron, causing it to precipitate as insoluble ferric compounds, which can be removed readily by filtration. Manganese, especially if it is present in appreciable quantities, is not so easily removed as iron. Aeration and filtration usually are not adequate for removal of manganese. Iron and manganese can be precipitated from aerated water effectively by the addition of lime. Manganese requires more lime for its precipitation than iron. A pH of 9.4 is usually adequate for the precipitation of manganese. The reduction of carbon dioxide by aeration decreases the cost of softening water if lime is used as the reagent. One objection to the aeration of oxygen-free groundwater is that it becomes oxygenated and under some circumstances may be more corrosive.

Aeration is expensive. Fortunately, many water supplies do not need to be aerated.

CAUSES AND TREATMENT OF HARD WATER

The hardness of water is caused by the presence of calcium and magnesium salts. Water hardness reduces the effectiveness of soap and detergents in wash water and may preclude direct use of the water in certain food and beverage operations.

One method for determining the hardness of water is based on the ability of soap to form insoluble precipitates with calcium and magnesium salts. In the usual test procedure, a standard solution of soap is added to a sample of water until a permanent lather or foam is formed, which serves as the end point in calculating the hardness. Hardness is expressed in grains per gallon or parts per million of calcium carbonate. Water can contain temporary or permanent hardness or both. However, the type of hardness present determines the treatment procedure, as described later.

Temporary (carbonate) hardness is that part of the total hardness that can be removed by boiling. It is caused by bicarbonates of calcium or magnesium in the water. This type of hardness gives an alkaline reaction when titrated with an acid. Because of the slight solubility of calcium and magnesium carbonates formed when a solution of their bicarbonates is boiled, not quite all of the temporary hardness is removed.

Permanent (noncarbonate) hardness cannot be removed by boiling and is caused by sulfates and chlorides of calcium and magnesium. If the temporary hardness is removed by boiling, the permanent hardness (i.e., calcium and magnesium salts) remaining in solution can be determined by adding standard soda reagent (equal parts of NaOH and Na_2CO_3), which precipitates the magnesium as hydroxide and the calcium as carbonate. The amount of standard soda reagent remaining represents the amount consumed as calcium and magnesium when the reagent is added in excess and titrated with a standard acid.

The total hardness of a water supply may be determined by first neutralizing a sample with H_2SO_4, and then determining the permanent hardness with soda reagent.

Obviously, soft water is highly desirable because of its inherent cleaning properties. As water increases in hardness, especially permanent hardness, soap consumption and waste is greater. Hardness also alters the chemical composition of detergents and reduces the efficiency of these products. Hence, the calcium and magnesium sulfates and chlorides, which are soluble in the wash water, tend to form insoluble carbonates, phosphates, and silicates, while the soluble sodium compounds form the basis of any cleaner being used.

Acid cleaners tend to increase the solubility of calcium and magnesium sulfates. Their use largely prevents water stone deposits from hard water. Although acid cleaners tend to be corrosive, this problem can be overcome by the use of corrosion inhibitors.

Softening Water

Procedures for removing or reducing water hardness are called *softening*. In the softening of water, a chemical reaction takes place that either converts the hardness-causing compounds to harmless dissolved salts or causes them to precipitate as scale, which can be removed. In hard water, the calcium and magnesium salts are present in solution and, therefore, cannot be removed by filtration.

A common procedure for softening water involves treatment with lime ($Ca(OH)_2$) and soda ash (sodium carbonate, $NaCO_3$). The effectiveness of this treatment is due to the low solubility of calcium carbonate ($CaCO_3$) and magnesium hydroxide ($Mg(OH)_2$). If sufficient quantities of lime and soda are added to hard water and then stirred, these insoluble compounds eventually form and settle out upon standing. The supernatant water is soft. Aluminum sulfate can be added to facilitate the precipitation of $CaCO_3$ and $Mg(OH)_2$, which otherwise settle very slowly.

Rapid sand filters are used to remove and prevent suspended salts from forming scale deposits.

Temporary (Carbonate) Hardness

In the presence of sufficient carbon dioxide, calcium and magnesium carbonates in water will form bicarbonates, which are soluble.

$$H_2O + CO_2 \rightarrow H_2CO_3$$
$$H_2CO_3 + CaCO_3 \rightarrow Ca(HCO_3)_2$$
$$H_2CO_3 + MgCO_3 \rightarrow Mg(HCO_3)_2$$

If such water is boiled, the carbon dioxide is driven off and most of the carbonates precipitate, thus largely removing the hardness from the water (hence the name temporary).

$$Ca(HCO_3)_2 \rightarrow CaCO_3 + CO_2 + H_2O$$
$$Mg(HCO_3)_2 \rightarrow MgCO_3 + CO_2 + H_2O$$

If hardness is caused by the presence of calcium bicarbonate only, it can be removed by precipitation with exactly the required quantity of lime.

$$Ca(HCO_3)_2 + Ca(OH)_2 \rightarrow 2CaCO_3 + 2H_2O$$

Since calcium carbonate is extremely insoluble, it settles as a sediment and can be removed.

Generally, magnesium bicarbonate accounts for about 60% of the hardness in water. To remove magnesium bicarbonate, excess lime must be added to form magnesium hydroxide, which is very insoluble. Because magnesium carbonate is slightly soluble, it does not entirely precipitate. Thus, in the absence of excess lime, some hardness would remain.

$$Mg(HCO_3)_2 + Ca(OH)_2 \rightarrow MgCO_3 + CaCO_3 + 2H_2O$$
$$MgCO_3 + Ca(OH)_2 \rightarrow Mg(OH)_2 + CaCO_3$$

Adding, $$Mg(HCO_3)_2 + 2Ca(OH)_2 \rightarrow Mg(OH)_2 + 2CaCO_3 + 2H_2O$$

Generally, aluminum sulfate (alum) is added. In the presence of basic substances (e.g., sodium carbonate or calcium hydroxide), this will form insoluble aluminum hydroxide, a flocculent precipitate, which causes the calcium carbonate and magnesium hydroxide to settle more quickly.

$$Al_2(SO_4)_3 + 6Na_2CO_3 + 3H_2O \rightarrow 2Al(OH)_3 + 3Na_2SO_4 + 6NaHCO_3$$
$$Al_2(SO_4)_3 + 3Ca(OH)_2 \rightarrow 2Al(OH)_3 + 3CaSO_4$$

In the latter case, the slightly soluble calcium sulfate would remain as permanent hardness, which can be removed as discussed in the next section.

The settling of any precipitate, especially a flocculent precipitate has the incidental but very valuable effect of sweeping down and thus removing nearly all microorganisms originally present in the water. Hence, removing hardness purifies water biologically as well.

Permanent (Noncarbonate) Hardness

The presence of noncarbonate salts of calcium and magnesium, such as the sulfates and chlorides, causes permanent hardness, which cannot be removed by simple boiling. Calcium salts are removed with soda ash (sodium carbonate).

$$CaSO_4 + Na_2CO_3 \rightarrow CaCO_3 + Na_2SO_4$$

Noncarbonate magnesium salts are treated with lime and soda ash.

$$MgSO_4 + Ca(OH)_2 \rightarrow Mg(OH)_2 + CaSO_4$$
$$CaSO_4 + Na_2CO_3 \rightarrow CaCO_3 + Na_2SO_4$$

Adding, $MgSO_4 + Ca(OH)_2 + Na_2CO_3 \rightarrow Mg(OH)_2 + CaCO_3 + Na_2SO_4$

Survival of Bacteria in Lime-Treated Water

As the previous sections indicate, most hard water is treated with lime to soften it. Reihl *et al.* (1952) studied the survival of bacteria in lime-treated water when a high pH was maintained. They reported that *Escherichia coli, Salmonella typhi,* and *Salmonella montivideo* did not survive for prolonged periods in water when high pH levels (11.0–11.5) were maintained by the addition of excess lime, at a temperature of 59°F (15°C), and with a holding period of slightly longer than 4 hr. This method was effective in destroying many of the test organisms. Freshly isolated strains of the test bacteria were more resistant than the same species propagated for several months on culture media.

WASTE TREATMENT

At the present time, sewage treatment is based largely on natural physical, chemical, and biological processes. However, treatment systems are designed to provide optimum conditions so that these natural processes can operate at an accelerated rate.

There are two principal types of waste treatment systems in use: primary and secondary treatment. The former consists of processes that mechanically remove solids from liquid wastes. Primary treatment may involve screening, aeration, grit settling, plain settling, and the precipitation of colloidal particles by the addition of chemicals. The solids remaining are known as *sludge*. In secondary treatment, sludge is acted upon by bacteria, which reduce it to an inert humuslike residue.

The liquid portion of sewage may be discharged into a stream, providing there is sufficient flow of water in the stream to dilute the effluent so that it will not be a nuisance. In other words, there must be sufficient dissolved oxygen in the water to oxidize any suspended solids in the effluent. A more satisfactory method is to discharge the liquid onto a filter bed, thus allowing desirable oxidizing bacteria to digest the suspended solids. The remaining liquid is sparkling clear after it passes through the filter bed. This is the most efficient method, but it may not be economical.

In many instances, primary treatment may be sufficient to handle industrial wastes providing the discharge stream flow is large enough to dilute the wastes. In this case, the normal biological action that takes place in the stream is sufficient to handle the extra load of wastes discharged into the water. However, secondary treatment is necessary to supplement primary treatment. This is especially true during a drought when the stream is not able to absorb the pollution load.

Various primary and secondary treatment systems are in use at the present time. Regardless of their design, they should include the following components for efficient and safe operation:

1. A screen of sufficient size to remove the larger floating or suspended solids
2. A grit chamber to remove inert materials, thus allowing them to settle out and ultimately be removed
3. A sedimentation tank, where the lighter organic and inorganic solids are allowed to settle and thus be removed as sludge and scum
4. A sludge digestion tank and drying bed. The sludge is removed from the sedimentation tank and placed in a sludge digester where bacterial action takes place

The biological action is caused by anaerobic bacteria, which break down the solids, forming gases, liquids, and inert solids. The gases, largely methane gas, are a valuable by-product, which can be used as fuel in combustion engines to produce electrical power and heat for plant operation. The inert solids are placed on drying beds and allowed to dry. They may be used as landfill, burned as fuel, or combined with other plant nutrients and sold as commercial fertilizer. The liquid remaining after digestion is passed through gravel filters, where aerobic oxidation stablizes it, and then discharged into a stream.

There are several variations in the primary treatment processes. The chemical treatment of incoming raw sewage by coagulation and flocculation to assure a more rapid and complete precipitation of suspended solids is receiving considerable attention. Chemical treatment is often used as an intermediate step between primary and secondary treatments.

Secondary treatment of sewage is used only after some form of primary treatment. Biological action is necessary to destroy the organic material in solution. The trickling filter and the activated sludge method are the two principal processes used at the present time.

The *trickling filter* consists of a bed of coarse stone, provided with an underdrain. After primary treatment, sewage is sprayed over the bed and allowed to percolate down through it. Biological life accumulates on the surface of the stones, and this mixed microflora called *schmutsdecke,* meaning groundcover, oxidizes the organic materials to produce an effluent from which about 85% of the pollution load of the original raw sewage has been removed.

The *activated sludge* process works on the same biological principle, except the mixed microflora are cultivated in the sludge. Tanks are constructed of sufficient capacity to allow the sewage to flow through at a slow rate. Large volumes of air are blown in at the bottom of the tank, thus providing stirring and oxygen. The incoming raw sewage is inoculated with a complex biological life by returning at least one-fifth of the sludge already removed from sewage previously treated. A thorough mixing of biological life with the raw sewage, along with sufficient air, provides purification in much the same manner as a trickling filter. The activated sludge process accelerates the settling of the solids in the raw sewage, thus saving approximately two-thirds of the time required for the natural settling of the solids in primary treatment.

Effects of Industrial Wastes

During the past few years, a marked increase in industrial activity has resulted in many industrial wastes being dumped into municipal sewage systems. The outcome of this practice, in many cases, is an upset of the normal biological activity in municipal sewage treatment plants. Obviously, the character or composition of industrial wastes will determine the biological action of the ordinary sewage when these wastes are dumped into it.

Milk product factory wastes are common in municipal sewage. Since dairy products undergo an acid fermentation, the wastes are usually acid. If too much milk waste enters the sewage, it becomes acid and may interrupt the normal biological activity of the sewage.

In many cases, industrial wastes must be pretreated before they enter a municipal sewage system. At present, a great deal of attention is given to such pretreatment, which is considered as an integral part of waste treatment. Each waste must be treated by the most economical and practical method. Federal and state laws prohibit the discharge of municipal raw sewage and industrial wastes of any kind into a surface water supply without some preliminary treatment to render it less objectionable for a good quality water supply.

Many food processing plants have their own treatment plant to take care of their wastes. This is especially true in the canning industry. The use of lagoons, covering large areas of land, has been very successful. The biological changes involved have been effective in transforming the wastes into stable products with little odor and a maximum reduction in the volume of wastes. However, food wastes from processing plants vary widely in composition. Some of the constituents are easily oxidized, while others are more complex and may be difficult to decompose by microbial action.

It takes about 50 tons of pure oxygen a day to process the sewage in a city with a population of around 500,000. This is about 250 tons of air or about 4 million cubic feet. Every molecule of oxygen goes through a bacterial cell. When oil cracking plants, plastic factories, and industries discharge their wastes into sewage, carbonic acid, phenols, and other compounds that are poisonous to bacteria prevent the organisms from oxidizing or stabilizing the sewage.

The ever-increasing use of detergents in the home and elsewhere also may cause the biological system in sewage plants to function abnormally. Studies by Malaney *et al.* (1960) demonstrated the effect of different concentrations of detergents on the biological processes normally associated with sewage treatment.

CHARACTERISTICS OF WASTES

Knowledge of the characteristics and components of a waste is important in assessing the kind of treatment required and the effect of the waste material on a treatment plant. Likewise, the effluents remaining after treatment should meet certain standards before being discharged into surface waters.

Biological Oxygen Demand

Fresh surface water supplies may contain up to 10 ppm of free oxygen. Obviously, the addition of organic wastes results in action by microorganisms until the free oxygen is used up. The term *biochemical oxygen demand* (BOD) is used to indicate the quantity of oxygen required to completely oxidize the organic wastes present. When the BOD of surface waters rises above 4 ppm, aquatic life ceases to exist. Then anaerobic conditions (without free oxygen) begin to operate, resulting in hydrolysis, putrefaction, and fermentation. Under anaerobic conditions the end products are not completely oxidized, giving rise to offensive odors. Such waters are not suitable for many uses. A major objective of any waste treatment operation is to reduce the BOD of the waste material to the point that the effluent will not cause deterioration of the body of water into which it is discharged.

A BOD analysis is an attempt to simulate the effect a waste will have on the dissolved oxygen of a stream by a laboratory test. It has been the most widely used method for estimating the strength of domestic or other biodegradable wastes. It must be applied with greater caution to many industrial wastes since the presence of certain compounds can inhibit the analysis. Although it is a useful measurement in characterizing industrial wastes, its use in monitoring is limited because 5 days are required to run the test. It is expected, however, that BOD will continue to be a standard for regulatory agencies for many years. Therefore, an understanding of this parameter is essential.

The BOD test gives an indication of the amount of oxygen needed to stablize or biologically oxidize a waste. The analysis determines the biodegradable organic carbon and, under certain conditions, the oxidizable nitrogen present in a waste. The measurement of oxidizable nitrogen may be avoided by adding inhibitors for the nitrifying bacteria. However, the ammonia content of the waste should be measured separately because it also affects the oxygen balance in a stream. After the carbon has been oxidized, nitrifying bacteria begin using oxygen to oxidize the ammonia (4.56 mg O_2/mg NH_4^+).

The advantage of the BOD test is that it measures only the organics that are oxidized by bacteria. The disadvantages of the BOD test are the time lag between sampling and results of the analysis (5 days for BOD_5) and the difficulty in obtaining consistent repetitive values. It is possible that organics not degraded in a BOD bottle will be oxidized in the environment by bacteria that are acclimatized to that environment. Normally, the BOD bottle is not shaken and the CO_2 produced accumulates in the bottle. Both shaking and CO_2 accumulation influence the test results. A further disadvantage of BOD is the poor reproducibility of the test. The BOD of the same sample analyzed by two different laboratories seldom agrees within 10%.

Manometric methods used to determine BOD are more reproducible than the bottle method. The mixture is usually stirred and the CO_2 is absorbed by a strong basic solution. For individual BOD analyses, however, the BOD bottle method is the only economically feasible method. If BOD is to be continuously monitored, the use of the Hach type of apparatus or the electrolysis BOD device should be considered. The use of the Warburg apparatus is justified for research purposes but probably not for routine measurement of BOD.

Acids and Alkalies

The pH of a waste material is important because a sudden change in pH can cause serious damage in surface waters or treatment plants. The measurement used to determine the required dosage of neutralizing agent, either $Ca(OH)_2$ or H_2SO_4, is termed acidity or alkalinity, respectively.

Acidity refers to the capacity to donate protons. Acidity is attributable to the unionized portions of weakly mineral acids, hydrolyzing salts, and mineral acids. In most wastes, mineral acids are probably the most significant group. It is difficult to predict neutralization requirements when diverse forms of mineral acidity are prevalent. Microbial systems may reduce acidity in some instances through biological degradation of organic acids.

Alkalinity, or the ability of wastewater to accept protons, is significant in the same general way as acidity, although the biological degradation process does offer some buffer capacity by furnishing carbon dioxide as a degradation end product. Alkalinity can be due to the presence of HCO_3^-, CO_3^{2-}, or OH^-. It has been estimated that approximately 0.5 lb of alkalinity (as $CaCO_3$) is neutralized for each pound of BOD removed. Excess alkalinity often has to be removed by neutralization.

Suspended Solids

Suspended solids represent the undissolved substances in a wastewater retained on a 0.45-μm filter. The residue retained on the filter is dried in an oven at $105°C$. Nonhomogeneous particulate matter should be excluded from the sample. Analysis for suspended solids should begin as soon as possible since preservation of the sample is not practical.

The use of glass filters has increased considerably and these filters appear to give comparable results to millipore filters. Glass fiber filters have one advantage over combustible materials, such as the millipore. Since the glass fibers are noncombustible at the temperatures used for determination of volatile suspended solids, the same crucible used for suspended solids determinations can be employed directly for determining volatile content. The combustible materials must be placed in a crucible and the final weight must be corrected to account for combustion of the filter.

Settleable Solids

The term *settleable solids* refers to solids in suspension that will settle under quiescent conditions. Only the coarser suspended solids with a specific gravity greater than that of water will settle. The test for settleable solids is conducted in a 1-hr settling time. Samples should be at room temperature and the test conducted in a location away from direct sunlight. The settled solids volume is measured and reported in terms of milliliters or settleable solids per liter.

The settleable solids test is important since it is the principal means to establish the need for and assist in the design of sedimentation facilities. This test is widely used in sewage and industrial waste treatment plant operation to determine the efficiency of sedimentation units.

Oil, Grease, and Immiscible Liquids

Oil, grease, and immiscible liquids can produce unsightly conditions. In most cases the quantities permitted in wastewater are restricted by regulatory agenices. In sewer systems the presence of oils and immiscible liquids, such as naphthene and ether, may cause explosive conditions. Wastes from the meat-packing industry, particularly those containing fats from the slaughtering of sheep and cattle, have resulted in serious decreases in the capacity of sewers. In treatment plants, wastewater with a high grease content may cause trouble with aerobic biological treatment.

The term grease refers to a wide variety of organic substances that may be extracted from aqueous solution or suspension by hexane. Hydrocarbons, esters, oils, fats, waxes, and high-molecular-weight fatty acids are the major materials dissolved by hexane. These materials have a greasy feel and are associated with problems in aerobic waste treatment.

Fats and oils are esters of the trihydroxy alcohol glycerol, while waxes are esters of long-chain monohydroxy alcohols. The glycerides of fatty acids that are liquid at ordinary temperatures are called oils, and those that are solids are called fats. The term oil also refers to a wide variety of hydrocarbons of mineral origin, spanning the range from gasoline through heavy fuel to lubricating oils.

Oils and greases of vegetable and animal origin are generally biodegradable and, in an emulsified form, can be successfully treated by a biological treatment facility. On the other hand, oils and greases of mineral origin may be relatively resistant to biodegradation and will require removal by methods other than biological treatment. Unfortunately, a satisfactory method of distinguishing between oils and greases of vegetable and animal origin and those of mineral origin is not readily available.

Pathogenic Bacteria

Wastewaters that contain pathogenic bacteria can originate from livestock production (cattle, poultry, swine, lab animals), tanneries, pharmaceutical manufacturers, and food processing plants (Fig. 19.1). Pathogenic bacteria in wastewaters may be destroyed by the process of chlorination.

The bacteriological safety of wastewater is normally measured by the number of fecal coliform bacteria present. Coliform bacteria are not pathogenic but are an indication of the probability that pathogenic bacteria are present. Examples of pathogenic bacteria are *Salmonella, Shigella, Leptospira,* and *Vibrio.* To this group of undesirable pathogens also can be added the enteric viruses and parasites, such as *Endamoeba histolytica.*

Toxic Materials and Heavy Metal Ions

For biological waste treatment plants, the maximum tolerable concentrations of toxic materials have been reported for many materials. In general, the threshold toxicity levels for biological treatment systems are higher than the allowable standards for surface waters. Establishing maximum concentrations for toxicants in biological treatment plants is useful only if the amount of toxicant is reduced during the treatment, as is the case with phenols. Often, it is necessary to decrease the concentration of the toxic material by pretreatment.

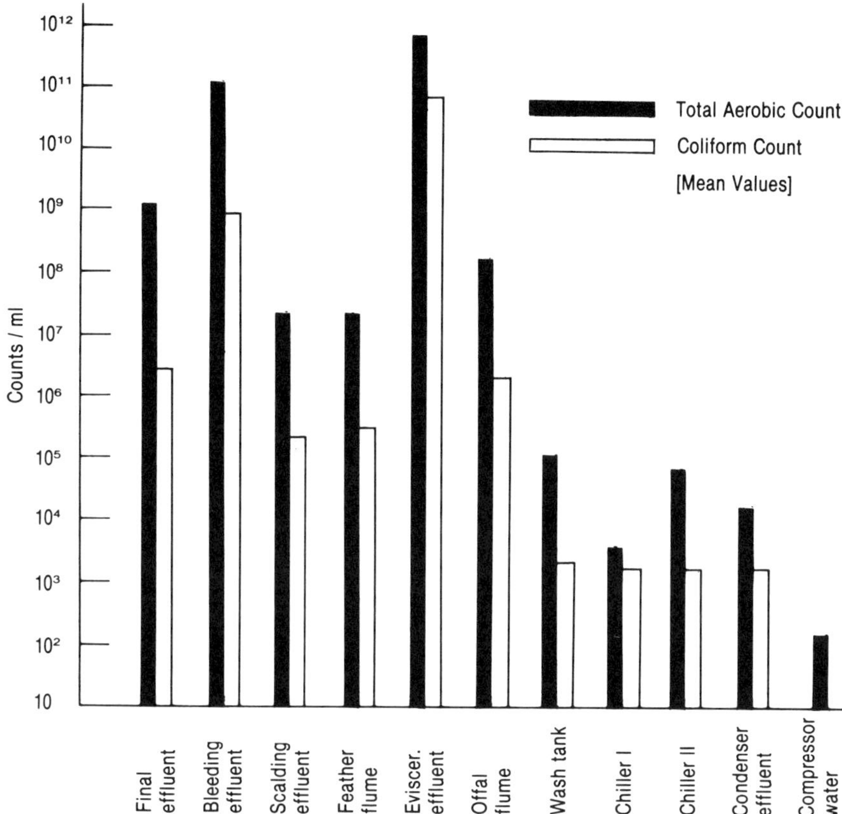

Fig. 19.1. Bacterial counts of poultry processing wastewater. Courtesy of U.S. Environmental Protection Agency.

It is necessary to guard against the synergistic effects of certain materials. One industrial plant may be allowed to discharge zinc below the toxic level, while another plant may be allowed to do the same with copper. The resulting combination of both discharges will have a synergistic effect and may cause biological deterioration in the receiving stream or the municipal treatment system.

Ammonia nitrogen is present in many natural waters in relatively low concentrations, while industrial streams often contain exceedingly high concentrations of ammonia. Nitrogen in excess of 1600 mg/liter has proven to be inhibitory to many microorganisms present in activated sludge basins. Sulfides also are present in many wastewaters as a mixture of HS^--H_2S (depending on pH), sulfonated organic compounds, or metallic sulfides.

The influence of heavy metals on biological unit processes has been the subject of many investigations. Toxic thresholds for copper, zinc, and cadmium have been established at approximately 1 mg/liter.

Nitrogen and Phosphorus

When effluents are discharged into lakes, ponds, and surface streams, the presence of nitrogen and phosphorus is particularly undesirable since these nutrients enhance eutrophication and stimulate undesirable algae growth. On the other hand, industrial wastes containing insufficient nitrogen and phosphorus for biological development in waste treatment systems require addition of these nutrients in forms such as anhydrous ammonia and phosphoric acid.

The chemical form in which nutrients are present in wastewaters vary with the degree of treatment. Nitrogen can be present as ammonia, nitrate, and nitrite, and organic nitrogen in the form of proteins, urea, and amino acids. Phosphorus can be present as inorganic phosphates or phosphorus-containing organic compounds.

Color and Turbidity

Color and turbidity present aesthetic problems. Low concentrations of compounds such as lignins and tannins impart color to natural waters and may be intensified when combined with other materials. An example of this is iron and tannin, which combine to form iron tannate, a common base of blue-black ink.

It is possible to differentiate between the true and apparent color of a sample. True color is due to matter that is in true solution, while apparent color includes the effects of matter in the suspended and colloidal states as well. Examples of true color constituents are soluble dyes used in industry. Constituents that cause apparent color are usually finely divided metal hydroxide particles.

Total Dissolved Solids

Total dissolved solids can be obtained by evaporating a sample of filtrate on a water bath. After the sample is evaporated, the dish is dried in an oven at a temperature of 105°C or 180°C. When total solids are measured, large floating particles should be removed from the sample. Oil and grease present in the sample should be included and dispersed by blending before evaporating.

Some materials, such as metallic hydroxides, will retain an associated water of hydration at 105°C, resulting in an apparent higher measurement of solids. At 180°C, organic matter can be reduced by volatilization but is not completely destroyed, some chloride and nitrate salts may be lost, and bicarbonate may be converted to carbonate, which may be partially decomposed to oxide or to basic salts. In general, samples containing considerable organic matter, or those with pH greater than 9, should be dried at the higher temperature. The report should indicate the drying temperature used in the analysis.

SELECTED REFERENCES

Anon. 1973. Handbook for Monitoring Industrial Wastewater. U.S. Environmental Protection Agency, Washington, DC.

Crosswhite, W. M., Carawan, R. E., and Macon, J. A. 1971. Waste and waste management in poultry processing. *Proc. 2nd National Symp. on Food Processing Wastes,* pp. 323–335. U.S. Environmental Protection Agency, Corvallis, Oregon.

Gainey, P. L., and Lord, T. H. 1952. "Microbiology of Water and Sewage." Prentice-Hall, New York.

Hoover, C. P. 1951. "Water Supply and Treatment." National Lime Assoc., Washington, DC.

Phelps, E. B. 1944. "Stream Sanitation." John Wiley, New York.

Prescott, S. C., Winslow, C. A., and McCrady, M. H. 1946. "Water Bacteriology," 6th ed. John Wiley, New York.

Reihl, M. L., Weiser, H. H., and Rheins, B. T. 1952. Effect of lime treated water upon survival of bacteria. *J. Am. Water Works Assoc.* 44 (5), 466–470.

20
Food Spoilage

Microorganisms and Food Spoilage 299
  Bacteria 299
  Yeasts 300
  Molds 300
Spoilage in Canned Foods 300
  Types of Canned Food Spoilage 301
  External Appearance of Canned Foods 305
  Determining Whether Canned Foods Are Safe 305
Spoilage in Meat and Other Unprocessed Foods 306
Microbiological Analysis of Foods 308
Selected References 311

Although food spoilage can be caused by chemical or biological processes, the most significant types of food spoilage result from biological changes caused by enzymatic activity and the growth of microorganisms.

Certain enzymes naturally present in a food may alter its texture, flavor, and odor. If the action of these enzymes is not checked, marked deteriorative biochemical changes occur in the food, a condition usually associated with food spoilage. For example, the enzyme lipase in raw, fresh milk will, under certain conditions, cause a series of undesirable changes, particularly rancidity, in the milk fat. In commercial practice, the action of enzymes can be minimized by cooking or canning a food shortly after it is harvested. The cooking or canning process, if continued long enough, completely inactivates enzymes. If a food must be stored prior to processing, it should be kept at a low enough temperature to retard enzyme action.

The presence and growth of bacteria, molds, and yeasts in foods is far more likely to cause spoilage than the action of endogenous enzymes. Microorganisms are present everywhere—in the air and water, on most foods, and on people's hands.

Usually there are many kinds of microorganisms present in and on raw food before it is preserved, and under certain conditions the method of food preservation may actually create a favorable environment for their growth. A good example is anaerobic bacteria, which grow in the absence of atmospheric oxygen. Many of these organisms form spores, which may not be destroyed by the heat treatment involved in canning. Since canned foods are sealed under vacuum with air excluded, once any spores present germinate, the conditions favor rapid growth of the anaerobic bacteria. Molds and yeasts are usually associated with canned products in which the air seal is not secure, because they usually require atmospheric oxygen for growth.

According to Kaufman (1947), spoilage of all edible foods would eventually take place if man did not do something to stop or at least partially slow up normal biological processes in foods. Various methods of food preservation have been used by man for many centuries (Table 20.1) to prevent spoilage and to keep surplus food for future use. Nowadays, the convenience and diversity of preserved and processed foods also is valued by consumers.

Although less important than enzymic activity or the growth of microorganisms, foods may undergo deteriorative chemical changes independent of any biological factors. The chemical composition of a food

Table 20.1. Food Preservation Methods

Asepsis
Additives
  Salt or sugar, or both
  Spices
  Acids (vinegar, sour milk, etc.)
  Smoking
  Inhibitors
Refrigeration
  Chilling
  Freezing
Heat Processing
  Pasteurization
  Boiling
  Canning
  Pressure processing
Fermentation
  Acid
  Alcohol
Removal of moisture (dehydration)
Filtration
Pressure
Gases (controlled atmosphere)
Radiation
  Ultraviolet
  Ionizing radiations
Fungicides (yeasts and mold inhibitors)
  Sorbic acid
  Sodium benzoate

determines its susceptibility to this kind of spoilage. For example, foods high in fat may become rancid when exposed to atmospheric oxygen. Exposure to sunlight causes some foods to undergo undesirable changes. In other foods, the formation of hydrogen gas results from a chemical reaction of the food constituents. In canned foods where oxygen has not been completely exhausted from tin-lined cans after sealing, oxygen initiates a chemical reaction causing detinning of the can.

MICROORGANISMS AND FOOD SPOILAGE

As noted already, the most important cause of food spoilage is the growth of bacteria, molds, and yeasts. Fortunately, most organisms can be destroyed or controlled by proper processing and handling of food products.

Bacteria

Bacteria exist in both active and resting forms. The resting form is known as a spore, which is quite resistant to heat. In the active or vegetative stage, bacteria are destroyed at boiling temperature.

The presence of acids in foods materially shortens the time required to render a product sterile during processing. In foods with high acid content (all fruits, tomatoes, and pickles), both spores and vegetative cells are killed easily at the boiling point of water. With low-acid foods such as meats, corn, peas, beans, and practically all vegetables except tomatoes, a 6-hr heating at the boiling point of water may be required to destroy spores. However, spores in low-acid foods are killed more quickly at 240°F (115.5°C), the temperature obtained by steam under 10 lb pressure. The temperature and treatment time required to destroy spores varies with each kind of food.

The types of bacteria present on raw foods depends on the type of food, the time of year, the locality, and the conditions under which the food is produced. Because some of the most heat-resistant bacteria are in the soil, special care is required in the preparation and processing of foods such as spinach and green beans.

Bacterial Spoilage of Protein Foods

Usually aerobic bacteria break down proteins to simple substances by a series of oxidation reactions. The end products are completely oxidized and no odoriferous compounds are noticeable.

In contrast, the breakdown of proteins by anaerobic bacteria results in end products that are not completely oxidized to stable compounds. Termed *putrefaction,* the anaerobic breakdown of proteins generally is accompanied by offensive odors due to the formation of indole, mercaptans, hydrogen sulfide, and ammonia, especially in canned meats and vegetables with low acidity.

Yeasts

Yeasts are responsible for much of the fermentation in fruits and fruit products, which eventually causes the food material to become sour. Yeasts are easily killed by heat during preheating or processing. A few heat-resistant varieties of yeasts have been found in canned orange juice and tomato products.

The acid tolerance of certain yeasts makes them important in *foamy cream,* a term used to describe the gassy condition in sour cream due primarily to the growth of *Candida pseudotrophicalis (Torula cremoris).* Foaminess in cream is caused by a mixed fermentation, the yeasts producing gas and some other species, perhaps bacteria, causing protein coagulation by acid production.

Molds

Molds will grow on many kinds of food, especially where temperature, air, and humidity are favorable for their growth. Although mold growth may be visible only on the surface of a food, it often changes the flavor and quality of the entire contents of a jar.

Molds are easily killed by moist heat. A temperature of 160°–180°F (71°–82°C) for 60 min is usually sufficient to kill most vegetative molds. Experimental results show that a temperature of 212°F (100°C) for several minutes will destroy a large percentage of mold spores. Thus, mold spores are unable to survive the temperature used in the processing of most foods.

SPOILAGE IN CANNED FOODS

Underprocessing of canned foods may result in microbial spoilage, especially if highly heat-resistant spores are present. Heat-resistant spores generally are associated with foods having a pH above 4.5. Sporeforming organisms commonly present in canned foods include thermophilic types such as *Bacillus stearothermophilus,* an aerobic flat-sour organism; an aerogenic anaerobe, *Clostridium thermosaccharolyticum;* and the nonaerogenic hydrogen sulfide-producing anaerobe *Clostridium nigrificans.* Mesophilic organisms include the putrefactive anaerobes and possibly a few aerobic sporeformers. Aciduric bacteria and yeasts may be present in acid foods that receive only a short heat treatment, which these organisms can survive.

Types of Canned Food Spoilage

Spoilage in canned foods can be classified accoridng to the nature of the changes involved (e.g., formation of gas or acid) or by the causative agent (e.g., aerobic or anaerobic bacteria). In some cases, the same type of change (e.g., gas production resulting in swelling of cans) may be caused by different bacteria or even by chemical reactions.

In this section, spoilage in canned foods is classified into the following types: flat-sour spoilage; putrefaction and sulfide spoilage; swells and other changes caused by aerobic organisms; and chemical and physical swells and other defects.

Flat-Sour Spoilage

Flat-sour spoilage is characterized by the production of acid and no gas. The causative bacteria are usually facultative anaerobes, although obligate thermophiles may be involved.

Low-acid foods such as peas, corn, and snap beans are particularly susceptible to flat-sour spoilage. The food has a sour taste and a slight disagreeable odor may be present, which may be the only indication that any change has taken place. As a rule, there is no obvious change in the physical appearance of the food.

The bacteria causing flat-sour spoilage grow best at temperatures of about 130°–140°F (54°–60°C). This type of spoilage frequently occurs when foods are not cooled quickly after canning or are held at too high storage temperatures before and after canning. The organisms causing this type of spoilage are fairly widely distributed in nature, so their introduction into canned food products may not be unusual.

Microscopic examination of a food with flat-sour spoilage reveals rod-shaped bacteria characteristic of this defect. Spores are usually present, but they are not formed in the presence of acid. Thus, flat sours is suspected in a food if rod-shaped bacteria are present that produce acid without gas in glucose broth incubated at 130°F (54°C) and if spores are produced on a neutral culture medium.

A flat-sour condition in beets, caused by a small group of mesophilic bacteria, imparts a distinct medicinal flavor to the product and a black color when small amounts of dissolved iron are present. A combination of causal organism, beet pigment, and iron in solution cause the black color, which may extend throughout the beet.

Bacillus coagulans, sometimes referred to as *Bacillus thermoacidurans,* is responsible for flat-sour spoilage in tomato juice. This is an exception because spoilage of this kind is not common in acid foods. As the organism grows in the tomato juice a phenolic-like flavor develops and the pH of the juice drops from 4.5 to 3.5. A microscopic examination shows large vegetative rods, although the acid environment will destroy the organisms rapidly. As a rule no spores are formed in the product. The causative organism responsible for this type of spoilage may be identified by microscopic examination, a change in pH of the product, and a characteristic off-flavor.

Putrefaction and Sulfide Spoilage

Putrefaction is most likely to occur in meats and low-acid vegetables, especially asparagus. This spoilage is easily recognized by a very bad odor, presence of gas (bulged lid), and the softening and darkening of the canned food.

Gas-forming thermophilic anaerobes cause putrefaction, which often is accompanied by sulfide spoilage. These organisms produce acid and gas, which is largely hydrogen. An affected can may swell until the seam is ruptured. The food is sour and many times has a distinct rancid odor caused by butyric acid.

Bacillus stearothermophilus is a common cause of putrefaction in spinach and asparagus (semi-acid foods) and in low-acid foods. Microscopic examination of affected foods reveals long vegetative rods. When cultured in a liver–agar shake tube, the organisms form gas. The organisms will form spores when cultured in a neutral liver–agar culture medium.

Sulfide spoilage is characterized by the formation of hydrogen sulfide and blackening of the food. It is most common in low-acid foods such as corn and peas. The causative organisms produce hydrogen sulfide from the decomposition of proteins containing sulfur and thus impart a "rotten egg" odor to the food. This defect can be easily detected by the odor and black appearance of the product caused by the iron sulfide formed.

The activity of anaerobic bacteria in canned food products is indicated by the following:

- Development of offensive odors
- Formation of black sediment or residue

- Reduction in amount of dissolved or free O_2
- Reduction in amount of available O_2
- Increase in carbonaceous (or oxidizable) matter

The putrefactive anaerobe group also includes several *Clostridium* species. From a public health standpoint, *C. botulinum* is the most significant. This organism is a typical sporeformer and occurs in nearly all types of soil. During growth, it produces a powerful exotoxin. Ingestion of toxin-containing foods causes botulism in humans. The symptoms of botulism and ways to guard against it are discussed in Chapter 21.

Clostridium sporogenes and *C. putrefaciens* also are common anaerobic bacteria widely distributed in the soil. Their presence in canned vegetables and meat products in large numbers usually is evident by the production of gas and acid formation in some instances, as well as by offensive odors and undesirable biochemical changes in the product.

Swells and Other Changes Caused by Aerobes

Spoilage by gas-forming aerobic organisms causes cans to swell. An affected container tends to swell at the ends or in the middle, especially along the seams where liquor from the food is forced out by internal pressure.

A slight defect or weak point in the seam of a can is often an ideal place for aerobic spoilage organisms to enter. If such a defect is present, non-sporeforming bacteria, molds, or yeasts may enter the container in air or water sucked in during cooling. Therefore, seals on jars and tin cans should be carefully checked. It is recommended that sample cans be taken off the line right after sealing for examination, which should include "tear down" of the can.

Coliform bacteria may be involved in aerobic spoilage, provided sufficient oxygen is present. When conditions are favorable for their growth, such as the presence of carbohydrates, proteins, or both, acid and gas formation is possible in sufficient amounts to bulge the container.

Various changes may occur in foods contaminated with aerobic organisms. The food may have a mold mat on the surface. It may be slimy or it may be frothy. Sometimes a pungent, sour odor is evident. A swelling of cans caused by gas formation is very common; usually the food contents will leak out through the defective seam. In glass jars, bubbles of gas are evident (Fig. 20.1), and the contents may leak out around the lid and rubber gasket. Any product showing signs of such leakage should not be consumed.

The aerobic organisms involved in this type of spoilage usually are

Fig. 20.1. Cloudy liquid and bubbles of gas (*right*) indicate spoilage of food. Normal jar appears at left.

fairly heat sensitive and normally are destroyed during processing. In general, then, these types of changes are associated with contamination after processing.

Chemical and Physical Swells and Other Defects

In some cases, hydrogen gas may be produced in canned products by a chemical reaction between the metal of the container and biological acid present in the food. A pH of 4.0 is most favorable for such reactions. If sufficient gas is formed, the can will swell.

Swelling also may result from strictly physical processes, under the following conditions:

- Overfilling the can at too low a temperature causing the container to bulge when room temperature is reached.
- Filling the can under a reduced pressure such as at high altitude. This would tend to cause collapse rather than swelling.
- Freezing the liquid portion of the food and thus causing an expansion of the can.

When oxygen is not completely exhausted from a can before sealing, it causes an oxidation reaction that eventually detins the can. Marked corrosion on the interior of containers is likely if sulfides are present in high-acid foods.

External Appearance of Canned Foods

Certain abnormalities in the external appearance of a canned food indicate that a microbiological or chemical reaction has occurred in the product. Several specialized terms are used for these defects. In addition to these, any mechanical injury to a can may lead to biological activity in the contents. Consequently, all foods in damaged containers should be suspected of having some type of food spoilage.

Breather is a container that is bacteriologically sealed but not completely airtight. This condition results in a slow interchange of atmospheric air with the food contents in the can. Naturally, spoilage, if it does occur, takes place after a long time.

Springers are cans with a marked bulging at one or both ends. Gas-producing bacteria, chemical swells, or not properly exhausting the air from the can during the filling and sealing process may be responsible for this condition.

Flippers usually result from excessive mechanical pressure exerted on the can externally. As a rule, the food in the container does not undergo any spoilage.

Buckles, which occur most often in large cans, are caused by formation of a vacuum set up inside the container. The condition is most serious when it causes the seam to open, thus providing entry for aerobic spoilage organisms.

Determining Whether Canned Foods Are Safe

Since canned foods, like all foods, are subject to various kinds of spoilage, they should be examined before use to be sure they are wholesome. By following the procedure described below, one can determine whether a canned food is safe to eat:

1. Inspect the can or jar before opening. Both ends of tin cans should be flat or curved slightly inward; all seams should be tight with no trace of leakage. In the case of glass jars, metal lids should be firm and flat or curved slightly inward; there should be no signs of leakage around the rubber ring or elsewhere.

2. As the can or jar is opened, notice whether there is an inrush or

an outrush of air. Spoilage is indicated when air rushes out or the liquid spurts.

3. Smell the contents at once. The odor should be characteristic of the food. An "off-odor" probably means spoilage.

4. Examine the food carefully to see that it appears sound and natural in texture and color. The broth over canned meat and chicken may or may not be jellied. Liquids in all foods should be clean. Any change from the natural texture and color indicates spoilage.

5. If the can is tin, notice the appearance of the inside. It should be smooth and clean or well lacquered and not corroded. Foods may be left in a tin can after the can is open provided the food is kept covered and cool. The same care should be given to food stored in an opened can as is given to any other cooked food.

6. If foods have passed the above tests, *boil all low-acid vegetables and meats for 10 min before using or tasting.* Boiling for at least 10 min removes the danger of botulism.

7. Discard or destroy all food showing signs of spoilage. Spoiled low-acid foods should be burned.

SPOILAGE IN MEAT AND OTHER UNPROCESSED FOODS

Nearly all raw foods are subject to microbial spoilage under certain conditions. Creating an unfavorable environment for microbial growth or survival is the basic principle behind various food preservation methods.

Drying, one of the oldest methods for preserving foods, is effective because a lack of moisture is not favorable for microbial growth. Salting, an accepted method for preserving meat and fish, is effective because most microorganisms are sensitive to high salt concentrations. Perhaps the most widely used preservation method is canning, even though the high temperatures required to kill organisms often cause changes in the flavor and texture of food products. Refrigeration helps to preserve food by retarding the growth of microorganisms, but it does not destroy them.

When meat is placed in cold storage, mold growth usually begins at 38°F (3.3°C) when the relative humidity is high. Usually, the meat surface becomes musty in odor and later develops a characteristic flavor. Mold growth on meat can be trimmed off, but this represents an economic waste. Bacteria involved in meat spoilage usually grow at temperatures above 40°F (4.4°C). A slimy film on the surface of the meat is a good indication that bacterial spoilage is present. Figure 20.2 illus-

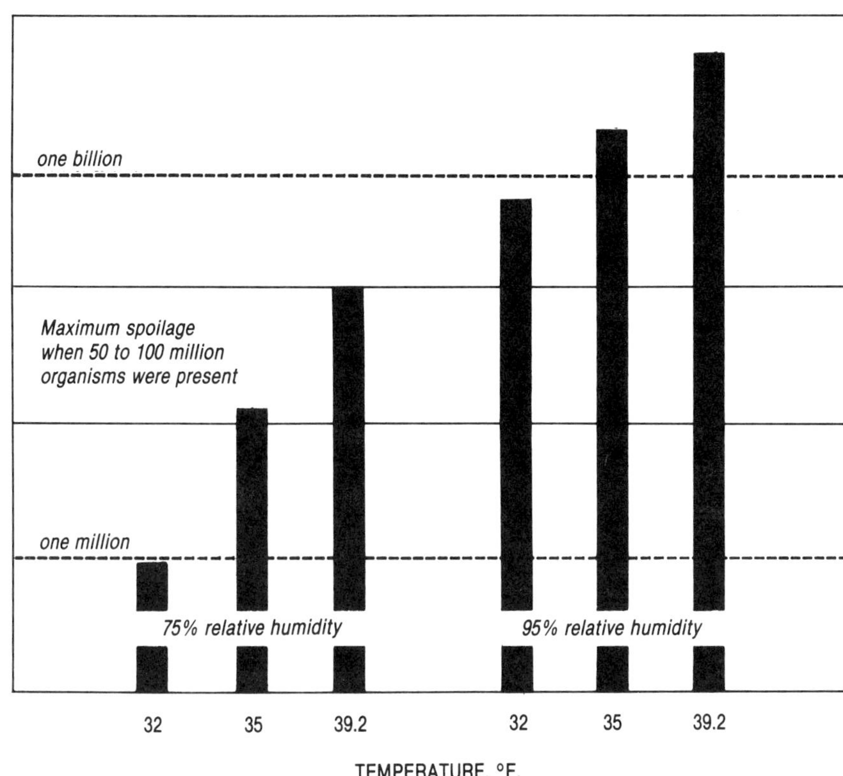

Fig. 20.2. Effect of temperature and relative humidity on growth of microorganisms in meat.

trates the effect of temperature and relative humidity on microbial growth in meat.

Spoilage in various types of foods is discussed in earlier chapters on specific foods. Refer to these chapters for information on the causative organisms and conditions favoring spoilage in different types of food products.

The physical appearance of food often indicates if spoilage or other contaminants are present. The following guidelines are helpful for detecting foods that should not be consumed and for handling foods safely:

Fruits and Vegetables: The appearance of white or grayish powder around stems of fruit and at the junction of leaves and stems of cabbage, cauliflower, celery, and lettuce indicates spray residues. All fruits and vegetables must be washed before eaten or cooked. Cooking will not destroy the toxicity of spray chemicals.

Cereals: Spread the cereal on brown paper. If insects are present, they can be easily observed. If one insect is seen, the entire batch of cereal should be destroyed. Insects are not necessarily dangerous, but neither are they appetizing.

Salads: Chicken salad, tuna and other fish salads, and nonacid potato salad must be refrigerated at all times. Refrigeration will keep infection from increasing. Spoilage is often impossible to detect until foods are completely spoiled. Salads should be served immediately after they are removed from the refrigerator.

Frozen Foods: Frozen foods should not be removed from the cold locker until they are ready to be served. Allowing frozen foods to stand after thawing favors growth of organisms that may induce spoilage in the product. Frozen vegetables should be thoroughly cooked before serving.

Leftover Foods: Discoloration, off-odor, and mold growth are common signs of spoilage in leftover foods. All highly perishable foods, even if cooked, that have not been refrigerated below 45°F (7°C) may be dangerous to eat, especially if the food is discolored. Abnormal odors in spoiled food are not always apparent. Do not eat leftover food stored for more than 36 hr unless it is cooked again.

Fish: Common signs of spoilage in fish are off-odor, gray or greenish gills, sunken eyes, flesh that easily pulls away from the bones; soft flesh in which a fingernail indentation remains, lack of rigidity.

Shrimp: A pink color on the upper fins and near the tail and ammonia-like off-odor are signs of spoilage in raw shrimp. Some types of shrimp are naturally pink; cooked shrimp are also pink. Both are wholesome if the odor is not abnormal.

Beef and Pork: Spoilage in beef and pork is commonly accompanied by off-odors and a slimy feel. Beef usually spoils first on the surface. Pork spoils first at the meeting point of bone and flesh in the inner portions. To test for spoiled pork use a pointed knife to reach the interior of the meat. An off-odor on the knife usually indicates spoilage.

Dressed Poultry: Spoilage in dressed poultry is indicated by off-odor under the wings, at the point where legs and body join, and on the upper surface of the tail. Darkening of the wing tips is another common sign of spoilage. Dressed poultry should be washed thoroughly with clean water before cooking. Wash your hands after handling.

MICROBIOLOGICAL ANALYSIS OF FOODS

Selective groups of microorganisms are omnipresent in many kinds of foods regardless of the manner in which they are processed. It is essential to recognize the sources of contaminating organisms and also to understand the selective effect of temperture, moisture, pH, and processing procedures.

Humans may be the most common source of pathogenic organisms in foods, although *Salmonella* are naturally present in some foods such as frozen eggs and meats. Fortunately, the low incidence of enteric disease is the result of cooking of foods prior to eating. Federal inspection of meats eliminates many potential infections in man caused by contaminated meat.

The coliform tests used in water analysis are not applicable to foods, except where these organisms are found in foods previously heated to temperatures sufficient to destroy vegetative cells. Total bacterial count does not measure safety of foods but should be a part of the quality control program.

There is no relationship between the bacterial counts in processed foods and those processed by the consumer. High counts may be caused by improper storage temperatures. For example, research on chicken pies stored at $-13°F$ ($-25°C$) for 3 months showed no increase in psychrophilic or mesophilic counts. However, when pies were stored at $32°F$ ($0°C$), the bacterial counts increased appreciably. At $41°F$ ($5°C$) the psychrophilic count increased slowly for the first 5 days, then increased very rapidly. At $68°F$ ($20°C$) similar products spoiled in 2 days, with mesophilic bacteria predominating. When 2.8 billion *Escherichia coli*, 15 billion *Streptococcus faecalis*, and 15 million *Staphylococcus aureus* were inoculated into chicken pies, there were no appreciable increases in the counts after freezing for 24 hr.

Selective culture media may give erroneous results in determining the quality of food from the bacteriological standpoint. The selective culture medium used in water analysis for detecting lactose-fermenting organisms cannot be used with foods. Ice cream and citrus juices containing sucrose or dextrose will give a large number of false positive tests. Samples of food tested for *Salmonella* may appear negative when large quantities of food material are inoculated into selenite F, tetrathionate, or other selective media, yet the results may be positive if small inocula are used. The removal of soluble materials in the supernatant by centrifuging is an erroneous factor interfering with the specificity of the selective medium.

A few common foods may contain *Salmonella, Micrococcus,* or *Clostridium* species, which can produce typical gastrointestinal syndromes if certain numerical levels are reached. These three genera are introduced by (1) human carriers, (b) live animal carriers, and (3) animal products used as foodstuffs.

Human carrier rates of *Salmonella* are not known, but outbreaks

from this source occur each year; this in turn increases the number of
carriers since salmonella organisms may be harbored for from several
weeks to several years. Coagulase-positive staphylococci are present in
the nasal tracts of 50–60% and on hands of 15–20% of the general
population. Approximately 10–20% are the bacteriophage and serological types associated with food poisoning. *Clostridium perfringens*
associated with food-poisoning outbreaks is carried by only 2–5% of
the normal population. Strains of *C. perfringens* are present in 15–20%
of dogs, cats, and hogs. It is well to measure total anaerobic as well as
aerobic flora. *Clostridium perfringens* poisoning may take place in the
presence of low aerobic and high anaerobic counts. Animal products
such as milk, eggs, and meat are likely to contain *Salmonella,* which
may be associated in some way with manufacturing processes. Staphylococci are more intimately associated with the flora of the normal and
abnormal udder.

One should keep in mind that foods containing certain pathogens
or indicator species do not necessarily cause disease. Food containing
Salmonella, C. perfringens, E. coli, fecal streptococci and staphylococci
may be found on the market with no associated outbreaks. However,
foods of this type are usually dirty and are more likely to cause disease
than clean foods.

Microbiological analyses of products in food-processing plants
should be conducted routinely to determine the types and levels of contamination. Based on the results, corrective measures can be taken to
improve hygiene and sanitation in a plant. Some of the more common
bacteriological procedures that should be performed include the following:

- Aerobic plate counts in tryptone–dextrose–yeast extract agar
- Proteolytic aerobic counts by drop plates on tryptone–yeast ex-
  tract–gelatin agar
- Anaerobic counts by two methods
- Coliform counts in crystal violet–neutral red bile–lactose–mannitol
  agar or in McConkey's bile–lactose–peptone water with subsequent
  confirmation
- Fecal streptococci in Packer's crystal violet–sodium azide–blood
  agar at 103°F (39°C) with subsequent confirmation
- *Salmonella* by enrichment in both selenite and tetrathionite broths
  followed by confirmation
- *Staphylococcus aureus* on 7.5% NaCl agar followed by confirmation
- Detection of illegally added preservatives and antibiotics by the *Saccharomyces cerevisiae* fermentation test

Errors of sampling and failure to revive organisms injured by processing must be recognized. Interpretation of data can only be made after careful study of all factors concerned with survival of organisms in foodstuffs.

SELECTED REFERENCES

Anon. 1968. "Laboratory Manual for Food Canners and Processors," 3rd ed. Vol. I, Microbiology and Processing. AVI Publishing Co., Westport, Connecticut.

Anon. 1985. An Evaluation of the Role of Microbiological Criteria for Foods and Food Ingredients. National Academy Press, Washington, DC.

Goldblith, S. A., Joslyn, M. A., and Nickerson, J. R. 1961. "An Introduction to the Thermal Processing of Foods." AVI Publishing Co., Westport, Connecticut.

Jacobs, M. B. 1951. "Spoilage of Canned Foods. The Chemistry and Technology of Food and Food Products," 2nd ed. Interscience Publishers, New York.

Kaufman, A. W. 1947. Deteriorative changes in some food specialties. Proc. Subsistence Research and Development Laboratory. Quartermaster General, U.S. Army, Chicago, Illinois.

Lopez, A. 1981. "A Complete Course in Canning," Book 1-Basic Information on Canning. 11th ed. Canning Trade, Inc., Baltimore, Maryland.

Pivnick, H., and Bird, H. 1965. Toxigenesis by *Clostridium botulinum* types a and e in perishable cooked meats vacuumed packed in plastic pouches. *Food Technol.* **19**, 132–140.

Splittoesser, D. F., and Wettergreen, W. P. 1964. The significance of coliforms in frozen vegetables. *Food Technol.* 18 (3), 134–136.

21
Foodborne Illness

Etiological Agents 313
Foodborne Infections 315
  Salmonella Gastroenteritis 315
  Streptococcal Gastroenteritis 317
  Minor Types of Food Poisoning 317
  Communicable Diseases Transmitted through Foods 318
Foodborne Intoxications 321
  Staphylococcal Intoxication 323
  Botulism 324
  Other Food Intoxications 326
Chemical Food Poisoning 327
Epidemiology of Foodborne Illness 327
  Preventing Foodborne Illness 329
Selected References 335

The general term *foodborne illness* refers to all diseases and other disorders that can result from consumption of food. Foodborne illnesses can be categorized into four types based on the causative (etiological) agent: food idiosyncrasies, foodborne infections, foodborne intoxications, and chemical food poisoning. The term *food poisoning,* which is commonly used, generally refers only to foodborne infections and foodborne intoxications, which are caused by microorganisms.

ETIOLOGICAL AGENTS

Many people suffer from allergies, which are hypersensitivities or reactions that some people show toward specific substances, called allergens. When an allergen is taken in with food, a hypersensitive response is called a *food idiosyncrasy*. A variety of foods are known to cause a hypersensitive response. The symptoms will vary with each individual from a slight reaction to a severe shock. Food idiosyncrasies are due to *anaphylaxis,* which is a hypersensitivity produced in an animal sensitized to an allergen in such a way that when the allergen is introduced later a violent reaction takes place, sometimes resulting in death. Many individuals are sensitive to different foods. The reactions may occur in the form of skin rashes, bronchial asthma, or intestinal disturbances.

Such foods as whole wheat bread, eggs, strawberries, cheese, tomatoes, and peanuts are often involved in food idiosyncrasies. The only remedy for this unpleasant discomfort is to avoid eating foods known to contain an allergic factor. In some cases, individuals allergic to eating or handling wheat flour may be desensitized by being given small injections of wheat protein. Eventually, the person may eat modest amounts of wheat bread daily. This same phenomenon occurs with eggs, strawberries, cheese, milk, and other foods. If an allergen is an essential food, the problem of desensitization is more complicated. Desensitization is absolutely necessary if the food cannot be eliminated from the diet.

The two types of foodborne illness commonly called food poisoning are caused by microorganisms. The first type, *foodborne infection*, occurs when food containing certain microorganisms is eaten. These organisms enter the alimentary tract where they establish an infection. Sometimes organisms enter the bloodstream and are carried to other organs in the body, although ordinarily this does not occur. Salmonella gastroenteritis, dysentery, scarlet fever, brucellosis, hepatitis, and typhoid fever are typical foodborne infections. Usually, no symptoms appear until the organisms have had time to localize and grow in the host. This is called the *incubation period.* After a period of incubation, symptoms may appear, which may serve to identify the type of food poisoning involved.

The second type of bacterial food poisoning is known as *foodborne intoxication*, which occurs when toxin-containing food is consumed. During growth of certain bacteria, an exotoxin (poison) is formed and liberated in the food. The toxin, not the bacteria themselves, causes the symptoms of illness. In order for sufficient toxin to cause symptoms to be present, the bacteria must grow and actively produce toxin for some time in a food. However, once the food is ingested, symptoms occur in a very short time if sufficient toxin is present. Herein lies a marked distinction between foodborne infection and intoxication. The former depends upon a varying period of incubation of the organisms in the human host before any characteristic symptoms are observed, while symptoms of the latter appear quite rapidly after ingestion of affected food because the toxin is already present.

Certain foods contain toxic substances that may cause so-called *chemical food poisoning.* These toxic substances may be natural metabolites of a plant or animal food, or they may be added by human activities. Examples of the first type include clams during some seasons, certain toadstools (often confused with edible mushrooms), and milk from cows that have ingested certain weeds, which impart a poison to the milk but do not affect the cow. The most common example of added toxic substances is pesticide residues. If pesticides are not properly used, they may remain on crops in sufficient amounts to cause various symptoms of poisoning after the food is ingested.

FOODBORNE INFECTIONS

The etiology and symptoms of the common foodborne infections are listed in Table 21.1. Generally, the term food poisoning refers to salmonella or streptococcal gastroenteritis, which are diseases accompanied by vomiting, diarrhea, nausea, and abdominal pains. In addition, a number of communicable diseases (e.g., cholera, dysentery, scarlet fever, trichinosis) may be transmitted by contaminated foods. The symptoms of these diseases usually differ, at least in some respects, from those of common food poisoning, and for many of them the primary mode of transmission is not via foods.

Salmonella Gastroenteritis

Salmonella species have been implicated as the causative agents in many outbreaks of food poisoning. Probably all serotypes of *Salmonella* are pathogenic for man, some for animals, and a few for both man and animals. Over 1200 serotypes are known.

Although salmonella gastroenteritis is an infection, its incubation time after ingestion of contaminated food is quite short. The reason for this is that salmonella bacteria are capable of multiplying very rapidly in certain kinds of food, so their numbers in contaminated foods is likely to be high. Thus, once contaminated food is ingested and the organisms find their way into the intestinal tract, the typical symptoms become evident in a short time. The mechanism of food poisoning by *Salmonella* is not well understood.

The onset of symptoms of food poisoning after ingestion (the incubation period) varies from 12 to 72 hr, depending on the number of salmonella organisms present in the ingested food and the susceptibility of the individual. The greater the number of organisms, the more severe the infection and the shorter the incubation time.

Salmonella food infection is usually characterized by severe headache followed by chills. As the symptoms progress, vomiting, abdomi-

Table 21.1. Characteristics of Common Foodborne Infections

Illness	Causative agent	Food usually involved	Incubation period	Symptoms
Shigellosis (bacillary dysentery)	*Shigella* spp.	Moist prepared foods; milk and other dairy products contaminated with excreta	Usually 2–3 days	Diarrhea, bloody stools, fever in severe cases
Cholera	*Vibrio comma*	Fecally contaminated food and water	2–5 days	Nausea, vomiting, diarrhea, and abdominal cramps
Brucellosis (undulant fever)	*Brucella abortus, B. melitensis,* or *B. suis*	Raw milk or dairy products contaminated with raw milk, animal contact (meat)	3–21 days; sometimes several months	Chills, sweats, weakness, malaise, headache, fever, muscle and joint pains, and loss of weight
Diphtheria	*Corynebacterium diphtheriae*	Milk contaminated from human sources	3–7 days	Insidious onset, inflammation of throat and nose
Scarlet fever and septic sore throat	Beta-hemolytic streptococci	Food contaminated with nasal or oral discharges and milk from cows having udder infections	1–7 days	Fever, sore throat, sometimes rash
Salmonella gastroenteritis	*S. typhimurium* *S. enteritis* *S. cholera suis* *S. newport* and others	Meat, poultry salads, and egg products	12–72 hr	Abdominal pain, diarrhea, chills, fever, vomiting, and prostration
Streptococcal gastroenteritis	*Streptococcus fecalis;* other enterococci	Food contaminated with excreta or human carrier	2–18 days	Nausea, vomiting, pains, and diarrhea
Typhoid fever	*Salmonella typhi*	Any food contaminated with excreta from human case or carrier	Usually 7–21 days	Malaise, lack of appetite, headache, fever
Paratyphoid A.	*S. paratyphi* A.	Same as for typhoid fever	1–10 days	Same as for typhoid fever
Tuberculosis	*Mycobacterium tuberculosis* (human and bovine types A and B)	Raw contaminated milk and other dairy products	Variable	Depends on part of body affected
Tularemia	*Pasteurella tularensis*	Wild game animals	3–10 days	Similar to typhoid fever
Trichinosis	*Trichinella spiralis*	Raw pork or similar products	36–72 hr	Sudden onset, headache, chills, body pains, fever, vomiting, swollen lymph glands, and loss of appetite

nal pain, diarrhea, and rise in temperature usually follow. The symptoms and effects are much more pronounced in young children than in adults. Fortunately, with proper medical care the infection lasts only about a week. However, weakness may persist for 2–3 weeks.

The human element plays an important role in the transmission of salmonella food poisoning. Food handlers, in particular, may be carriers of and transmit the usual causative species, *S. typhimurium, S. enteritis, S. newport, and S. choleraesuis,* all of which are frequently present in the intestinal tract. Unclean personal habits, especially after using a toilet, can result in contamination of food during its preparation, handling, and serving. Infected rats and mice also may be involved, as well as cockroaches and flies, in transmitting these organisms. Salads, especially those containing eggs, pastries, and nonfat dried milk solids are frequently the foods involved in salmonella food poisoning.

Salmonellae can be destroyed by thorough cooking. The greatest danger involves uncooked foods and prepared food contaminated by food handlers. A lack of refrigeration of prepared foods enables the organisms to increase in large numbers. High counts of *Salmonella* in foods are necessary to induce a severe case of food poisoning.

Streptococcal Gastroenteritis

Streptococcus faecalis (alpha type) also cases an infection-type food poisoning when large numbers of organisms are ingested. The incubation period ranges from 2 to 18; symptoms include nausea, abdominal pains, diarrhea, and vomiting, but no appreciable rise in temperature.

Streptococcus faecalis grows at 55°–115°F (13°–46°C), shows a high degree of salt tolerance, and causes a greenish discoloration when grown on blood agar. Pasteurizing temperatures will destroy a high percentage of these organisms.

The usual sources of *S. faecalis* are the intestinal contents of man, mammals, and birds. Poulty dressings that have not been properly cooked after being stuffed into the bird have been involved (Jensen 1954). Adequate heating of food, followed by refrigeration, is necessary in order to destroy the organisms. Foodborne outbreaks are due largely to infected food handlers.

Minor Types of Food Poisoning

Proteus vulgaris and *P. mirabilis* have been isolated in cases of food poisoning traced to fish, ham, and sausage. Escherichia–Aerobacter microorganisms also have been suspected in some food-poisoning outbreaks. Although they have been isolated and identified from characteristic cases of food poisoning, the role they play is not well understood and there is no convincing evidence that they are directly responsible for food poisoning.

Communicable Diseases Transmitted Through Foods

During harvesting, processing, handling, and distribution, foods may become contaminated with various pathogenic bacteria from intestinal discharges of infected persons or even carriers. Typhoid fever and dysentery are typical examples of bacterial diseases transmitted through food contaminated by humans. Milk and other dairy products may be involved in the transmission of pathogenic bacteria from an infected cow or human being, or from carriers who handle the food product before it reaches consumers. Cows may harbor certain species of streptococci that may be largely responsible for septic sore throat, scarlet fever, and other streptococcal infections in man.

Certain strains of *Mycobacterium tuberculosis,* especially the bovine type, may be responsible for a high incidence of tuberculosis in children, since children are large consumers of milk. Fortunately, the detection and elimination of infected cows and the universal pasteurization of milk, have practically eliminated the dairy cow as a source of this disease.

Causative organisms of brucellosis belong to the genus *Brucella. Brucella abortus* causes contagious abortion in cattle and swine, and undulant fever in man; *B. suis* infects swine, chickens, horses, dogs, cows, monkeys, and man; and *B. melitensis* infects goats and man. However, the symptoms of brucellosis in various animals are different.

Like tuberculosis in cattle, brucellosis can be detected in dairy cows, and thus infected animals may be eliminated. The brucella organisms are easily destroyed when exposed to sunlight. Studies show that these organisms may remain viable in cream stored at 50°F (10°C) for several days. They also resist drying and remain alive in dust for long intervals of time. The pasteurization of milk makes it safe from brucellosis.

There is evidence to show that the causative organisms of scarlet fever, tuberculosis (human type), typhoid fever, and certain kinds of dysenteries may come from human sources and thus be a potential source of danger in foods from a public health standpoint.

Tularemia, an infection produced by *Pasteurella tularensis,* can be transmitted to man by direct contact or by the ingestion of uncooked rabbit and squirrel meat. Infection through the alimentary tract is not common and it produces a syndrome similar to typhoid fever without localized lesions.

Many human intestinal infections may come from water polluted with raw sewage. Oysters, clams, and many kinds of shell fish may become contaminated when they grow in polluted waters. Insects, rodents, cats, and dogs sometimes are involved in the contamination of foods.

Bacillary dysentery, cholera, amoebic dysentery, and trichinosis, all of which may be transmitted through foods, have symptoms rather similar to typical food poisoning caused by *Salmonella* and *Streptococcus faecalis*. For this reason, it is very important to identify the causative organisms in any outbreak of foodborne infection. What may be suspected as typical food poisoning may in fact be something entirely different.

Bacillary Dysentery

Little is known about the natural habitat of the *Shigella* species that cause bacillary dysentery. However, in areas where typhoid fever and cholera are controlled, bacillary dysentery remains endemic. The infection appears to be common where people are closely associated (e.g., prisons, army camps, and summer resorts).

The isolation and identification of dysentery bacilli from feces is difficult primarily because of the rapid disappearance of the organisms from the stools during convalescence. Undoubtedly many individuals are undetected carriers. Bacillary dysentery may be spread by food contaminated with feces from carriers or from those suffering from the disease.

The incubation period for bacillary dysentery is usually 2–3 days. The recovery and identification of the bacilli from stools is usually sufficient to differentiate the disease from typical food poisoning. Fortunately, the incidence of bacillary dysentery caused by contaminated foods is quite low, according to public health reports.

Cholera

In man cholera is acquired by ingestion of the spirillum *Vibrio comma*, usually found in food or water contaminated by fecal material. The ingested organisms multiply in the small intestine and after 2–5 days cause a sudden onset of nausea, vomiting, diarrhea, and abdominal cramps.

Amoebic Dysentery

Another disease that may be confused with food poisoning is amoebic dysentery. Its insidious and chronic nature usually differentiates it from typical food poisoning. The incubation period after the ingestion of the amoeba varies from a few days to 3–4 months. The symptoms of amoebic dysentery in man vary from a mild diarrhea to an acute bloody discharge from the bowels.

Endamoeba histolytica, a protozoan, is the causative agent of amoebic dysentery. It is good laboratory procedure to test for the protozoan and exclude it before making a final diagnosis of food poisoning. Its presence in stools, however, should be interpreted with some reservation because about 10% of the population is known to be infected, but many of these persons suffer no noticeable symptoms.

The disease usually occurs sporadically and may continue to occur over a long time. The Chicago epidemic in 1933 was waterborne and was due primarily to faulty plumbing and back siphonage of sewage, which occurred in water lines in certain hotels. However, the disease usually is endemic and spread by intimate personal contact.

Although amoebic dysentery, like bacillary dysentery, may be transmitted through foods, this is not a common mode of transmission. Where good sanitary practices are followed in processing and handling food, little fear should arise from the dangers of foodborne amoebic dysentery.

Trichinosis

Trichinosis is another disease that may be confused with typical food poisoning. The causative agent is *Trichinella spiralis,* a nematode, which is usually obtained by man from poorly cooked and infected pork (see Chapter 4). When the encysted larvae are eaten, the cysts are digested in the stomach and the larvae enter the duodenum, where they localize in the duodenal and jejunal mucosa. The symptoms include nausea, vomiting, diarrhea, dysentery, colic, and profuse sweating.

It is important that the infection be recognized abut 48 hr after the ingestion of the larvae and be differentiated from typical food poisoning. A case history of eating uncooked pork and the finding of encysted nematodes in the food is important in diagnosis. Dack (1956) states that eosinophilia is suggestive of trichinosis. During the intestinal phase, adult worms are usually recovered from the feces. During and after larval migration, the larvae localize in muscle strips removed at biopsy.

Apparently there is no way to exclude trichinella from pork at the time of slaughter. The one sure safeguard against this disease is thorough cooking of pork.

Viral Diseases

Although infectious hepatitis can be transmitted in foods, information is meager concerning foodborne transmission of the virus causing this disease. Fecal–oral routes have been suspected, especially since persons are known to be fecal carriers. The incubation period may range from 10 to 40 days. The symptoms are usually headache, general malaise, fatigue, nausea, chills, vomiting, and jaundice in some cases.

No effective therapeutic measures are available for infectious hepatitis. However, since the symptoms are quite similar to those of salmonella and streptococcal gastroenteritis, it is recommended that the hepatitis agent be tested for in any outbreak of food poisoning. Unfortunately, the laboratory procedure for doing so is difficult and inconclusive.

Many authorities believe that hepatitis, poliomyelitis, coxsackie, herpes simplex, mumps, influenza, and perhaps other viruses are present in the alimentary tract of healthy individuals. One group affects the nose, throat, and lungs and is known either as the ARD (acute respiratory disease) group or the ADC (adenoidopharyngo-conjunctival) group.

Another group of viruses are called ECHO (enteric cytopathogenic human orphan) viruses. They are viruses of unknown classification and unknown reactions, hence orphans. Little is known about the extent to which any of these viruses is transmitted in foods.

FOODBORNE INTOXICATIONS

A second major type of food poisoning is foodborne intoxications. As noted earlier, these are caused by ingestion of foods containing bacterial toxins. The most common species responsible for toxin formation are *Staphylococcus aureus, Clostridum botulinum,* and *C. perfringens* (Table 21.2).

The staphylococci are aerobic and non-sporeformers, while the clostridia are anaerobic sporeformers. Toxins produced by anaerobic bacteria (especially *C. botulinum*) are extremely potent and may even cause death, whereas aerobic bacterial toxins are much less potent and rarely cause death. On the other hand, anaerobic toxins are more easily destroyed by heat than aerobic toxins. Since the aerobic toxin-producing

Table 21.2. Characteristics of Common Foodborne Intoxications

Illness	Causative agent	Foods usually involved	Incubation period	Symptoms
Staphylococcal intoxication	Enterotoxin from *S. aureus*	Meats, food rich in carbohydrates, salads, and warmed-over foods	2–11 hr	Nausea, vomiting, diarrhea, and abdominal cramps
Botulism	Exotoxin from *Clostridium botulinum* and *C. parabotulinum*	Home-processed foods and contaminated canned foods with pH over 4.5	12 hr to 6 days	Dizziness, double vision, muscular weakness, difficulty in swallowing, speech and respiration
Clostridium perfringens	Type A exotoxin from *C. perfringens*	Cold and reheated meats, water, milk, salt rising bread. Found in intestinal tract of man and animals	8–22 hr (variable)	Acute abdominal pains, diarrhea, nausea, and vomiting rare

species are all non-sporeformers, they generally are destroyed by the heat treatments commonly employed in canning. In contrast, the anaerobic species produce heat-resistant spores, which may survive the canning process if it is not done carefully.

Although it is generally assumed that staphylococcal and botulinum toxins are relatively sensitive to heat, some investigators have questioned this. Since the heat stability of all bacterial toxins has not been accurately determined and may even vary depending on conditions, any suspected food should be discarded. Preparing and cooking foods far in advance of consumption and storing them at room temperature should be avoided in all cases. Such practices are conducive to growth of many food-poisoning organisms.

Staphylococcal Intoxication

Many cases of food poisoning are caused by *Staphylococcus* species, particulary *S. aureus.* The toxin liberated by these bacteria is called an *enterotoxin* (*entero* = intestine) because it can produce a marked irritating effect on the gastrointestinal tract. It is still quite difficult to detect the different strains of enterotoxin-producing organisms even by cultural, biochemical, or serological tests. Enterotoxin formed by these organisms is solely responsible for staphylococcal food poisoning. It has been determined that staphylococcus enterotoxin B is a simple protein with a molecular weight of about 35,000, which is very soluble in water and dilute salt solutions.

Perhaps one unique feature of staphylococcal food poisoning is the short incubation period; usually about 3 hr, after infected food is consumed, symptoms appear. The typical symptoms are nausea, vomiting followed by abdominal cramps, and severe diarrhea. Fortunately, the symptoms are of short duration, and in most instances recovery occurs within 20–48 hr. Because of the quick onset and short duration of staphylococcal intoxication, it can be readily distinguished from infection-type food poisoning even though many of the symptoms are similar.

Food rich in carbohydrates serves as an excellent medium for the growth of staphylococci, provided no acid-producing bacteria are present, as acid may inhibit their growth. Bakery products such as eclairs, cream puffs, custard-filled doughnuts, and cream and custard pies are examples of such food.

Meat products of various kinds are potentially dangerous, especially leftover roast turkey and chicken that is improperly refrigerated. Precooked hams also can be involved in this type of food poisoning. Various dairy products such as milk, cream, cheese, and ice cream have been reported as causing staphylococcal infection if not properly handled. If foods to be frozen are not kept cold enough before freezing, staphylococci can multiply and form toxins before freezing. Thawing frozen food and holding it at room temperature for several hours also should be avoided, since staphylococci will grow well at such temperatures.

Many outbreaks of staphylococcal food poisoning are caused by unrefrigerated foods, especially custard-filled bakery products. Food held at temperatures below 45°F (7°C) is usually safe because enterotoxin is not formed. The presence of enterotoxin in food cannot be detected easily, since it does not impart an unusual odor or taste to food.

In one study, more than 1000 food handlers were examined to detect the presence of staphylococci in throat and nasal passages. Approximately 60% of the individuals examined harbored staphylococci, and 20–25% of these carried coagulase-positive staphylococci (a pathogenic strain). These results emphasize the danger of staphylococcal contamination by food handlers.

Botulism

From a public health standpoint, botulism may be the potentially most serious type of food poisoning for several reasons. The causative organism is *Clostridium botulinum*, an anaerobic sporeformer that produces a powerful exotoxin called *botulin*.

The spores of *C. botulinum* are extremely heat resistant and may remain viable after exposure to boiling water for several hours. Esty and Meyer (1922) reported that *C. botulinum* spores were destroyed by exposure to the following temperatures for the indicated times:

$$4 \text{ min at } 248°F(120°C)$$
$$10 \text{ min at } 239°F(115°C)$$
$$32 \text{ min at } 230°F(110°C)$$
$$100 \text{ min at } 221°F(105°C)$$
$$330 \text{ min at } 212°F(100°C)$$

Young spores of *C. botulinum* are more resistant to heat than old spores. This may be true of all bacterial spores.

If the spores are ingested in a food, they will not reproduce in humans. Likewise, the vegetative cells of *C. botulinum* will not multiply in the intestinal tract nor cause any ill effect in humans.

Properties of Botulin

The exotoxin formed by *C. botulinum* is absorbed from the intestinal tract into the bloodstream. It has a specific affinity for the nervous system and, thus, is classified as a *neurotoxin*.

At least six types of exotoxin may be produced by *C. botulinum*, designated as A, B, C, D, E, and F. Types A and B toxins are chiefly responsible for human cases of botulism, while types C and D affect cattle, horses, sheep, and goats. It appears that type C is responsible for "limber neck," a term used to designate botulism in domestic fowl and wild ducks. Biochemists have obtained type A toxin in the form of pure white needle-shaped crystals. One gram of the crystal form of this toxin would furnish about 8,000,000 lethal doses. About 68% of humans ingesting botulin die.

Fortunately, in view of its lethality, botulin can be inactivated by exposure to 175°F (79°C) for 30 min or to 212°F (100°C) for a few minutes. Unfortunately, the presence of the toxin generally is not indicated by any obvious signs of spoilage. Sanitarians have recommended heating any foods suspected of containing botulin for 10–20 min at boiling temperatures with frequent stirring.

Distribution of C. botulinum *in Foods*

Clostridium botulinum is widely distributed in soil and, thus, is commonly present on fruits and vegetables grown in or near soil. Although the number of organisms will vary with different types of soil and climatic conditions, soil in the Great Plains area contain type A toxin-producing *C. botulinum*, whereas in the Great Lakes region and the Atlantic states, type B organisms are the principal form.

Botulinum organisms can grow in canned foods providing the pH is near the neutral point. Low-acid foods such as green beans, corn, beets, and meats serve as excellent material for the growth of *C. botulinum*.

Botulinum organisms usually do not grow in foods having a pH as low as 4.5, such as tomatoes, fruits, pickles, and sauerkraut. Any spores present in such foods remain viable for long periods. If molds or acid-tolerant bacteria proliferate, using acids as a source of energy, the pH may increase enough so that spores germinate and vegetative cells grow and liberate the toxin.

There is very little danger of botulism from fresh foods, cooked or raw, because the spores and vegetative cells are harmless when present in food. *Clostridium botulinum* can grow and produce exotoxin only under anaerobic conditions or reduced oxygen tension such as are found in sealed containers.

Symptoms of Botulism

Symptoms occur 12 hours to several days after ingestion of botulin. Usually fatigue and muscular weakness are the first symptoms followed by double vision (diplopia), drooping of the upper eye lids, dilated pupils, dryness of the mouth, swelling of the tongue, persistent constipation and, finally, difficulty in swallowing and speaking. There are usually no gastrointestinal disturbances, and death results from a paralysis of the respiratory muscles. Consequently, in fatal cases the victim suffocates because of an inability to breathe.

Precautions to Avoid Botulism

Most cases of botulism in the United States can be traced to inadequately processed home-canned foods that are consumed without cooking after removal from the jar.

All home-canned meats and low-acid vegetables (green beans, corn, beets) should be removed from the can, brought to the boiling point, and boiled actively for at least 10 min before any portion is eaten. Do not even taste a small amount before boiling it. One should be sure that all parts of the mass of food in the vessel have been thoroughly cooked. Such treatment destroys any toxin present. This type of heating does not, however, destroy the vegetative cells of *C. botulinum.* So any food left over should be reboiled for another 10-min period before being used.

Although botulism is relatively rare, considering the amounts of canned food that are consumed, all possible safety precautions should be taken because of the very serious consequences of this type of food poisoning.

Other Food Intoxications

Clostridium perfringens also produces an exotoxin, which causes symptoms quite similar to those of streptococcal infection. Although not as serious as botulism, *C. perfringens* intoxication is much more common.

Bacillus subtilis has been reported to form a toxin when grown in meat. In the mid-1950s *Bacillus cereus* caused four outbreaks of food poisoning involving 600 persons who ate custard made from commercial cornstarch containing large numbers of the organism.

The fungus *Claviceps purpurea,* commonly known as ergot, produces several toxic alkaloids. Ergotism results from eating bread made of rye on which the fungus has grown. Symptoms are itching, muscle cramps, gangrene, and convulsions. Toxins produced by fungi are called mycotoxins. They are discussed in the next chapter.

CHEMICAL FOOD POISONING

As mentioned at the beginning of this chapter, some foods contain substances that can cause various symptoms of poisoning when such foods are consumed. Common examples of naturally occurring toxic foods are as follows:

- Inedible mushrooms produce a poisonous substance called muscarine or phallin. Symptoms are nausea, vomiting, thirst, and later convulsions.
- Oxalic acid, which occurs in rhubarb leaves, causes poisoning when the leaves are eaten as cooked greens.
- Milk sickness, or trembles, follows consumption of milk from cows that have grazed on white snake root. Symptoms are vomiting, constipation, profound weakness, and later alkalosis.
- Shellfish poison is probably produced by a protozoan (flagellate) on which shellfish feed. Symptoms are muscle weakness and paralysis, including respiratory paralysis. Paralytic shellfish poisoning is discussed in detail in Chapter 12.

The addition of toxic substances to foods also may result in chemical food poisoning. Several chemical agents involved and their most common sources are listed in Table 21.3. Obviously, any toxic substance deliberately or accidentally added to a food may cause poisoning symptoms after the food is ingested.

EPIDEMIOLOGY OF FOODBORNE ILLNESS

Although mild cases of foodborne illness probably often are unrecognized as such or unreported, more serious cases, especially if associated with outbreaks involving many individuals, are documented by public health authorities. Table 21.4 contains data on the etiologic agents involved in outbreaks of foodborne illness in 1982. In these years, as is generally true, *C. perfringens* and *Salmonella* were the leading causes of foodborne illness, both in terms of number of outbreaks and number of individual cases.

Table 21.3. Toxic Substances That Cause Chemical Food Poisoning

Chemical	Source
Arsenic	Deliberately added to food
Antimony	Food prepared in poor-quality gray enamel pans
Cadmium	Acid foods cooked in cadmium-plated utensils
Zinc	Acid foods cooked in galvanized utensils
Lead	Same as arsenic
Sodium fluoride	Cockroach powder spilled into foods or mistaken for baking powder or powdered milk
Germicides, herbicides, fungicides	Residues on foods after improper use or accidentally added
Antibiotics	Antibiotics used in livestock production

Table 21.4. Confirmed Foodborne Disease Outbreaks, Cases, and Deaths, by Etiologic Agents, United States, 1982

Etiologic agent	Outbreaks No.	Outbreaks (%)	Cases No.	Cases (%)	Deaths No.	Deaths (%)
BACTERIAL						
Bacillus cereus	8	(3.6)	200	(1.8)	0	(0.0)
Brucella	1	(0.5)	3	(<0.1)	0	(0.0)
Campylobacter jejuni	2	(0.9)	31	(0.3)	0	(0.0)
Clostridium botulinum	21	(9.5)	30	(0.3)	5	(20.9)
Clostridiium perfringens	22	(10.0)	1,189	(10.8)	0	(0.0)
Escherichia coli	2	(0.9)	47	(0.4)	0	(0.0)
Salmonella	55	(25.0)	2,056	(18.6)	8	(33.3)
Shigella	4	(1.8)	116	(1.1)	0	(0.0)
Staphylococcus aureus	28	(12.7)	669	(6.0)	0	(0.0)
Streptococcus Group A	1	(0.5)	34	(0.3)	0	(0.0)
Vibrio cholerae O1	1	(0.5)	892	(8.0)	11	(45.8)
Vibrio cholerae non-O1	1	(0.5)	7	(0.1)	0	(0.0)
Vibrio parahaemolyticus	3	(1.4)	39	(0.4)	0	(0.0)
Yersinia enterocolitica	2	(0.9)	188	(1.7)	0	(0.0)
Total	151	(68.7)	5,501	(49.9)	24	(100.0)
CHEMICAL						
Ciguatoxin	8	(3.6)	37	(0.3)	0	(0.0)
Heavy metals	5	(2.3)	26	(0.2)	0	(0.0)
Monosodium glutamate	3	(1.4)	10	(0.1)	0	(0.0)
Mushrooms	4	(1.8)	9	(0.1)	0	(0.0)
Scombrotoxin	18	(8.2)	58	(0.5)	0	(0.0)
Shellfish	1	(0.5)	5	(<0.1)	0	(0.0)
Other	8	(3.6)	75	(0.7)	0	(0.0)
Total	47	(21.4)	220	(1.9)	0	(0.0)
PARASITIC						
Trichinella spiralis	1	(0.5)	4	(<0.1)	0	(0.0)
Total	1	(0.5)	4	(<0.1)	0	(0.0)
VIRAL						
Hepatitis A	19	(8.5)	325	(2.9)	0	(0.0)
Norwalk virus	2	(0.9)	5,000	(45.2)	0	(0.0)
Total	21	(9.4)	5,325	(48.1)	0	(0.0)
CONFIRMED TOTAL	**220**	**(100.0)**	**11,050**	**(100.0)**	**24**	**(100.0)**

Table 21.5 shows an analysis of the contributing factors associated with outbreaks of foodborne illness in 1982. Improper holding temperatures were the most frequent contributing factor. As shown in Table 21.6, outbreaks of foodborne illness in 1982 were acquired most often in restaurants, homes, or schools.

As emphasized throughout this chapter, most cases of foodborne illness ultimately can be traced to direct contamination of foods by humans or to improper handling of food. Few cases result from inadequate processing procedures (e.g., canning, pasteurization), with the exception of home canning. Some poor food handling practices, which can result in transmission of foodborne illnesses, are summarized in Table 21.7.

In any significant outbreak of foodborne illness, public health authorities attempt to determine the source and causative agent involved. This information can serve as a basis for corrective measures to prevent future outbreaks. The steps involved in the detection and diagnosis of bacterial food poisoning are outlined in Table 21.8.

Preventing Foodborne Illness

Throughout this chapter, various precautionary measures that help to prevent foodborne illnesses have been mentioned. A summary of these guidelines follows:

1. Food handlers should observe strict standards of personal cleanliness and follow sanitary procedures in the use of utensils and preparation of foods.
2. A known carrier of enteric pathogens should not be allowed to handle or prepare food for human consumption.
3. Any person suffering from an upper respiratory infection should keep away from foodservice operations.
4. Foods should be kept at suitable temperatures in order to ensure a minimum of bacterial growth.
5. All foods should be protected from insects, rodents, and vermin.
6. Cream-filled pastries and custards should not be eaten in public eating places unless they are refrigerated.
7. Home-canned low-acid foods should be processed under steam pressure.
8. Canned foods that have an off-flavor should not be used unless they are first boiled for at least 12–15 min.
9. All fresh fruit and vegetables should be washed thoroughly before eating.

Table 21.5. Foodborne Disease Outbreaks by Etiologic Agent and Contributing Factors, United States, 1982

Etiologic agent	Number of reported out-breaks	Out-breaks in which factors reported	Improper holding tempera-tures	Inade-quate cooking	Contami-nated equip-ment	Food from unsafe source	Poor personal hygiene	Other
BACTERIAL								
B. cereus	8	5	5	—	—	—	1	—
Brucella	1	—	—	—	—	—	—	—
C. jejuni	2	2	—	—	—	1	—	1
C. botulinum	21	4	1	—	—	—	—	3
C. perfringens	22	20	19	8	3	—	—	2
E. coli	2	1	1	—	—	—	—	—
Salmonella	55	34	16	18	6	6	7	4
Shigella	4	4	1	—	—	—	4	—
S. aureus	28	21	20	3	4	1	9	1
Streptococcus Group A	1	1	1	—	—	—	1	—
Streptococcus Group D	1	—	—	—	—	—	—	—
V. cholerae O1	1	—	—	—	—	—	—	—
V. cholerae non-O1	1	1	—	1	—	—	—	—
V. parahaemolyticus	3	2	—	—	—	2	—	—
Y. enterocolitica	2	—	—	—	—	—	—	—
Total	151	95	64	30	13	10	22	11

330

CHEMICAL								
Ciguatoxin	8	1	—	—	—	1	—	—
Heavy metals	5	4	—	—	2	—	1	2
Monosodium glutamate	3	2	—	—	—	—	—	2
Mushrooms	4	2	—	—	—	2	—	—
Shellfish	1	1	—	—	—	1	—	—
Scrombotoxin	18	9	5	—	—	5	—	—
Other	8	6	—	—	3	—	—	3
Total	47	25	5	—	5	9	1	7
PARASITIC								
Trichinella spiralis	1	1	—	1	—	1	—	—
Total	1	1	—	1	—	1	—	—
VIRAL								
Hepatitis A	19	14	2	2	—	6	9	3
Norwalk virus	2	2	—	—	—	—	2	—
Total	21	16	2	2	—	6	11	3
CONFIRMED TOTAL	220	137	71	33	18	26	34	21
UNKNOWN	436	225	87	73	35	94	59	16
TOTAL 1982	**656**	**362**	**158**	**106**	**53**	**120**	**93**	**37**

Table 21.6. Foodborne Disease Outbreaks, by Specific Etiologic Agent and Place Where Food Was Eaten, United States, 1982

Etiologic agent	Home	Delica- tessen, cafeteria, or restau- rant	School	Picnic	Church	Camp	Other	Un- known	Total
BACTERIAL									
B. cereus	—	3	2	—	—	1	2	—	8
Brucella	1	—	—	—	—	—	—	—	1
C. jejuni	1	—	—	—	—	—	1	—	2
C. botulinum	21	—	—	—	—	—	—	—	21
C. perfringens	—	9	1	—	—	—	12	—	22
E. coli	—	—	—	—	—	—	2	—	2
Salmonella	12	17	6	3	4	1	12	—	55
Shigella	1	2	—	—	1	—	—	—	4
S. aureus	4	7	2	2	2	1	10	—	28
Streptococcus Group A	—	—	—	—	—	—	1	—	1
V. cholerae O1	1	—	—	—	—	—	—	—	1
V. cholerae non-O1	1	—	—	—	—	—	—	—	1
V. parahaemolyticus	1	—	—	1	—	—	1	—	3
Y. enterocolitica	1	—	—	—	—	—	1	—	2
Total	44	38	11	6	7	3	42	—	151

CHEMICAL									
Ciguatoxin	3	5	—	—	—	—	—	—	8
Heavy metals	1	3	—	—	—	—	1	—	5
Monosodium glutamate	1	2	—	—	—	—	—	—	3
Mushroom	3	—	—	—	—	—	1	—	4
Shellfish	1	—	—	—	—	—	—	—	1
Scrombotoxin	7	9	—	—	—	—	2	—	18
Other chemical	4	4	—	—	—	—	—	—	8
Total	20	23	—	—	—	—	4	—	47
PARASITIC									
Trichinella spiralis	1	—	—	—	—	—	—	—	1
Total	1	—	—	—	—	—	—	—	1
Viral									
Hepatitis A	5	6	—	1	—	1	6	—	19
Norwalk virus	—	1	—	—	—	—	1	—	2
Total	5	7	—	1	—	1	7	—	21
CONFIRMED TOTAL	70	68	11	7	7	4	53	—	220
UNKNOWN	127	221	17	16	2	1	49	3	436
TOTAL 1982	197	289	28	23	9	5	102	3	656

Table 21.7. Food Handling and Public Health[a]

(a) Unsanitized glasses, dishes, and silverware. Exposure of food to coughing, sneezing, spitting:
    Common cold
    Diphtheria
    Encephalitis
    Measles
    Mumps
    Pneumonia
    Poliomyelitis
    Scarlet fever
    Septic sore throat
    Tuberculosis
    Whooping cough
	(b) Use of common cup, towels, and toilet articles by employees:
	    Common cold
	    Influenza
	    Septic sore throat
	    Syphilis
	    Tuberculosis
	    Typhoid fever

(c) Failure of employees to wash hands thoroughly after each visit to toilet:
    Dysentery, bacillary
    Dysentery, amoebic
    Food infection and poisoning
    Paratyphoid fever
	(d) Employment of people who are carriers[b]
	    Dysentery, bacillary
	    Dysentery, amoebic
	    Paratyphoid fever
	    Tuberculosis
	    Typhoid fever

(e) Improper washing, preparation, and refrigeration of foods:
    Botulism
    Dysentery, bacillary
    Dysentery, amoebic
    Food infection and poisoning
    Paratyphoid fever
    Typhoid fever
	(f) Employment of personnel in preparation of foods with sores and boils on hands or arms:
	    Food infection and poisoning

(g) Use of ungraded milk and milk products:
    Diphtheria
    Dysentery, bacillary
    Paratyphoid fever
    Scarlet fever
    Septic sore throat
    Tuberculosis
    Typhoid fever
    Undulant fever (brucellosis)
	(h) Use of uncertified oysters:
	    Paratyphoid fever
	    Typhoid fever
	    Infectious hepatitis

(i) Use of contaminated drinking water and home-made drinks:
    Dysentery, bacillary
    Dysentery, amoebic
    Paratyphoid fever
    Typhoid fever
	(j) Lack of screening, exposure of food to flies:
	    Dysentery, bacillary
	    Dysentery, amoebic
	    Typhoid fever
	    Paratyphoid fever

(k) Improper storage of food, lack of rat control:
    Paratyphoid fever
    Typhus fever
    Infectious jaundice
    Food infection and poisoning
	(l) Plumbing and water supply incorrectly connected:
	    Dysentery, amoebic
	    Dysentery, bacillary
	    Paratyphoid fever
	    Typhoid fever

[a] From Food Handler's Training Program, Texas State Board of Health.
[b] Carrier: A person who, without symptoms of a communicable disease, harbors and disseminates the specific microorganisms.

Table 21.8. Detection and Diagnosis of Bacterial Food Poisoning

Primarily Clinical
1. Secure complete list of cases involved.
2. Obtain details and complete history in each individual case, if possible.
3. Ascertain vehicle of infection by appropriate laboratory techniques.
4. Study the history of implicated food.
5. Obtain evidence as to the manner in which the food was contaminated.

Laboratory Examination
1. If possible, obtain samples of infected food consumed.
2. Examine blood, liver, spleen, and intestinal contents of fatal cases.
3. Examine feces and blood of suspected carriers.
4. If the bacteriological findings are negative, then an examination for the presence of harmful chemicals should be made.

Epidemiology of Foods
1. Prepare a spot map of the area involved.
2. Complete bacteriological examination of the milk supply in the area.
3. Bacteriological examination of the water supply, especially shallow wells and wells not properly protected from surface water.
4. Record age groups of persons, also incidence and number of cases involved.

SELECTED REFERENCES

Anon. 1950. Ordinance and Code Regulating Eating and Drinking Establishments. U.S. Public Health Service Publ. *27.*

Dack, G. M. 1956. "Food Poisoning." 3rd ed. University of Chicago Press, Chicago, Illinois.

Esty, J. R., and Meyer, K. F. 1922. The heat resistence of the spores of *B. botulinus* and allied anaerobes. *J. Infect. Dis.* **31,** 650–653.

Guthrie, R. K. 1980. "Food Sanitation," 2nd ed. AVI Publishing Co., Westport, Connecticut.

Jensen, L. B. 1954. "Microbiology of Meats," 3rd ed. Gerrard Press, Champaign, Illinois.

Lewis, K. H., and Angelotti, R. 1964. Examination of Foods for Enteropathogenic and Indicator Bacteria. U.S. Dept. of Health, Education, and Welfare, Washington, DC.

22
Mycotoxins

Specific Mycotoxins 338
   Aspergillus Toxins 338
   Penicillium Toxins 340
   Fusarium Toxins 340
   Other Mycotoxins 340
Human Mycotoxins 340
Selected References 343

The term *mycotoxin* is derived from the two Greek words, *mykes* (mushroom) and *toxikon* (poison), and is used to describe the group of fungal metabolites that are toxic to animals and in most cases probably to humans. It has now been established that 148 or more species of microfungi produce mycotoxins. Only a few mycotoxins have actually been implicated as causative agents in human diseases, but it is believed that many more are toxic.

Mycotoxicosis is the general name given to diseases caused by ingesting mycotoxins; the term *mycosis* refers to diseases in which mold itself actually grows in the organism. Although most human outbreaks of mycotoxicosis come from direct consumption of mold-contaminated foods, toxins in feeds eaten by livestock can be transmitted through the animal tissues as well as through their milk and eggs. Table 22.1 lists a number of mycotoxins, and Table 22.2 lists the mold sources for some of these.

A number of human epidemics in the past are now known to have been caused by consumption of foods contaminated with mycotoxins. In the Middle Ages, a disease referred to as St. Anthony's Fire was caused by the fungus *Claviceps purpurea,* commonly known as ergot, which parasitizes rye and other grasses. An outbreak of alimentary toxic aleukia (ATA) in the Orenburg Province of Russia during World War II was caused by the human population consuming grain that had become moldy because it was left in the field until spring. In this outbreak, up to 10% of the population was affected, and mortality rates were as high as 60%. The molds *Fusarium poae, Fusarium sporotri-*

Table 22.1. Mycotoxins Classifed by Degree of Toxicity[a]

High	Medium	Low
*Aflatoxins	Aspergillic acid	Fumagillin
*Citreoviridin	*Aspertoxin	Fusaric acid
*Cyclochlorotine	*Butenolide	Gentisic acid
Diacetoxyscirpenol	Citrinin	Geodin
Ergot alkaloids	Emodin	Glauconic acid
*Fusarenone-X	Gliotoxin	Griseofulvin
*Luteoskyrin	Iridoskyrin	Kojic acid
Maltoryzin	β-Nitropropionic acid	Mycophenolic acid
Nivalenol	Patulin	Oxalic acid
*Nidulotoxin	Penicillic acid	Trichodermin
Ochratoxin	Roridins	Xanthocillin
Rubratoxin	*Sterigmatocystin	Zearalenone
Sporidesmim	Slaframine	
Stachybotryotoxins	Verrucarins	
T-2 toxin		
Tremorgens		

[a]Asterisk indicates those mycotoxins known to cause chronic toxicity.

choides, and several species of *Cladosporium* were later identified as the causative agents. In epidemiological studies in Thailand, a close correlation was observed between incidence of liver cancer and consumption of food contaminated with mycotoxins from the mold *Aspergillus flavus* and other fungal toxins. A similar study in India suggested aflatoxins as the cause of infantile liver cirrhosis. Mycotoxins from different sources have been observed to be hemorrhagic, hepatoxic, carcinogenic, neurotoxic, and uterotrophic.

Fungi that contaminate foods are generally divided into two groups: parasites of field fungi and saprophytic or storage fungi. Because field fungi grow on the seeds before harvest and require high-moisture atmospheres for growth, they are more common during years of heavy rainfall and in moist tropical areas. Storage fungi, on the other hand, can grow even under low moisture conditions.

SPECIFIC MYCOTOXINS

Those mycotoxins currently considered to be the most important ones found in feed and food are discussed in this section.

Aspergillus Toxins

Aflatoxins are a group of closely related toxic compounds produced by the genus *Aspergillus.* Most are produced by *Aspergillus flavus,*

Table 22.2. Known Mold Sources of Selected Mycotoxins

Mycotoxin	Mold species		
Citrinin	*Penicillium ci-trinum*	*P. citreoviride*	*P. canescens*
	P. viridicatum	*P. notatum*	*P. velutinum*
	P. lividum	*P. steckii*	
	P. fellutanum	*P. palitans*	*Aspergillus niveus*
	P. implicatum	*P. claviforme*	*A. terreus*
	P. jenseni	*P. expansum*	*A. candidus*
Ochratoxin	*A. ochraceus*	*A. sclerotiorum*	*A. cyclopium*
	A. alliaceus	*A. sulphureus*	*P. variable*
	A. melleus		*P. purpurescens*
	A. ostianus	*P. viridicatum*	*P. palitans*
	A. petrakii	*P. commune*	
Patulin	*P. patulum*	*P. urticae*	*P. leucopus*
	P. claviforme	*P. melinii*	*P. novae-zealan-diae*
	P. expansum	*P. divergens*	*Byssochlamus fulva*
	P. cyclopium	*P. lapidosum*	*A. clavatus*
	P. griseofulvum	*P. equinum*	*A. giganteus*
			A. terreus
Penicillic Acid	*P. martensii*	*P. thomii*	*A. ochraceus*
	P. puberulum	*P. viridicatum*	*A. ostianus*
	P. cyclopium	*P. janthinellum*	*A. sulphureus*
	P. roqueforti		
Rubratoxin	*P. rubrum*		
	P. purpurogenum		
Sterigmatocystin	*Aspergillus versi-color*	*A. parasiticus*	*A. amstelodami*
	A. nidulans	*A. chevalieri*	*Bipolaris* (?)
	A. flavus	*A. ruber*	*P. luteum*
T-2 Toxin	*Fusarium tricinc-tum (F. poae)*	*F. roseum* "grami-nearum"	*(G. zeae)*
	F. solani	*F. rigidosum*	*Trichothecium vir-ide*
		F. lateritium	
	F. sporotrichioides	*(Gibberella zeae)*	*T. lignorum*
Zearalenone	*F. roseum* "grami-nearum"		*F. culmorum*
	F. moniliforme		*F. equiseti*
	F. tricinctum		*F. gibbosum*

which is found widely distributed in soils, forage, decaying vegetation, and stored foods and seeds. Cotton seed, peanuts, coconuts, olives, castor beans, and sunflower seeds provide ideal growth media for *A. flavus* because of their high oil content.

Ochratoxins, a group of toxins produced by *A. ochraceus, A. sulfureus,* and *A. melleus,* are common in soils and decaying vegetation but can also grow on stored grains. Ochratoxins are found on wheat, corn, cottonseeds, legumes, peppers, apples, onions, and pears.

Sterigmatocystin is a toxin produced by *A. versicolor*. It is not as toxic as the aflatoxins and ochratoxins. Several other substances, toxic to animals but of lesser concern, are produced by other *Aspergillus* molds.

Penicillium Toxins

Patulin is produced by seven species of *Penicillium*. One, *P. expansum*, causes storage rot in apples. Another patulin-producing mold, *P. patulum*, has been isolated from the stubble of mulched grain fields and from mixed feeds.

Eleven species of penicillium have been reported to produce penicillic acid. *Penicillium martensi* produces a blue-eyed mold and toxin in corn. Some data suggest that penicillic acid is a carcinogen.

Penicillium purpurogenum and *P. rubrum*, both produce rubratoxin. They have been isolated from cereals and legumes.

Fusarium Toxins

Zeralenone, a common mycotoxin contaminant in foods and feeds, causes vulvovaginitis in swine. *Fusarium graminarum* (*Gibrella zeae*), *F. trincinctum*, and *F. roseum culmorum* are three species that produce fusarium toxins.

Other Mycotoxins

Tremorgens and cyclopiazonic acid are uncharacterized toxins that produce acute neurotoxic side effects in the form of trembling and convulsions. They have been isolated from animal feeds.

Rice toxins are found in stored rice and usually are produced by organisms in Aspergillus and Penicillium.

The group of toxins consisting of 12,13-epoxy,9-tricothecenes causes severe skin irritation and hemorrhages on the lips, mouth, throat, and gastrointestinal tract.

HUMAN MYCOTOXICOSIS

Even though there have been some severe human food-poisoning epidemics from mycotoxins, the risk of large human epidemics of acute mycotoxicosis is relatively low. When mold growth on a food is enough to produce sufficient toxin to cause immediate pathogenic symptoms, the food is already unpalatable to humans. In the past, many of the epidemics of mycotoxicoses have occurred in populations weakened by war, stress, starvation, and malnutrition. Under normal circumstances, the contaminated foods would never have been consumed.

The effects of ingesting low levels of mycotoxins over extended periods are a matter of considerable concern, especially since many mycotoxins have been shown to be carcinogenic to animals. A number of foods have been shown to contain low levels of these toxins. Figure 22.1 presents the FDA program strategy for mycotoxins in foods.

Contamination of cereals has been demonstrated in several surveys. Shipments of nuts and oilseeds, especially those that have been mishandled during harvesting, processing, or storage, have repeatedly contained mycotoxins. The probability of mycotoxin contamination from these sources is greatly reduced by using stringent testing and selection procedures.

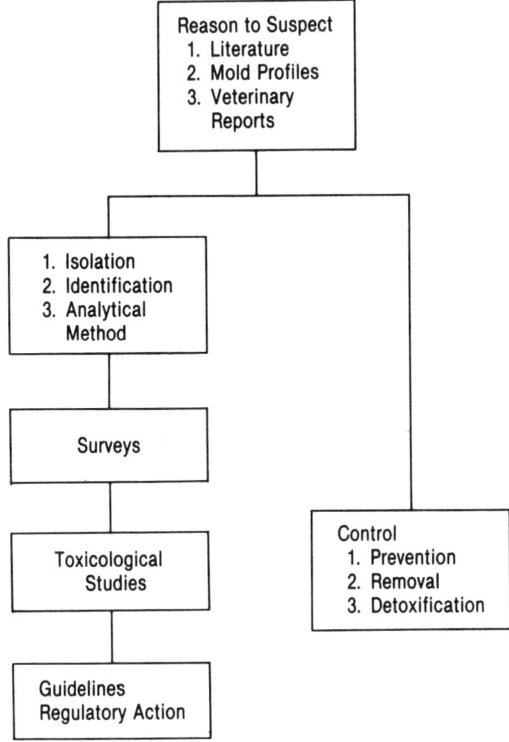

Fig. 22.1. FDA program strategy for controlling mycotoxins in foods.

Soybeans occasionally become contaminated with aflatoxins. When the beans are processed, the aflatoxin stays with the meal portion. The toxins are not destroyed when soybeans are fermented. Aflatoxins have also been found in peanut butter samples. Figure 22.2 is a scheme for aflatoxin control in groundnuts for direct human consumption.

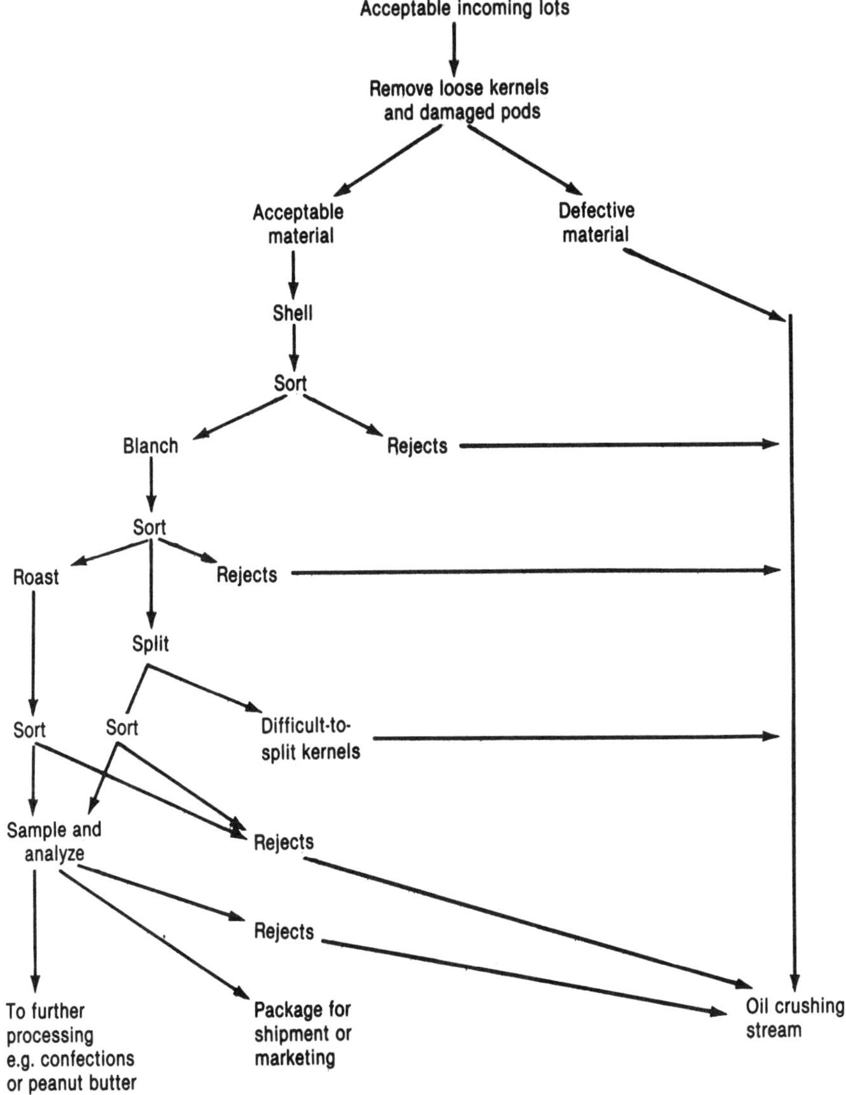

Fig. 22.2.   Scheme for aflatoxin control in ground nuts for direct human consumption.

Mold-contaminated fruits are immediately rejected by consumers, but there is the possibility that toxins could get into processed fruit products. Since low-quality fruits are frequently used for cider and other fruit juices there is an increased chance for contamination.

Animal products can also contain low levels of mycotoxins. Lactating animals fed moldy feed secrete modified aflatoxins in their milk. Aflatoxins can also be found in the meat from pigs, beef, and poultry fed contaminated feed. Fortunately, the strains of *Penicillium* used for starter organisms in cheese do not produce aflatoxins.

SELECTED REFERENCES

Bullerman, L. B. 1978. Significance of mycotoxins to food safety and human health. *J. Food Protect.* 42 (1), 65–86.

Campbell, T. C., and Stoloff, L. 1974. Implications of mycotoxins for human health. *J. Agric. Food Chem.* 22, 1006.

Davis, N. D., and Deiner, U. L. 1987. Mycotoxins. *In* "Food and Beverage Mycology," 2nd ed. L. R. Beuchat (Editor). Van Nostrand Reinhold Co., New York.

Jarvis, B. 1976. Mycotoxins in food. *In* "Food Agriculture, Fisheries, and Food," F. A. Skinner and J. A. Carr (Editors). Academic Press, New York.

Purchase, I. F. H. (Editor). 1974. "Mycotoxins." American Elsevier Publishing Co., New York.

Rodricks, J. V., and Lovett, J. 1976. Toxigenic fungi. *In* "Compendium of Methods for the Microbiological Examination of Foods." Am. Public Health Assoc., Washington, DC.

Index

Acetobacter
  acetic, 243
  ascendens, 243
  xylinum, 243
Achromabacter, 2, 3, 21, 51, 158,
    161, 164, 167
Achromobacteriaceae, 50
Acidified foods, 81
Acid in foods, 229
Acidophilus milk, 201
Acids
  acetic, 121
  benzoic, 122
  boric, 123
  citric, 123
  formic, 124
  germicidal effects of, 120
  hydrochloric, 125
  lactic, 124
  monochloroacetic, 126
  oxalic, 327
  propionic, 126
  pyroligneous, 125
  salicylic, 124
  sorbic, 127
  sulfurous, 125
  vanillic, 127
Aerobacter, 167
  aerogenes, 52, 193, 246, 262
Aerobes, 301, 303
Alcaligenes, 2, 108, 167
  viscosus, 187
  visolactics, 51
Algae, 61
Alkaline germicidal agents, 128
Alkali-producing organisms, 127
Alternaria tenuis, 58
Amoebic dysentery, 320
Anaerobic bacteria, 301
Anaphylaxis, 313

Aspergillus, 116, 165
  flavus, 338, 339
  glaucus, 57
  niger, 57, 137, 247
  ochraceous, 57
  ripens, 57
Ascomycetes, 61
Ascorbic acid, 114

Bacillaceae, 51
Bacillary dysentery, 319
Bacillus, 2, 3
  anthacis, 88, 89, 103, 218
  calidolactic, 188
  cereus, 46, 187, 326
  coagulans, 51, 302
  mesentericus, 46, 152, 261, 264
  polymyxa, 51
  stearothermophilus, 51, 300, 302
  subtilis, 51, 87, 246, 326
  thermoacidurans, 78, 256, 257
Bacteria, classification of, 49
  Achromobacteriaceae, 50
  Bacillaceae, 51
  Brevibacteraceae, 51
  Corynebacteriaceae, 51
  Enterobacteriaceae, 54
  Lactobacillaceae, 54
  Proprionibacteriaceae, 54
  Pseudomonadaceae, 50
  Streptococcae, 55
  thermoduric, 94
  thermophilic, 94
Balantidium coli, 66
Benzoic acid, 122
Blanching, 111
BOD, 289
Botrytis, 2
Botulism, 321, 324, 325, 326

346   Index

Breadmaking, 263
  moldy, 265
  red, 264
  ropy, 264
  sour, 264
Brevibacteraceae, 51
Brevibacterium lineus, 51
  erythrogenes, 51
Brucella, 188
  abortus, 103, 318
  melitensis, 318
  mycoides, 138
  suis, 318
Byssochlamys fulva, 2

Candida
  krusei, 61
  mycoderma, 61
  pseudotrophicalis, 300
Canned foods
  breather in, 305
  buckles in, 305
  flippers in, 305
  safety of, 305
  springers in, 305
Carbohydrates
  biological role, 9
  classification, 6
Catalase, 41
Catsup, 257
Cellulase, 38
Cellulomanas, 262
Cellulytic, 3
Cheese, 205
  classification, 209
  processing, 205
Chemical food poisoning, 312, 327
Cholera, 319
Chromobacterium, 161
Citric acid, 247
Cladosporium herbarum, 214
Claviceps puparea, 337
Clostridium, 2, 3
  botulinum, 2, 51, 77, 88, 90, 103,
    132–33, 172, 321, 324
  butyricum, 51, 138, 246

distribution, 325
heat resistance, 324
nigrificans, 301
perfringes, 2, 51, 263, 310, 321,
  326
precautions, 326
properties, 325
putrefaciens, 2, 303
sporogenes, 51, 188, 303
symptoms, 326
tetani, 253
thermosaccharolyticum, 51, 301
types, 325
tyrobutyricum, 188
Corynebacteriaceae, 51
Corynebacterium
  diphtheriae, 51
  pyogenes, 51
Cryptococcus, 167
  kefyr, 61
  utilis, 61
Cultures, butter, 192

Dairy products. *See* Milk

Eberthella, 167
Eggs and egg products
  antibacterial substance, 217
  bacteriostatic substance, 219
  dried, 223
  enzymes, 218
  microbial content, 213
  microflora, 215
  pasteurization, 222
  storage, 220
  structure, 212
Endamoeba histolytica, 65, 320
Enterobacteriaceae, 52
Enterobacter levans, 263
Enzyme, activity, 29, 112
  autolytic, 31
  carbohydrases, 35–36
  carboxylase, 43
  catalytic properties, 43
  chlorophyllase, 35

classification, 25
exoenzymes, 26
hydrolases, 33
hydrolytic, 33
industrial use, 31
lipases, 26, 34
oxidation-reduction, 40
peroxidase, 29
phosphatase, 35, 45
proteases, 26
proteolytic, 39, 124
Epiphytic, 261
Erwinia carotovora, 52, 254
Escherichia coli, 3, 52, 121, 154,
167, 174, 176, 177, 188, 250,
253, 258, 263, 285
Etiological agents, 313

Fats. *See* Lipids
Fermentations, 229
Fish and shellfish
lobsters, 173
microbiology of, 170
mollusks, 177
oysters, 174
preservation, 169
Flat sour, 301
Flavobacterium, 2, 108, 167, 262
Food composition, 5
Foodborne infection, 314
Foodborne illness, 313, 327
chemical poisoning, 314
detection, 335
and food handling, 334
outbreaks, 330
prevention, 328
Foodborne intoxication, 314, 321
Food spoilage
aerobic, 303
bacteria, 299
beef and pork, 308
canned foods, 300
cereals, 308
chemicals, 304
enzymatic, 30
fish, 308

flat sour, 301
fresh fruits and vegetables, 307
frozen foods, 308
leftovers, 308
lysozyme activity, 219
meats, 306
molds, 300
poultry, 308
protein foods, 299
salads, 308
sulfide, 302
swells, 303
yeasts, 300
Freezing
effects of, 99
eggs, 109
fish, 109
milk products, 111
poultry, 110
process, 104
Frozen foods
egg products, 107
fruits and vegetables, 106
meat products, 107
quality of, 111
refreezing, 115
thawing, 115
Fruit color, 112
Fruits
dried, 252
microflora, 250
natural protection, 249
spoilage, 251, 253
sulfuring, 252
Fruit juices, 253
Fungi imperfecti, 61
Fusarium, 254
poae, 337
sporotrichoides, 338

Geotrichum, 2
candidum, 58
Giardia lamblia, 66

Gonyaulax catenella, 178
Grains, microflora, 261, 262

Halophilic organisms, 2, 135

Irradiation, 145

Lactobacillaceae, 54
Lactobacillus acidophilus, 54, 101, 202
  arabinosus, 84
  brevis, 54, 230
  bulgaricus, 54, 61, 200, 201, 202, 203, 204, 263
  casei, 54
  cucumeris, 230, 231, 233, 237
  lactis, 204
  leichmannii, 54
  lycopersici, 256
  pentoaceticus, 54, 230, 231
  plantarum, 230, 231, 245
  thermophilus, 54
  viridescens, 163
Leuconostoc, 163
  citrororum, 56
  cremoris, 191
  dextranicum, 56, 138
  mesenteroides, 56, 138, 230, 231, 232, 233, 237, 245, 253
Lipase, 34
Lipids (fats and oils), 17
  in foods, 19
  oxidation of, 20
  saturated, 23
  unsaturated, 23
Lipolytic, 3

Meat
  microbial content, 158
  microorganisms, 158
  proteolytic enzymes, 164
  spoilage, 161
  tenderization of, 164

Meningococcus, 218
Mesophilic, 2
Micrococcus, 161
  halodenitic ficans, 136
Microbial activity factors
  inhibitors, 81
  moisture, 69, 78, 79, 80
  nutritional requirements, 75
  oxygen, 71
  pH, 76, 78, 79, 80
  temperature, 73
Microbiological analysis of food, 308, 310
Microbiology, science of, 1
Microorganisms, classification of, 2
  amylytic, 3
  halophilic, 2
  lipolytic, 3
  osmophilic, 2
  pectinolytic, 3
  proteolytic, 3
  psychrotropic, 2
  saccharolytic, 3
  thermophilic, 2
Microorganisms, temperature effect, 83, 84, 87, 89, 90, 91, 92
Microwave heat, 116
Milk
  butter, 204
  fermentations, 190
    cultures, 192
    starters, 191
  fermented milk
    acidophilus, 201
    Bulgarian, 200
    kefir, 203
    kumiss, 203
    ripened, 200
    taetta, 203
    yogurt, 202
  microbial inhibitors, 198
  pathogens, 188
  processing, 184, 189
  spoilage, 185
Minerals, 23
Molds, 57
  meat, 158

spoilage, 350
Monila nigra, 188
Mycobacterium tuberculosis, 45,
    103, 318
Mycodermi vini, 244
Mucor pyriformis, 247
Mycotoxicosis, 337
Mycotoxins, 261, 337

Neurospora sitophila, 58
Nitrates, 13
Nitrogen cycle, 12
Nitrosomonas, 13

Off-flavors, 113
Olives, 215
    green, 245
    ripe, 246
    spoilage, 246
Oospora, 167
    lactic, 188
Osmophilic, 2
Osmotic pressure, 131
Oxidases, 40
Oxidoreductases, 40

Parasitic organisms, 63
Pasteurella tularensis, 318
Pathogenic organisms, 3
Pectinases, 38
Pectinolytic, 3
Penicillium, 117, 167
    camemberti, 57
    expansum, 251, 252
    luterim, 247
    roqueforti, 57, 187
Peroxidase, 29
pH, 6, 9
Physalospora malorum, 251
Phythium debaryanum, 254
Pickles, 236
    defects in, 237
    microbial growth in, 237
    processing, 237

Plasmoptysis, 131, 136
Poultry, microbiology of, 165
Preservation, 119
Propionibacteriaceae, 54
Propionibacterium shermanii, 54
Proteolytic, 3
Proteins, 10
    classification of, 11
    decomposition of, 14
    in milk, 15
Proteus, 53, 108, 164, 167
    mirabilis, 317
    vulgaris, 53, 317
Pseudomonadaceae, 50
Pseudomonus, 108, 109, 164, 167
    aeruginosa, 121, 124
    fluorescens, 50, 108
    fragi, 188
    ichthyosmia, 188
    nigrificans, 50
    putrefaciens, 188
Ptomaines, 16
Putrefaction, 302

Radiation
    effects of, 144
    history of, 141
    sources of, 143
Rhizopus, 117
Rhodotorula, 167
    glutinus, 61

Saccharolytic, 138
Saccharomyces
    carlbergensis, 35
    cerevisiae, 35, 122, 137, 243, 263,
        310
    ellipsoideus, 154, 243, 270
    mellis, 2
    rouxii, 2
Salmonella, 53, 90, 165, 221, 222,
        226, 309, 310, 315
    aetrycke, 124
    choleraesuis, 317
    enteritis, 53, 152, 215, 317
    gastroenteritis, 315

Salmonella (cont.)
  montivideo, 285
  newport, 317
  paratyphi, 53, 198, 215
  pullorum, 53, 212, 213
  schottmuelleri, 53, 198
  typhi, 53, 92, 103, 121, 124, 177,
    198, 218, 253, 254, 285
  typhimurium, 53, 317
Salt
  calcium, 135
  magnesium, 135
  preservation, 132
Sarcina, 167
Sauerkraut, 230
  effect of salt, 232
      starters, 233
      temperature, 232
  microbial growth, 230
  spoilage, 234
      dark, 235
      pink, 236
      soft, 234
Serratia, 52
  marcescens, 188, 262, 264
Shellfish poisoning, 178
Shigella, 54
  dysenteriae, 54, 198
  paradysenteriae, 198, 250
Sodium benzoate, 257
Spices
  aromatic properties of, 269
  classification of, 268
  flavoring properties of, 269
  germicidal, 270
  "hot" properties of, 269
  storage of, 271
Staphylococcal intoxication, 323
Staphyloccus, 218, 262
  aureus, 90, 100, 102, 103, 107,
    121, 124, 152, 188, 310, 321,
    323
Sterols, 22
Streptococcus
  agalactiae, 55, 188
  citrovorus, 193

  cremoris, 55, 63, 191, 193
  faecalis, 55, 56, 187, 253, 317
  gastroenteritis, 317
  lactis, 55, 56, 63, 187, 191, 193,
    196, 203, 233
  liquefaciens, 55, 56, 187
  paracitrovorus, 191, 193
  pyogenes, 55
  thermophilus, 63, 202
Streptomyces, 167
Sugar, use of, 136
Sugars, microorganisms on, 8

Thermoduric, 2, 94
Thermophylic, 2, 94
Tomatoes and tomato products
  bacterial swell, 256
  canned, 256
  flat sour, 256
Torula glutinis, 132, 188, 233, 264
Toxoplasma gondii, 66
Trichinella spiralis, 65, 91, 102, 320
Trichinosis, 320

Ultraviolet light, 148
Ultrasound, 153

Vegetables
  contamination of, 255
  microbiology of, 253
  microflora, 255
  spoilage, 254
Vibrio comma, 319
Vibrio para haemolyticus, 188
Vinegar, 238
  acetic acid, 239
  fringe, 242
  home process, 240
  modified orleans process, 241
  orleans process, 241
  processing, 243
  quick, 241
  spoilage, 244
  starters, 243

synthetic acetic acid, 242
Vitamins, 23, 114
Viruses, 63

Water
  bacteria in, 285
  ground, 273, 277
  hard, 282
  public, 274
  purification, 279
      aeration, 282
      filtration, 280
      sedimentation, 280
      storage, 281
  softening, 283
  surface, 273, 277
  uses, 274

Waste waters, characteristics of,
    288
  BOD, 289
  color, 294
  heavy ions, 292
  immiscible liquid, 291
  industrial uses, 286
  nitrogen, 294
  pathogenic bacteria, 292
  pH, 290
  phosphorus, 294
  settleable solids, 291
  suspended solids, 290
  total dissolved solids, 294
  toxic materials, 292
  treatment, 286
  turbidity, 294

Yeast, 60